Advances in Experimental Medicine and Biology

Cell Biology and Translational Medicine

Volume 1401

Cell Biology and Translational Medicine aims to publish articles that integrate the current advances in Cell Biology research with the latest developments in Translational Medicine. It is the latest subseries in the highly successful Advances in Experimental Medicine and Biology book series and provides a publication vehicle for articles focusing on new developments, methods and research, as well as opinions and principles. The Series will cover both basic and applied research of the cell and its organelles' structural and functional roles, physiology, signalling, cell stress, cell-cell communications, and its applications to the diagnosis and therapy of disease.

Individual volumes may include topics covering any aspect of life sciences and biomedicine e.g. cell biology, translational medicine, stem cell research, biochemistry, biophysics, regenerative medicine, immunology, molecular biology, and genetics. However, manuscripts will be selected on the basis of their contribution and advancement of our understanding of cell biology and its advancement in translational medicine. Each volume will focus on a specific topic as selected by the Editor. All submitted manuscripts shall be reviewed by the Editor provided they are related to the theme of the volume. Accepted articles will be published online no later than two months following acceptance.

The Cell Biology and Translational Medicine series is indexed in SCOPUS, Medline (PubMed), Journal Citation Reports/Science Edition, Science Citation Index Expanded (SciSearch, Web of Science), EMBASE, BIOSIS, Reaxys, EMBiology, the Chemical Abstracts Service (CAS), and Pathway Studio.

Kursad Turksen

Editor

Cell Biology and Translational Medicine, Volume 17

Stem Cells in Tissue Differentiation, Regulation and Disease

 Springer

Editor
Kursad Turksen (emeritus)
Ottawa Hospital Research Institute
Ottawa, ON, Canada

ISSN 0065-2598   ISSN 2214-8019   (electronic)
Advances in Experimental Medicine and Biology
ISSN 2522-090X   ISSN 2522-0918   (electronic)
Cell Biology and Translational Medicine
ISBN 978-3-031-20516-3   ISBN 978-3-031-20514-9   (eBook)
https://doi.org/10.1007/978-3-031-20514-9

This Springer imprint is published by the registered company Springer Nature Switzerland AG
The registered company address is: Gewerbestrasse 11, 6330 Cham, Switzerland

Preface

In this next volume in the Cell Biology and Translational Medicine series, we continue to explore the potential utility of stem cells in regenerative medicine. Amongst topics explored in this volume are repair and regenerative aspects of stem cells in different disease state. One goal of the series continues to be to highlight timely, often emerging, topics and novel approaches that can accelerate stem cell utility in regenerative medicine.

I remain very grateful to Gonzalo Cordova, the editor of the series, and wish to acknowledge his continued support.

A special thank you goes to Shanthi Ramamoorthy and Rathika Ramkumar for their outstanding efforts in the production of this volume.

Finally, sincere thanks to the contributors not only for their support of the series but also for their willingness to share their insights and all their efforts to capture both the advances and the remaining obstacles in their areas of research. I trust readers will find their contributions as interesting and helpful as I have.

Ottawa, ON, Canada   Kursad Turksen

Contents

Adv Exp Med Biol - Cell Biology and Translational Medicine (2022) 17: 1–22
https://doi.org/10.1007/5584_2022_712
© Springer Nature Switzerland AG 2022
Published online: 5 May 2022

Endothelialization and Inflammatory Reactions After Intracardiac Device Implantation

Christoph Edlinger, Vera Paar, Salma Haj Kheder,
Florian Krizanic, Eleni Lalou, Elke Boxhammer, Christian Butter,
Victoria Dworok, Marwin Bannehr, Uta C. Hoppe, Kristen Kopp,
and Michael Lichtenauer

Abstract

Background: Due to the advances in catheter-based interventional techniques, a wide range of heart diseases can now be treated with a purely interventional approach. Little is yet known regarding biological effects at the intracardiac implantation site or the effects on endothelialization and vascular inflammation in an in vivo environment. Detailed knowledge of ongoing vascular response, the process of endothelialization, and possible systemic inflammatory reactions after implantation is crucial for the clinical routine, since implants usually remain in the body for a lifetime.

Methods: For this narrative review, we conducted an extensive profound PubMed analysis of the current literature on the endothelialization processes of intracardially implanted devices, such as persistent foramen ovale (PFO) occluders, atrial septal defect (ASD) occluders, left atrial appendage (LAA) occluders, transcatheter aortic valve implantations (TAVIs), and leadless pacemakers.

Christoph Edlinger and Vera Paar have equally Contributed to this chapter.

The authors declare that there is no conflict of interests regarding the publication of this paper. All companies have granted image rights to depict their devices in this article.

C. Edlinger
Department of Cardiology, Heart Center Brandenburg, Bernau/Berlin, Germany

Brandenburg Medical School (MHB) "Theodor Fontane", Neuruppin, Germany

Department of Internal Medicine II, Division of Cardiology, Paracelsus Medical University of Salzburg, Salzburg, Austria

V. Paar, E. Boxhammer, U. C. Hoppe, K. Kopp, and M. Lichtenauer (✉)
Department of Internal Medicine II, Division of Cardiology, Paracelsus Medical University of Salzburg, Salzburg, Austria
e-mail: michael.lichtenauer@chello.at

S. H. Kheder
Brandenburg Medical School (MHB) "Theodor Fontane", Neuruppin, Germany

F. Krizanic
Department of Cardiology, Caritas Clinic Pankow, Berlin, Germany

E. Lalou, C. Butter, V. Dworok, and M. Bannehr
Department of Cardiology, Heart Center Brandenburg, Bernau/Berlin, Germany

Brandenburg Medical School (MHB) "Theodor Fontane", Neuruppin, Germany

Additionally, the known biological activities of common metallic and synthetic components of intracardiac devices in an "in vivo" setting have been evaluated.

Results: Nitinol, an alloy of nickel and titanium, is by far the most commonly used material found in intracardiac devices. Although allergies to both components are known, implantation can be performed safely in the vast majority of patients. Depending on the device used, endothelialization can be expected within a time frame of 3–6 months. For those patients with a known allergy, gold coating may be considered as a viable alternative.

Conclusion: Based on our analysis, we conclude that the vast majority of devices are made of a material that is both safe to implant and nontoxic in long-term treatment according to the current knowledge. The literature on the respective duration of endothelialization of individual devices however is highly divergent.

Keywords

Cardiac injury · Endothelialization · Endovascular inflammation

Abbreviations

ASD	Atrial Septal Defect
ASO	Amplatzer Septal Occluder
ATP	Adenosine Triphosphate
CMs	Cardiomyocytes
DAMPs	Damage-Associated Patterns
ECs	Endothelial Cells
ECM	Extracellular Matrix
EPCs	Endothelial Progenitor Cells
GSO	Gore Septal Occluder
HMGB1	High-mobility Group B1
HSPs	Heat Shock Proteins
ICAM-1	Intercellular Adhesion Molecule 1
IFN-γ	Interferon Gamma
LAA	Left Atrial Appendage
MERTK	Myeloid–Epithelial–Reproductive Tyrosine Kinase
MMPs	Matrix Metalloproteinases
MI	Myocardial Infarction
NOAC	New/"Non-Vitamin K" – Oral Anticoagulants
PCI	Percutaneous Coronary Intervention
PDGF	Platelet-Derived Growth Factor
PET	Polyethylene Terephthalate
PFO	Persistent Foramen Ovale
PRRs	Pattern Recognition Receptors
RAGE	Receptor for Advanced Glycation End Products
ROS	Reactive Oxygen Species
SMCs	Smooth Muscle Cells
TAVI	Transcatheter Aortic Valve Implantation
TGF-β	Transforming Growth Factor Beta
TLRs	Toll-Like Receptors
Treg	Regulatory T-cells
VEGF	Vascular Endothelial Growth Factor
VSD	Ventricular Septal Defect

1   Introduction

Interventional cardiology is a rapidly evolving field in modern clinical medicine. Although the enormous therapeutic potential was not immediately recognized after Forßmann carried out the first catheterization of the right heart in 1929 in a heroic self-experiment, the idea was taken up and further developed by Cournand et al., who are today regarded as founders of interventional cardiology (Forssmann-Falck 1997; Nicholls 2020).

Grüntzig's first successful percutaneous coronary intervention (PCI) in 1977 marked the dawn of a new era in clinical cardiology as catheter-based therapy changed from a purely diagnostic tool to an interventional treatment option for acute coronary syndrome (Ar et al. 1979).

During recent years, interventional cardiology has undergone an enormous transformation as catheter technology has developed rapidly and is used in many different cardiological diseases. Today, in addition to coronary heart disease, a wide range of valvular diseases or congenital heart defects, such as atrial septal defect (ASD) (Sievert et al. 1998) and persistent foramen ovale (PFO) can be successfully treated using a catheter-based interventional approach. Even the catheter-based implantation of cardiac pacemakers is widely used in countless clinics today.

Permanent intracardiac placement of devices represents an enormous challenge for product engineers, as all components must offer exceptional stability and durability combined with low weight and small size. In addition, any device exposed to blood flow in an "in vivo" environment must be safe in terms of hemostasiological interactions; more precisely, the device must not initiate thrombogenic or even hemolytic cascades. Furthermore, the materials components must have an extremely low allergic potential, both at the implantation site and, in case of systemic reactions, in the entire organism.

All the devices must undergo extensive testing on safety and durability for market approval in order to achieve CE certification. In addition, an optimally designed intracardiac device should also have acceptable biocompatibility, especially rapid endothelialization, to reduce the duration of anticoagulant use with the goal of reduced bleeding risk and to safely discontinue use of endocarditis prophylaxis.

The aim of this narrative review is to provide an overview of the current state of knowledge on endothelialization in common intracardiac devices as well as an overview of the known in vivo interactions, and the important components of different devices.

2 Intracardiac Devices

2.1 PFO Occluders

The patent foramen ovale (PFO) is an essential component of intrauterine circulation, allowing blood to bypass fetal lungs. Although spontaneous occlusion should occur shortly after birth, this physiological process is absent or incomplete in up to 27.3% of the population (Hagen et al. 1984).

In the vast majority of cases, a PFO does not have a clinical relevance. Nevertheless, it may play a crucial role in the genesis of migraine headaches or even cryptogenic stroke (Saver et al. 2017). In symptomatic patients, interventional treatment with an occluder can be considered to permanently close the opening. Recent studies (CLOSE, REDUCE, RESPECT LT, and DEFENSE PFO) showed that interventional PFO occlusion (Fig. 1a) was associated with a significant reduction in recurrent stroke compared to drug therapy (Mas et al. 2017; Saver et al. 2017; Søndergaard et al. 2017). Based on the long-term results of RESPECT and the results of the REDUCE study, two occluder devices, the AMPLATZER PFO closure and GSO (GORE Medical, Flagstaff, AZ, USA) received US Food and Drug Administration (FDA) approval for secondary stroke prevention in 2016 and 2018, respectively.

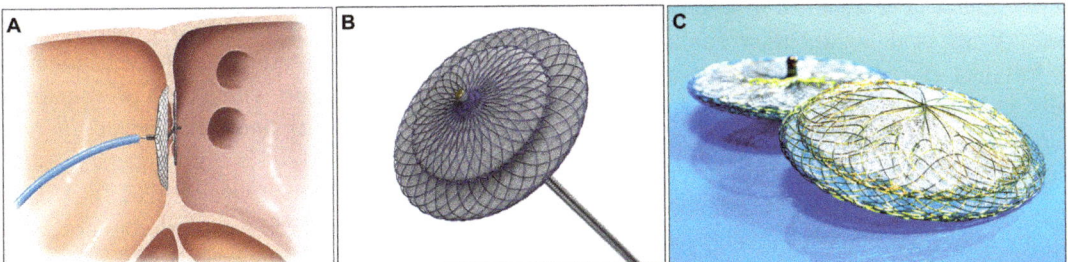

Fig. 1 PFO occluders. (**a**) Schematic representation of the regular position of a PFO occluder. (**b**) Amplatzer™ PFO Occluder. (**c**) Figulla Flex II® PFO Occluder. Pictures provided by Occlutech International AB

The later developed Flex II PFO occluder (Occlutech, Jena, Germany) received CE certification for clinical use in Europe in 2009. According to the manufacturer, more than 33,000 such devices have been delivered worldwide.

2.1.1 Amplatzer PFO Occluder

The Amplatzer occluder (Abbott Cardiovascular, North Chicago, Illinois, USA), first implanted in 1997, has been very well described elsewhere (Meier 2005; Madhkour et al. 2019). In brief, it is a double disc made of Nitinol mesh, the inside of which is made of polyester fabric.

A thin neck, which consists of the tightly woven wires of the discs, serves as a connector. The neck is rotated around its longitudinal axis so that it can be extended in principle (Fig. 1b). The two discs are sewn together with polyester fabric for better stabilization (Scalise et al. 2016).

The special feature is the so-called "shape memory," which means that the device returns to its original shape after being stretched through the guiding catheter.

The Amplatzer PFO Occluder is available in three different sizes (18 mm, 25 mm, 35 mm), whereby the size specification is based on the size of the right-sided disc. The most frequently implanted occluder is the medium size. The small version has its special value in the case of a small PFO with a largely stable septum primum, whereas the 35 mm version is used in the case of an extremely redundant septum primum and possibly in the case of an atrial septal aneurysm.

According to the current literature, the chances of success are very high, so that complete closure of the shunt can be assumed in well over 90% of cases (Bruch et al. 2002; Greutmann et al. 2009).

Residual shunts require surgical intervention only in rare cases, although the actual average duration until complete endothelialization "in vivo" is not completely clear. In a recent position paper, the German Cardiology Society recommends dual antiplatelet therapy with aspirin and clopidogrel for a period of 6 months. Should there be a concomitant indication for oral anticoagulation, this will be given as monotherapy, with NOACs being preferred.

2.1.2 Gore Septal Occluder

The Gore Septal Occluder (GSO) has been approved for the treatment of PFOs in Europe for 10 years now by means of CE certification.

In contrast to the more commonly implanted Amplatzer Occluder, it is intended to offer advantages in difficult anatomical conditions. The device consists of five nitinol wires formed into a left atrial and a right atrial disc. The outer frame is coated with polytetrafluoroethylene film. For implantation, the device comes already loaded on a 10 French introducer catheter. A special safety feature is the integrated retrieval cord, which can be used to retrieve the device if necessary. When fully deployed, two circular discs are formed, facing each other, which can be fixed in place by a locking mechanism in the center.

2.1.3 Flex II PFO Occluder

The Flex II PFO occluder (Occlutech, Jena, Germany) consists of two self-expanding woven nitinol discs (Fig. 1c). The special feature of this device is a central pin on the left atrial disc and also a ball socket joint connection. The complete closure is achieved by two biocompatible polyethylene terephthalate patches.

The superiority of this device is the flexibility and ability to angulate in order to achieve the maximal adaptation to the interatrial septum. This offers an advantage for the complex anatomical variations (Neuser et al. 2016).

2.2 ASD Occluders

Atrial septal defect (ASD) is a relatively common congenital heart defect with a birth prevalence of 1.43:1000 live births and an expected survival rate into adulthood of 97% (Anderson et al. 2002; Anderson 2016; Lee et al. 2018). ASD of the ostium secundum is the most common type, occurring in 70% of all patients with ASD, followed by ASD of the ostium primum (10%) and ASD of the sinus venosus (5–10%) (Moons et al. 2009). The "true atrial septum," that is, the tissue directly separating the atrial cavities, is restricted to the base of the oval fossa and the surrounding inferoanterior margins. Defects of the true atrial septum are called "secundum defects." Atrial septal defects are usually well tolerated in children, but can cause significant complications in adults (Campbell 1970). Early closure is therefore recommended and can be achieved using catheter deployment in the majority of cases. A symptomatic benefit can be seen at any age (Komar et al. 2014) Since left ventricular compliance decreases with age or in the presence of conditions that can increase left atrial pressure (e.g., high blood pressure, ischemic heart disease, cardiomyopathy, aortic and mitral valve disease), the left–right shunt can increase due to ASD (Le Gloan et al. 2018; Kumar et al. 2019). However, conditions that reduce right ventricular compliance (e.g., pulmonary arterial hypertension, pulmonary stenosis, right heart disease, tricuspid valve disease) may eventually reverse the shunt and cause cyanosis (Le Gloan et al. 2018).

A left–right shunt leads to right ventricular volume overload, which in turn results in right ventricular dilatation. It is well tolerated throughout childhood, despite a pulmonary–systemic flow ratio that can exceed 3:1. The pulmonary vascular system is also able to absorb the increased blood flow at low pulmonary artery pressure for many years. A persistent large left–right shunt leads to increased right atrial and right ventricular dilatation from late childhood onwards, which in some patients leads to arrhythmia and a progressive increase in pulmonary

vascular resistance (Le Gloan et al. 2018). Severe pulmonary vascular disease is rare (<5%), unless there are other associated factors (Nashat et al. 2018).

2.2.1 Amplatzer Septal Occluder

The Amplatzer® Septal Occluder (ASO) (Abbott Cardiovascular, North Chicago, Illinois, USA; former: St. Jude Medical, Inc., St. Paul, Minnesota) consists of nitinol–titanium memory wire mesh infused with polyester patches that facilitate occlusion and endothelialization (Fig. 2a). It consists of a smaller right and a larger left disc connected by a waist; the difference in size of both discs is 4 mm (Nassif et al. 2016).

ASO has been shown to be a practical, safe, and effective treatment option for ASD (Masura et al. 1997; Podnar et al. 2001; Masura et al. 2005; Cardoso et al. 2007; Knepp et al. 2010). Nevertheless, complications such as implant embolization, mispositioning, and fracture may occur in rare cases. Moreover, cases of erosion/perforation, cardiac arrhythmia, cardiac tamponades, and even infectious endocarditis have been reported (Sievert et al. 1998; Chessa et al. 2002; Fischer et al. 2003; Balasundaram et al. 2005; Sadiq et al. 2012).

2.2.2 Occlutech ASD Occluder

The Figulla Occlutech ASD closures (Occlutech, Jena, Germany) consist of individually braided, very thin (40–150 μm or 0.00157–0.00590 inches) nitinol strands. All strands end proximally

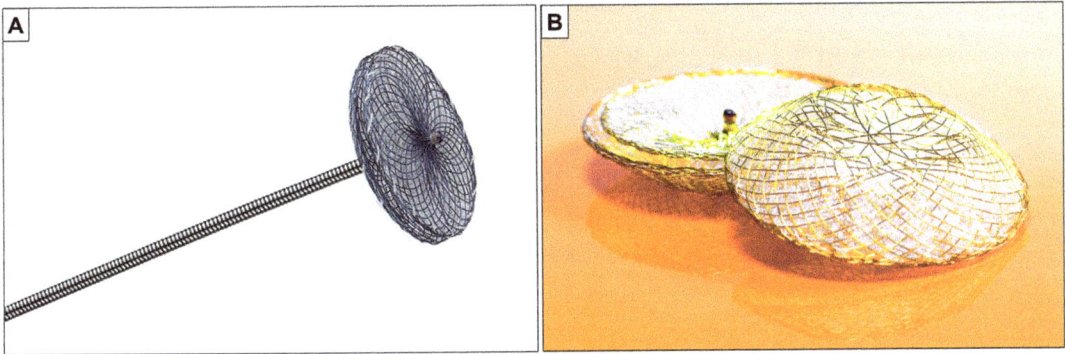

Fig. 2 ASD occluders. (**a**) Amplatzer™ Multi-Fenestrated Septal Occluder – "Cribriform". (**b**) Occlutech® Fenestrated Atrial Septal Defect Occluder. Pictures provided by Occlutech International AB

and therefore do not require clamping to the left disc (Fig. 2b). This results in a smaller amount of uncovered metallic material. The ultrathin fabrics made out of polyethylene terephthalate (PET) of the device promote the endothelial growth after implantation as well as the defect closure (Pedra et al. 2016). The design of the Flex II ASD occluder is intended to allow ideal alignment of the septum, which in turn should increase feasibility and patient safety during implantation. The device is made of Titanium oxide–covered nitinol, which should result in the lowest possible release of nickel.

2.3 Left Atrial Appendage Occluders

The term "LAA occluder" refers to devices that can be used to close the left atrial appendage (LAA). In patients with atrial fibrillation, the LAA is anatomically particularly important, as this is where the vast majority of cardio-embolic strokes originate (Alli and Holmes, 2015). In patients at high risk for bleeding complications, implantation of an occluder allows discontinuation of anticoagulants after the initial endothelialization phase. The most common product is the Watchman Device, which has been CE certified since 2005. Another product, CE certified in 2013, is the Amulet device, which is designed to provide benefits due to its wide range of available sizes, according to the manufacturer.

2.3.1 Watchman Device

The Watchman device (Boston Scientific, Marlborough, MA, USA) consists of a nitinol frame coated with a permeable 160 micron polyethylene terephthalate knit fabric on the left atrial surface (Fig. 3a) (Fountain et al. 2006). It is a parachute-shaped, self-expanding device (Kramer and Kesselheim, 2015). The PET knit fabric facilitates endothelialization over the device and serves as a filter for emboli that originate from the LAA pouch (Della Rocca et al. 2019). After femoral vein access and transseptal puncture, the Watchman device is delivered using a 12-Fr delivery catheter and is then deployed until its titanium dowel pin separates from the catheter (Fig. 3b). The Watchman device is affixed to the LAA wall by 10 fixation barbs, which are arranged around the mid-perimeter. To match different LAA orifice sizes, the device is manufactured in five sizes (21 mm, 24 mm, 27 mm, 30 mm, and 33 mm) to allow adequate placement. An adequate seal is defined as a leakage < 5 mm.

2.3.2 Amulet Occluder

The AMULET is a second-generation Amplatzer Cardiac Plug (Abbott Cardiovascular, North Chicago, Illinois, USA; former: St. Jude Medical, Inc., St. Paul, Minnesota). The self-expanding device is made of flexible, braided nitinol filled with polyester tissue. It consists of a proximal disc and a distal lobe shaped like a hockey puck

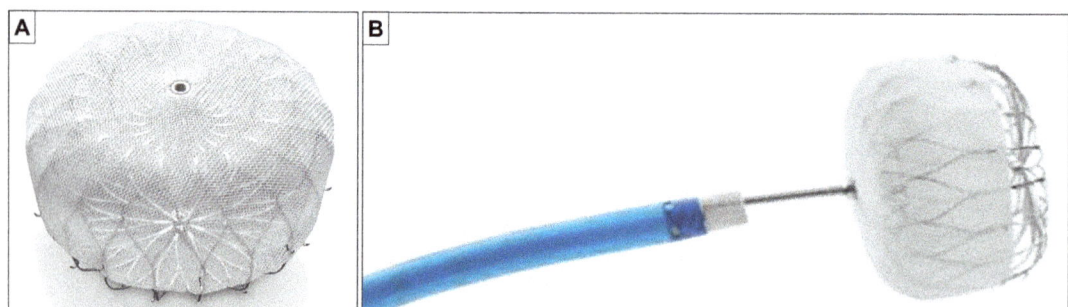

Fig. 3 Watchman Devices. (**a**) Watchman™. (**b**) Watchman™ on a 12 French delivery catheter. Pictures provided by Boston Scientific

connected by a flexible waist (Meerkin et al. 2013). The proximal disc covers the LAA orifice and the distal lobe with stabilization hooks that secure the engagement of the occluder to the LAA wall. The LAA Occluder is delivered by a 12-F or 14-F sheath into the left atrium after a transseptal puncture. It is available in eight sizes. Pre-interventional standard imaging, including transesophageal echocardiography and computerized tomography (CT) scan, is performed in order to determine the proper occluder size. The proximal disc is always slightly larger than the lobe and has a central screw.

2.4 Transcatheter Aortic Valve Implantation

Transfemoral aortic valve replacement, first performed in 2001 by Cribier et al., is beyond doubt one of the greatest achievements in interventional cardiology (Cribier et al. 2002). As alternative to conventional surgical aortic valve replacement, a minimally invasive transcatheter implantation of a valve prothesis is possible. The intervention was originally reserved for medium- to high-risk elderly patients, also due to material durability profiles. In recent years, its indication has been extended to younger patients age <75 years. Recently, transcatheter valve replacement has been approved by the FDA as the method of choice for all patients (Edlinger et al. 2020). The most important challenge is the material compatibility and the special required characteristics of the valve. Those consist of "durability, low thrombogenicity, hydrodynamics, hemocompatibility, low calcification susceptibility and crimping and deployment stability" (Rotman et al. 2018).

Various models have been developed over the years, whereby the self-expanding Medtronic valves (CoreValve, Evolut R) (Medtronic, Minneapolis, MN, USA) and the balloon-expanding Edwards valves (Sapien, Sapien XT, Sapien 3) (Edwards Lifesciences, Irvine, CA, USA) are the most widely implanted valve types (Chakos et al. 2017). The SAPIEN 3 and SAPIEN 3 Ultra are the latest generation of balloon-expanding Edwards valves. The Evolut R and Evolut PRO are the latest self-expanding valves from Medtronic (Renker and Kim, 2020). In several hospitals they consist more than 70% of the total transcatheter aortic valve implantation (TAVI) procedures.

Another important aspect concerns the valve-in-valve procedures. Here is the geometric orifice area that plays the most significant role. Therefore a valve-in-valve procedure should only be used after careful planning, because it can diminish the

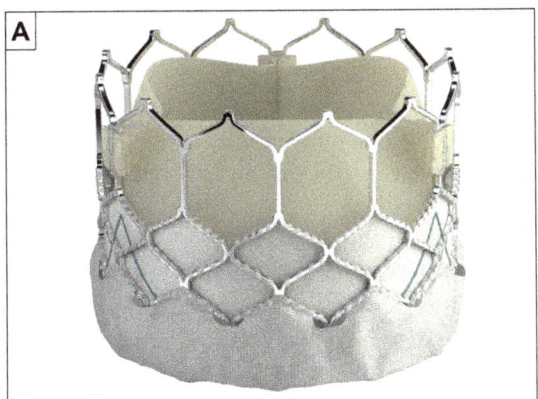
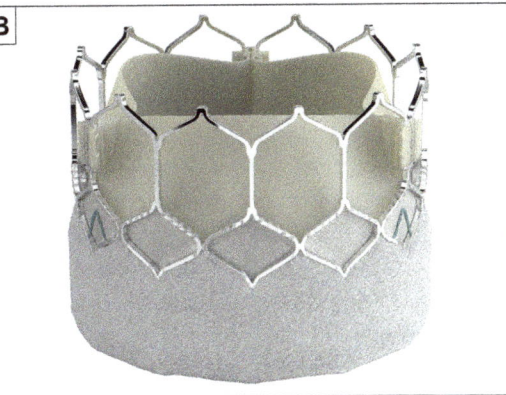
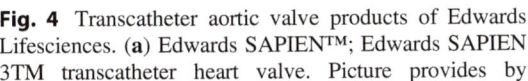

Fig. 4 Transcatheter aortic valve products of Edwards Lifesciences. (**a**) Edwards SAPIEN™; Edwards SAPIEN 3TM transcatheter heart valve. Picture provides by Edwards-Sapien. (**b**) Edwards SAPIEN™ 3 Ultra; Edwards SAPIEN 3 UltraTM transcatheter heart valve. Picture provided by Edwards-Sapien

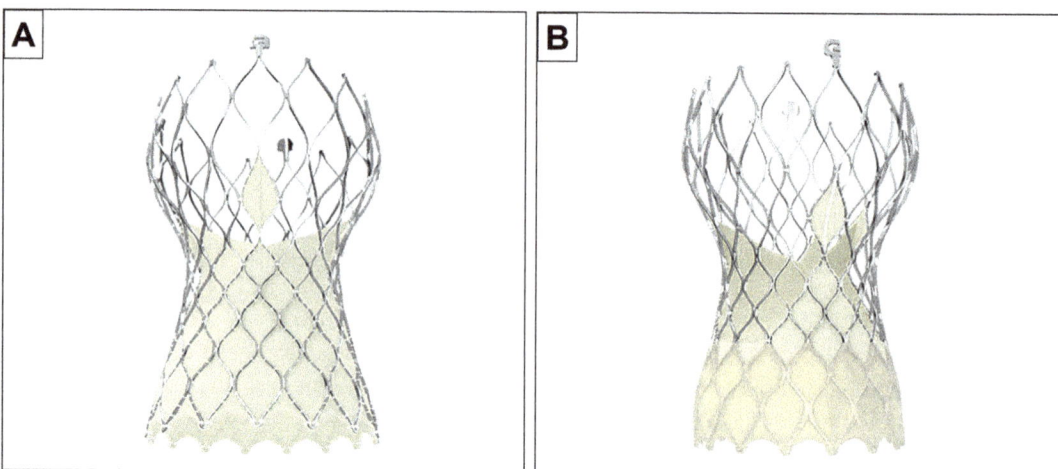

Fig. 5 Transcatheter aortic valve products of Medtronic. (**a**) CoreValve™ Evolut™ R. (**b**) CoreValve™ Evolut™ Pro. (Source: Medtronic GmbH)

hydrodynamic performance of the valve and significantly reduce the opening area (Rotman et al. 2018).

2.4.1 Edwards SAPIEN 3/SAPIEN 3 Ultra

The Edwards SAPIEN 3 (Fig. 4a) and SAPIEN 3 Ultra (Fig. 4b) consist of a cobalt–chromium stent frame, three leaflets of bovine pericardium, an inner and an outer skirt. The SAPIEN 3 has an internal skirt and an outer sealing skirt made of polyethylene terephthalate (PET) (Jose et al. 2015). In the SAPIEN 3 Ultra, the outer portion is textured PET material with a greater height in comparison to the SAPIEN 3 (Renker and Kim, 2020). The valves are manufactured in four sizes (20 mm, 23 mm, 26 mm, and 29 mm).

2.4.2 Medtronic Evolut R/Evolut PRO

Both valves consist of a self-expandable nitinol stent frame and three leaflets of porcine pericardium positioned in a supra-annular location. In contrast to the Evolut R (Fig. 5a), an external porcine pericardial sleeve has been added to the Evolut PRO (Fig. 5b) as a sealing sleeve. The Evolut R model is manufactured in four sizes: 23 mm, 26 mm, 29 mm, and 34 mm, while the Evolut PRO is only available in 23 mm, 26 mm, and 29 mm.

The valve size selection is of paramount significance for the physician and for the patient and is highly associated with the success of a TAVI procedure. In order to choose the right valve size, an MSCT Scan must be performed. Then the appropriate software should be used to quantify the aortic root.

The cut-off values of the manufacturer's sizing plan should be considered. There is often a difference between clinical valve selection and selection based on the computer software. The "device–host interaction" is very important in order to prevent complications such as obstruction of the coronary arteries, relevant aortic regurgitation after implantation as well as conduction disorders such as atrioventricular (AV) block (El Faquir et al. 2020).

A rare but sometimes life-threatening complication is the transcatheter heart valve migration into the outflow tract of the left ventricle. In this case is a balloon repositioning of the valve necessary as well as possibly a valve-in-valve procedure in order to prevent severe aortic regurgitation (Ito et al. 2017).

One of the deciding mortality as well as success parameters of a TAVI procedure – also long term – is the paravalvular regurgitation. This reinforces the importance of correct valve selection and the experience of the physician. In several publications, it is observed that in the second part of the study or cohort there are a greater number of good final results as the technique of

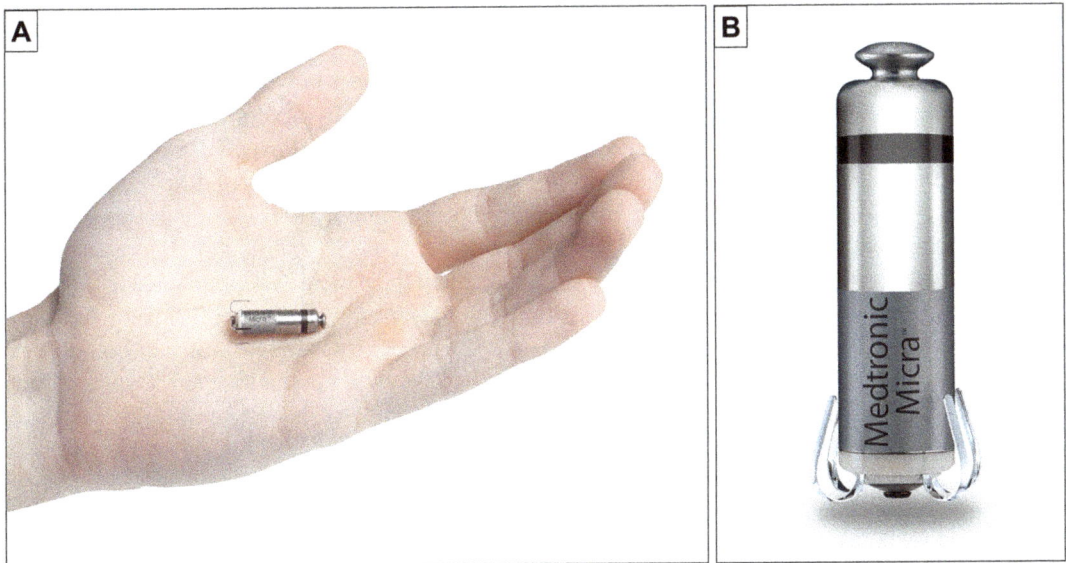

Fig. 6 Micra Pacemakers. (**a**) Illustration of Micra™ pacemaker size compared to a human hand. (**b**) Micra™. (Source: Medtronic GmbH)

the interventional cardiologist improves exponentially (Wang et al. 2021).

2.5 Leadless Pacemakers

Within the last decade, single-chamber leadless pacemaker devices have been developed, which are implanted to the inner side of the right ventricle via a steerable catheter insertion system (Reynolds et al. 2016). Currently, the most common device in clinical use is the MicraTM (Medtronic Inc., Minneapolis, MN, USA). Despite its small size of 0.8 m^3, a light capsule weight of 2.0 g, a length of 25.9 mm, and an outer diameter of 6.7 mm, it has all the features of a conventional single-chamber pacemaker system (Reddy et al. 2015). The Micra pacemaker itself consists of nitinol, gold, steel, titanium, and tungsten (Figs. 6a and 6b). The tines, with which the device is affixed to the endocardium of the right ventricle, consist entirely of nitinol.

There are now numerous clinical studies that prove both, the effectiveness and safety of the product. Medtronic postulates that the device would float in an in vivo environment at the inside of the right ventricle. According to the manufacturer, the only connection to the endothelium is through the pacemaker's anchoring system. As lead-free pacemaker technology is a relatively new topic, the effects on the intracardiac endothelium at the anchor point are not yet well understood. However, individual cases have also been published in which unexpected encapsulation was observed. . First data from autopsies show partial or even complete encapsulation (Tjong et al. 2015; Kypta et al. 2016). In the two published cases, complete endothelialization or even encapsulation is reported after 12 months and 19 months, respectively. Within one of our prior in vitro studies, we could identify a potential impact of the tungsten component in these processes of endothelialization (Edlinger et al. 2019).

3 Common Device Components

3.1 Nitinol

Nitinol, a material used in the majority of intracardiac devices, is a 55:45 nickel–titanium alloy. It is of enormous value for medical applications due to its thermal shape–memory effect and super elasticity (Ryhänen et al. 1998). Nitinol is not considered cytotoxic or thrombogenic, although its individual components, nickel and titanium, can be detected in peripheral blood samples (Shayan and Chun 2015).

It is assumed that the immunological effects are comparable to those of stainless steel without any toxic effects (Ryhänen et al. 1998). Nitinol is widely used, for example, in many PFO and ASD occluders, as well as the tines of the Micra pacemaker are made of this material.

3.2 Titanium

Due to its good in vivo compatibility, titanium is a widely used component for implants and alloys (Ungersboeck et al. 1995). Concentrations of 50–150 µg/l can be measured in peripheral blood samples, a level considered nontoxic (Ipach et al. 2012). Data from orthopedic studies could show an increase in classical inflammatory cytokines such as tumor necrosis factor alpha (TNF-α), interleukin-1beta (IL-1β) and IL-6 in patients with titanium implants. However, there is no evidence of toxicity to date (Sun et al. 2000; Östberg et al. 2015). In cardiology, for example, titanium is used in pacemaker generators with no long-term adverse effects. It is particularly important as a component of the nitinol alloy.

3.3 Tungsten

Tungsten is another component that is commonly used in medical devices. Certain publications point to toxic effects during long-term treatments (Witten et al. 2012). However, according to the current knowledge, no toxic effects are expected at normal corrosivity rates and slightly elevated serum levels are considered as normal (Peuster et al. 2003). It is considered to be a very strong and durable material as well as very resistant to corrosion. In cardiology, tungsten is widely used as a component of pacemaker probes. The Micra pacemaker also partly consists of this material.

3.4 Gold

Implants made of gold and its nanoparticles are known to be noncytotoxic and nonimmunogenic (Shukla et al. 2005). Cytokine elevations of interleukin 6 (IL-6), tumor necrosis factor-alpha (TNF-α), interleukin 1-beta (IL-1β), and mononuclear chemotactic protein 1 (MCP-1) are frequently found. However, these cytokine reactions are insignificant and have no influence on the cell viability (Zhang et al. 2011). For this reason, gold is of paramount value as a reserve material if the patient to be implanted suffers from a previously known allergy, for example to titanium. This is a rare allergy, which affects only the 0.6% of the total population. There are reports of patients who have been successfully implanted with a gold-plated leadless pacemaker for this indication. This was described in a 65-year-old man with a type IV allergy to titanium in the case report by Kypta et al. (Kypta et al. 2015; Goli et al. 2012).

3.5 Steel

Steel is considered to be a material of manufacturing precision, good hygiene, as well as high resistance against corrosion. It has been reported that the toxic effects of steel implants are higher than those of other components (Haynes et al. 1998). Lacey et al. observed a decrease in monocyte and macrophage survival in response to steel (Lacey et al. 2009), as well as a reduced leukocyte migration to the implant site or prosthetic implants. In addition, steel has been shown to induce abundant elevation of cytokines such as IL-1β (Haynes et al. 1998).

4 Cardiac Injury, Wound Healing, and Regeneration – A Brief Overview

Cardiac injury causes a severity-related damage of the myocardium, followed by cardiac repair or wound healing where damaged tissue is usually replaced by a fibrotic scar, as described in the literature (Deb and Ubil 2014; Talman and Ruskoaho 2016). Moreover, it has recently been reported that adult cardiomyocytes (CMs) have a slight ability to proliferate, raising the promise of promoting cardiac regeneration in humans

(Beltrami et al. 2001; Bergmann et al. 2009; Mollova et al. 2013). However, the feasibility of cardiac regeneration is largely dependent on the type and extent of immune responses, thus leading to an inflammatory response (Sattler and Rosenthal 2016; Cheng et al. 2017).

Although the implantation of an intracardiac device is necessary for the maintenance of proper cardiac pacing in various cardiac diseases, it is also accompanied by the injury of the cardiac tissue at the site of implantation. This inevitable injury initiates a complex series of tissue repair processes that comprise of the interaction and timely coordination of several cell types, cytokines, chemokines, and signaling cascades. Furthermore, there is a host reaction following the implantation of foreign biomaterial into the cardiac implantation site. This also includes blood–material interactions, inflammation, granulation, provisional matrix formation, and the fibrotic remodeling of the injured area (Gretzer et al. 2006; Luttikhuizen et al. 2006).

Generally, the immune response to cardiac injury is accomplished by the innate and the adaptive immune systems in synergy and can be divided into three phases: the pro-inflammatory phase, the proliferative phase, and the reparative phase (Lai et al. 2019).

4.1 Pro-inflammatory Phase

In the very early process after the implantation of a foreign biomaterial, a material–blood interaction occurs, whereby proteins from the blood adhere to the implant's surface. A provisional matrix based on the blood's components forms, that is, the initial thrombus or blood clot where further protein adsorption proceeds. This provisional matrix and the injured tissue are responsible for the recruitment of structural, biochemical, and cellular compartments that are essential for wound healing (Gristina 1994; Gretzer et al. 2006; Luttikhuizen et al. 2006). In this period, inflammatory cells are recruited to the site of injury to clear the damaged wound of dead cells and tissue, as well as to degrade the matrix debris. Furthermore, it initiates the processes necessary

to form the reparative scar. However, it has been described that prolonged or excessive inflammation is accompanied by a poor tissue remodeling and worse outcomes in patients or animal models with myocardial infarction (MI) (Timmers et al. 2008; Arslan et al. 2010; Frangogiannis 2012).

The initial immune response is driven by molecules released from necrotic cells, the so-called damage associated patterns (DAMPs) (Arslan et al. 2010). In addition, during tissue death, dying proteases, hydrolases, and mitochondrial reactive oxygen species (ROS) are also released into the extracellular space, generating further DAMPs that trigger the inflammatory response (Kono and Rock 2008). Subsequently, these DAMP molecules bind to pattern recognition receptors (PRRs), including toll-like receptors (TLRs) and the receptor for advanced glycation end products (RAGE), that are expressed by both tissue resident cells and recruited leukocytes (Muzio et al. 2000; Chavakis et al. 2004). Among other DAMPs present in cardiac inflammation, high-mobility group B1 (HMGB1) is one of the best characterized (Andrassy et al. 2008). HMGB1 is responsible for the initiation of inflammation in myocardial infarction (MI) and cardiac ischemia by promoting the migration of immune cells through its interaction with PRRs, such as TLR2/4 (most abundant TLRs in the heart) and RAGE (Nishimura and Naito 2005; Klune et al. 2008; Sims et al. 2009). Moreover, it induces tissue healing by changing the macrophages' phenotype, favoring neoangiogenesis and promoting stem cell activation and proliferation (Bianchi et al. 2017).

Physiologically, the extracellular matrix (ECM) is responsible for the support and the maintenance of the heart's structural integrity. However, during inflammation the ECM is degraded by matrix metalloproteinases (MMPs), activated by necrotic cells, neutrophils, and macrophages. This degraded ECM can in turn act as a DAMP, driving the inflammatory pathway forward (Dobaczewski et al. 2010a). In the context of cardiac injury, a switch to a transient fibrin-based ECM is achieved (González-Rosa et al. 2011; Frangogiannis 2017), further modulating and guiding inflammatory cells through TLRs (Corbett and Schwarzbauer 1998; Smiley et al. 2001; Flick et al. 2004) and promoting the proliferation of endothelial cells and fibroblasts (Frangogiannis 2017).

As mentioned above, dying cardiomyocytes release intracellular components, such as deoxyribonucleic acid (DNA) and ribonucleic acid (RNA), as well as intracellular components like adenosine triphosphate (ATP) and heat shock proteins (HSPs), that might accelerate the ongoing immune response (Arslan et al. 2011; Kono et al. 2014). Furthermore, reactive oxygen species (ROS), which stem from mitochondria of necrotic cells or are secreted by neutrophils, constitute a key player in the promotion of immune cells to infiltrate the injured tissue. ROS contributes to the onset of the nuclear factor kappa light chain enhancer of activated B-cells (NF-κB), a main chemotactic and pro-inflammatory protein complex (Thannickal and Fanburg 2000; Gloire et al. 2006), and directly activates the so-called inflammasome, as well as cardiac resident cells, such as fibroblasts and mast cells (Gilles et al. 2003; Kawaguchi et al. 2011). The inflammasome, a multiprotein complex of receptors and cytokines, in turn promotes the immune response and triggers the expression and activation of other cytokines (Latz et al. 2013).

After the immune response has been initiated by damage-associated molecular patterns (DAMPs) and related pattern recognition receptors (PRRs), resident immune cells and non-immune cells, such as resident macrophages, endothelial cells (ECs), and fibroblasts, drive the expression of pro-inflammatory cytokines and chemokines. In cardiac device implantation, the extent of immune responses is primarily mediated by the extent of injury that happened during the implantation procedure (Zdolsek et al. 2007; Tang et al. 1998). In the presence of DAMPs, cytokines, chemokines, activated platelets, and histamine, neutrophils are the first innate immune cells that are rapidly recruited to the injured tissue (Mcdonald et al. 2010; Soehnlein and Lindbom 2010). Contemporaneously, the cardiac endothelium is activated by pro-inflammatory cytokines, such as TNF-α, IL-1β, and histamine (Duperray

et al. 1995; Dewald et al. 2004; Debrunner et al. 2008). This ensemble of pro-inflammatory cytokines constitutes the inflammasome and facilitates the neutrophil transmigration between and through the endothelial wall to the site of tissue injury (Frangogiannis et al. 1998; Singh and Saini 2003). Furthermore, IL-6 seems to be a main mediator of tissue injury, since it is expressed by CMs and recruited neutrophils and macrophages (Youker et al. 1992). IL-6 in turn upregulates intercellular adhesion molecule 1 (ICAM-1) on CMs that mediates neutrophil binding and is associated with cytotoxic events (Entman et al. 1992; Youker et al. 1992).

4.2   Cell Proliferation Phase

The cellular proliferative phase is the second phase and is characterized by the expansion of neutrophils and macrophages that degrade dead cells and the matrix debris, further promoting the expression of cytokines and growth factors. Due to the pro-inflammatory and cytotoxic activity of neutrophils, excessive amounts or a prolonged presence of neutrophils have been associated with remodeling and a poor prognosis after MI (Mocatta et al. 2007; Akpek et al. 2012). On the other hand, they constitute a key factor in the resolution of inflammation and lead to a shift of the macrophages' phenotype to a reparative one (Čulić et al. 2002; Pase et al. 2012). Furthermore, they contribute to the initiation of angiogenesis during inflammation by expressing vascular endothelial growth factor (VEGF) (Gong and Koh 2010).

Generally, monocytes are a type of leukocytes that have the ability to differentiate into macrophages and dendritic cells. There are two different subpopulations of monocytes present in cardiac inflammation: the Ly6C^{high} and Ly6C^{low} (Hettinger et al. 2013; Yona et al. 2013). Ly6C^{high} monocytes belong to the primary subset that is recruited to the injured heart, driven by MCP-1. Therefore, Ly6C^{high} monocytes are commonly active in the early pro-inflammatory phase and are responsible for proteolytic and inflammatory processes. In contrast, Ly6C^{low} are sometimes known as resident monocytes due to their appearance of not being actively recruited into the injured myocardium (Geissmann et al. 2003; Nahrendorf et al. 2007). They have been shown to emerge later, in the resolution phase, demonstrating decreased inflammatory properties, as well as the expression of VEGF (Yao et al. 2012). It is not definitively clear if Ly6C^{low} arise from differentiation of Ly6C^{high} (Hanna et al. 2011; Yona et al. 2013), although it has been speculated that they arise from the same progenitor cells (Hettinger et al. 2013; Yona et al. 2013). In addition, two macrophage subsets (M1 and M2 macrophages) correspond with these different monocyte concentrations. M1 macrophages are present early after heart injury and are known to secrete pro-inflammatory cytokines, like IL-1β, TNF-α, IL-6, and IL-10 (Dewald et al. 2005), whereas M2 monocytes become active at the later stage of reparative heart tissue healing (Nahrendorf et al. 2010). However, the simple division of macrophages into two subsets should be considered due to the great variety of macrophage phenotypes (Martinez and Gordon 2014). The initial acute inflammatory response usually resolves within 1 week after device implantation, though it is also dependent on the extent of injury at the implant site (Gretzer et al.).

4.3 Endothelialization and Resolution of Inflammation

Finally, the conversion from inflammation to the repair phase is crucial for wound healing as a prolonged inflammatory response would lead to CM death, excessive fibrosis, cardiac remodeling, and damage. In cardiac device implantation, acute inflammation is often followed by a chronic inflammation period that is characterized by the presence of mononuclear cells, such as monocytes and lymphocytes. This chronic inflammation lasts for a short time of approximately 2 weeks and is strictly located to the site of implantation. The prolongation of the inflammation phase for greater than 3 weeks, usually a device infection is indicated (Luttikhuizen et al. 2006). Once the inflamed/injured area is cleared

of apoptotic cells, the repair process is initiated and a new ECM is produced (Frangogiannis 2014). The resolution phase is mainly characterized by the recruitment of lymphocytes, the activation of fibroblasts, and the proliferation of ECs, as well as the activation of smooth muscle cells (SMCs).

Originally, it was thought that circulating endothelial progenitor cells (EPCs), as progenitors of ECs, were a source of new endothelial cells, as first described by Asahara et al. in 1997 (Asahara et al. 1997). They originate from different hematopoietic progenitor cells located in the bone marrow, such as hematopoietic stem cells, myeloid precursors, and mesenchymal stem cells (Balistreri et al. 2015). However, EPCs also stem from different nonhematopoietic tissues, such as the umbilical cord etc. (Ingram et al. 2004; Mund et al. 2012; Chan et al. 2013). In response to tissue damage, they are released into the circulation and invade the site of injury attracted to inflammatory cytokines and chemoattractant proteins. As progenitor cells, EPCs constitute a source for ECs by differentiation and further promoting the proliferation of resident ECs (Buijs et al. 2004; Li et al. 2012). Furthermore, they release several growth factors, such as VEGF and angiopoietins, and other pro-endothelial factors that promote the healing process (MCP-1), stromal cell-derived factor 1, insulin-like growth factor 1, platelet-derived growth factor (PDGF), and macrophage inflammatory protein 1a (Rehman et al. 2003; Caiado et al. 2008). In turn, these factors stimulate ECM proteins and the proliferation of SMCs. An overview of the factors released by ECs, SMCs, and inflammatory cells was already provided by Welt and Rogers (Welt and Rogers 2002).

Both lymphocytes, B- and T-cells, comprise the main cellular components of the adaptive immune system. T-cells are further divided into CD8$^+$ and CD4$^+$ subsets, whereas CD4$^+$ T cells are the main actors in the healing process. According to their secreted cytokines they are further classified into Th1 (IL-2, TNF-α, and interferon gamma (IFN-γ); Th2 (IL-4, IL-4, IL-13); Th17 (IL-17, IL-21, IL-22); and regulatory T-cells (Treg) (transforming growth factor beta (TGF-β), IL-35) (Hofmann and Frantz 2015). More precisely, especially Tregs play a key role in the healing phase through suppressing the immune response in the damaged tissue, promoting revascularization, and initiating the shift to a reparative phenotype of the macrophages (M2 microphages; as mentioned above) (Zouggari et al. 2009; Dobaczewski et al. 2010b; Weirather et al. 2014). Particularly, TGF-β was shown to be mainly responsible for the deactivation of inflammatory macrophages in MI (Dobaczewski et al. 2011). These M2 macrophages then have the ability to express high amounts of several different MMPs and secrete anti-inflammatory cytokines, such as IL-4, IL-13, and mainly IL-10 (Frangogiannis et al. 2000). Consequently, the extent of IL-4 and IL-13 expression also determines the extent and duration of the inflammatory response. Furthermore, it was found that myeloid–epithelial–reproductive tyrosine kinase (MERTK) (Wan et al. 2013) and platelet-derived growth factor (PDGF) (Zymek et al. 2006) are crucial for the transition to a reparative status too. Furthermore, TGF-β signaling and the decline in pro-inflammatory cytokine signaling result in the activation of interstitial and perivascular fibroblasts, EC proliferation, followed by reparative myocardial fibrosis and angiogenesis (Chen and Frangogiannis 2013). Any failure of accurate regulation of Treg or TGF-β signaling may lead to excessive scar formation, an ongoing chronic inflammation (Kypta et al. 2016), such as described in the case report by Kypta et al. (Dobaczewski et al. 2011).

5   Discussion

In the last decades, an immense increase in the number of implanted intracardiac devices could be observed (Mond and Proclemer, 2011). At the same time, patients implanted with a device are getting older and therefore may live for decades with this foreign material embedded in the endocardium and exposed to blood flow (Leon et al. 2010; Proclemer et al. 2010). This is of relevance insofar as little is known about long-term toxic effects of implantable devices (Eliaz 2019; Nasakina et al. 2019). Furthermore, at present it is not known with certainty whether there is a "critical concentration" of the metallic components which, if exceeded, can be expected to cause consequential damage to health. Nitinol, by far the most commonly used alloy for intracardiac devices, appears to have a number of good properties especially during implantation and durability in long-term treatment. On the one hand, it is highly malleable, which is of enormous importance in the context of implantation (Stoeckel et al. 2004; Henderson et al. 2011; Maleckis et al. 2018); on the other hand, it is considered to be extremely durable with overall good tolerance (Eliaz 2019).

We know from numerous preliminary reports that as a result of intracardiac positioning, endothelialization on nitinol surfaces is expected to occur after only a few weeks (Zahn et al. 2001; Sigler et al. 2005a; Schwartz et al. 2010) and depends on numerous factors, such as the size of the device or, in the case of occluders, the primary interventional outcome (Granier et al. 2018). Incomplete endothelialization could lead to complications at site of the implantation, such as thrombus formation (Sigler et al. 2005b; Sellers et al. 2019). Moreover, patient-specific factors must be taken into consideration. For instance, there are known cases of patients with multiple allergies where excessive endothelialization was found in the autopsy (Kypta et al. 2016). Vice versa, it seems conceivable that endothelialization processes or the healing phase can be negatively influenced by the intake of immunosuppressive substances such as glucocorticoids (Radovsky et al. 1988) or TNF-α inhibitors (Sandberg et al. 2012). The same is conceivable for patients in whom cytotoxic substances or radiation therapies are used (Hopewell 1990). As a wide overall variation in anatomic conditions is to be expected in PFO, ASD, or LAA occlusions, positional control by transesophageal echocardiography appears essential (Krizanic et al. 2010; Saw et al. 2016).

Intracardiac pacemakers are a special case in this context, as no endothelial surface usually forms over the implanted cardiac device due to direct contact with the blood flow (Jana 2019). According to the manufacturers, the devices should only be anchored to the endocardium at their base, while the majority of the device should remain floating in the blood flow. However, there are now several published cases reporting the contrary. Namely, a complete or at least partial endothelialization/encapsulation of the device in the right ventricular wall (Candinas et al. 1999; Esposito et al. 2002; Tjong et al. 2015; Keiler et al. 2017). Interestingly, there are also reports from autopsies where histological processing has shown a clear evidence of inflammatory processes around the encapsulated pacemaker (Dvorak et al. 2012). Exact knowledge of any expected endothelialization is of enormous clinical relevance, since an influence on the stimulation threshold is at least conceivable through the encapsulation (Stokes et al. 1991). There are also issues of what to do in the event of battery exhaustion. An extraction, as originally intended by the developers, seems unlikely in the case of complete encapsulation. It remains to be seen whether the limited space at the surface of the inner heart is sufficient to safely implant an additional device. For the Micra pacemaker, it could be shown in an animal model that up to three devices can be implanted without any problems (Omdahl et al. 2016).

Another major uncertainty is the importance of allergies in long-term use. Nitinol is an alloy which consists of 45–50% nickel (Eliaz 2019), a relevant allergen. Nickel allergies are type IV allergies, that is, a contact allergy caused by long-term exposure, usually after 24 h to a few days (Tramontana et al. 2020), and are relatively common with a prevalence of approximately 8% to 19% in adults (Diepgen et al. 2016). Whether the allergenic potential within the blood flow is particularly high, or whether a weakening occurs once endothelialization has been achieved, remains completely unclear to date. Allergies are also known to occur with exposure to titanium, which is the other component of the nitinol alloy (Fage et al. 2016). However, the incidence is

significantly lower for titanium; consequently therefore the clinical relevance is probably of secondary importance (Grosse Meininghaus et al. 2020). In case of a confirmed allergy, it is possible to coat the device with a less/nonallergenic substance such as gold, which has already been done in individual cases (Kypta et al. 2015). The measurement of any metal released from cardiac devices has already been performed (Ries et al. 2003; Saylor et al. 2018). An open question for the future will be whether there are measurable parameters that can be used to estimate the degree of endothelialization. For example, it is conceivable that the metallic components could be measured as nanoparticles in peripheral blood, but their concentration would decrease during the healing phase. It may also be assumed, that with complete endothelialization achieved, the metal content might fall below the detection limit, which in turn provides important additional information for the estimated duration of the healing process.

6 Conclusion

In summary, we conclude that the vast majority of intracardiac devices meet very high safety standards from a hemostasiological point of view, and that there is currently no evidence of any therapy-limiting toxic effects in long-term treatment. Nitinol, as a component of many devices, is of particular importance in this context. However, there are currently gaps in knowledge for patients who are under immunosuppressive medication. Moreover, the impact of an optimal implantation technique on the initial healing phase and endothelialization phase has not been fully understood in many cases.

For these reasons, it seems indispensable to us that patients continue to be treated at the respective healthcare center after primary implantation and are followed up by means of imaging, so that individualized coagulation management can be determined if necessary.

7   Limitations

We could only include devices which are already in broad clinical use. Nevertheless, a number of less-established devices in the field of interventional cardiology might already have received market approval or might be in the preclinical testing phase. However, in our opinion, the current knowledge of the established devices provide a fundamental basis for the above review and the obtained conclusions.

Contributions C.E planned and coordinated the study, compiled and analyzed data, wrote the manuscript, and contributed in the final submission. V.P., E.B., S.H.K., and M.B. analyzed data, prepared figures, and contributed to manuscript preparation. F.K. analyzed data and contributed to final submission. V.D. contributed in data acquisition. K.K. carried out English-language editing of the paper. E.L., C.B., and U.C.H. revised the article critically for the content. M.L. planned the study and provided the final approval of the article.

Acknowledgments This research did not receive any specific grant from funding agencies in the public, commercial, or not-for-profit sectors. We would like to express our gratitude to the companies mentioned in the journal for permission to reprint the product images.

References

Akpek M, Kaya MG, Lam YY, Sahin O, Elcik D, Celik T, Ergin A, Gibson CM (2012) Relation of neutrophil/lymphocyte ratio to coronary flow to in-hospital major adverse cardiac events in patients with ST-elevated myocardial infarction undergoing primary coronary intervention. Am J Cardiol 110:621–627

Alli O, Holmes DR (2015) Left atrial appendage occlusion for stroke prevention. Curr Probl Cardiol 40:429–476

Andrassy M, Volz HC, Igwe JC, Funke B, Eichberger SN, Kaya Z, Buss S, Autschbach F, Pleger ST, Lukic IK (2008) High-mobility group box-1 in ischemia-reperfusion injury of the heart. Circulation 117:3216

Anderson RH (2016) Development of the atrial septum. Heart 102:481

Anderson RH, Brown NA, Webb S (2002) Development and structure of the atrial septum. Heart 88:104–110

Ar IG, Senning A, Siegenthaler WE (1979) Non-operative dilatation of coronary artery stenosis. N Engl J Med 301:61–68

Arslan F, De Kleijn DP, Pasterkamp G (2011) Innate immune signaling in cardiac ischemia. Nat Rev Cardiol 8:292

Arslan F, Smeets MB, O'Neill LA, Keogh B, Mcguirk P, Timmers L, Tersteeg C, Hoefer IE, Doevendans PA, Pasterkamp G, De Kleijn DP (2010) Myocardial ischemia/reperfusion injury is mediated by leukocytic toll-like receptor-2 and reduced by systemic administration of a novel anti-toll-like receptor-2 antibody. Circulation 121:80–90

Asahara T, Murohara T, Sullivan A, Silver M, Van Der Zee R, Li T, Witzenbichler B, Schatteman G, Isner JM (1997) Isolation of putative progenitor endothelial cells for angiogenesis. Science 275:964–966

Balasundaram RP, Anandaraja S, Juneja R et al (2005) Infective endocarditis following implantation of amplatzer atrial septal occluder. Indian Heart J 57:167–169

Balistreri CR, Buffa S, Pisano C, Lio D, Ruvolo G, Mazzesi G (2015) Are endothelial progenitor cells the real solution for cardiovascular diseases? Focus on controversies and perspectives. Biomed Res Int:2015

Beltrami AP, Urbanek K, Kajstura J, Yan S-M, Finato N, Bussani R, Nadal-Ginard B, Silvestri F, Leri A, Beltrami CA (2001) Evidence that human cardiac myocytes divide after myocardial infarction. N Engl J Med 344:1750–1757

Bergmann O, Bhardwaj RD, Bernard S, Zdunek S, Barnabé-Heider F, Walsh S, Zupicich J, Alkass K, Buchholz BA, Druid H (2009) Evidence for cardiomyocyte renewal in humans. Science 324:98–102

Bianchi ME, Crippa MP, Manfredi AA, Mezzapelle R, Rovere Querini P, Venereau E (2017) High-mobility group box 1 protein orchestrates responses to tissue damage via inflammation, innate and adaptive immunity, and tissue repair. Immunol Rev 280:74–82

Bruch L, Parsi A, Grad MO, Rux S, Burmeister T, Krebs H, Kleber FX (2002) Transcatheter closure of interatrial communications for secondary prevention of paradoxical embolism: single-center experience. Circulation 105:2845–2848

Buijs JOD, Musters M, Verrips T, Post JA, Braam B, Van Riel N (2004) Mathematical modeling of vascular endothelial layer maintenance: the role of endothelial cell division, progenitor cell homing, and telomere shortening. Am J Phys Heart Circ Phys 287:H2651–H2658

Caiado F, Real C, Carvalho T, Dias S (2008) Notch pathway modulation on bone marrow-derived vascular precursor cells regulates their angiogenic and wound healing potential. PLoS One 3:e3752

Campbell M (1970) Natural history of atrial septal defect. Heart 32:820–826

Candinas R, Duru F, Schneider J et al (1999) Postmortem analysis of encapsulation around long-term ventricular endocardial pacing leads. Mayo Clin Proc 74:120–125

Cardoso CO, Rossi Filho RI, Machado PR et al (2007) Effectiveness of the Amplatzer device for transcatheter closure of an ostium secundum atrial septal defect. Arq Bras Cardiol 88:384–389

Chakos A, Wilson-Smith A, Arora S et al (2017) Long term outcomes of transcatheter aortic valve implantation (TAVI): a systematic review of 5-year survival and beyond. Ann Cardiothorac Surg 6:432–443

Chan JKY, Patel J, Seppanen E, Chong MS, Yeo JS, Teo EY, Fisk NM, Khosrotehrani K (2013) Prospective surface marker-based isolation and expansion of fetal endothelial colony-forming cells from human term placenta. Stem Cells Transl Med 2(11):839–847

Chavakis T, Bierhaus A, Nawroth PP (2004) RAGE (receptor for advanced glycation end products): a central player in the inflammatory response. Microbes Infect 6:1219–1225

Chen W, Frangogiannis NG (2013) Fibroblasts in post-infarction inflammation and cardiac repair. Biochimica et Biophysica Acta (BBA)-Mol Cell Res 1833:945–953

Cheng B, Chen H, Chou I, Tang TW, Hsieh PC (2017) Harnessing the early post-injury inflammatory responses for cardiac regeneration. J Biomed Sci 24:1–9

Chessa M, Carminati M, Cao QL et al (2002) Transcatheter closure of congenital and acquired muscular ventricular septal defects using the Amplatzer device. J Invasive Cardiol 14:322–327

Corbett SA, Schwarzbauer JE (1998) Fibronectin–fibrin cross-linking: a regulator of cell behavior. Trends Cardiovasc Med 8:357–362

Cribier A, Eltchaninoff H, Bash A et al (2002) Percutaneous transcatheter implantation of an aortic valve prosthesis for calcific aortic stenosis: first human case description. Circulation 106:3006–3008

Čulić O, Eraković V, Čepelak I, Barišić K, Brajša K, Ferenčić Ž, Galović R, Glojnarić I, Manojlović Z, Munić V (2002) Azithromycin modulates neutrophil function and circulating inflammatory mediators in healthy human subjects. Eur J Pharmacol 450:277–289

Deb A, Ubil E (2014) Cardiac fibroblast in development and wound healing. J Mol Cell Cardiol 70:47–55

Debrunner M, Schuiki E, Minder E, Straumann E, Naegeli B, Mury R, Bertel O, Frielingsdorf J (2008) Proinflammatory cytokines in acute myocardial infarction with and without cardiogenic shock. Clin Res Cardiol 97:298–305

Della Rocca DG, Prete AD, Di Biase L, Horton RP, Al-Ahmad A, Bassiouny M, Mohanty S, Trivedi C, Romero J, Gianni C, Burkhardt JD, Gallinghouse GJ, Sanchez JE, Versaci F, Natale A (2019) Current endocardial approaches for left atrial appendage closure. Eur J Arrhythm Electrophysiol 5(1):40–46

Dewald O, Ren G, Duerr GD, Zoerlein M, Klemm C, Gersch C, Tincey S, Michael LH, Entman ML, Frangogiannis NG (2004) Of mice and dogs: species-specific differences in the inflammatory response following myocardial infarction. Am J Pathol 164:665–677

Dewald O, Zymek P, Winkelmann K, Koerting A, Ren G, Abou-Khamis T, Michael LH, Rollins BJ, Entman ML, Frangogiannis NG (2005) CCL2/Monocyte Chemoattractant Protein-1 regulates inflammatory

responses critical to healing myocardial infarcts. Circ Res 96:881–889

Diepgen TL, Ofenloch RF, Bruze M et al (2016) Prevalence of contact allergy in the general population in different European regions. Br J Dermatol 174:319–329

Dobaczewski M, Chen W, Frangogiannis NG (2011) Transforming growth factor (TGF)-β signaling in cardiac remodeling. J Mol Cell Cardiol 51:600–606

Dobaczewski M, Gonzalez-Quesada C, Frangogiannis NG (2010a) The extracellular matrix as a modulator of the inflammatory and reparative response following myocardial infarction. J Mol Cell Cardiol 48:504–511

Dobaczewski M, Xia Y, Bujak M, Gonzalez-Quesada C, Frangogiannis NG (2010b) CCR5 signaling suppresses inflammation and reduces adverse remodeling of the infarcted heart, mediating recruitment of regulatory T cells. Am J Pathol 176:2177–2187

Duperray A, Mantovani A, Introna M, Dejana E (1995) Endothelial cell regulation of leukocyte infiltration in inflammatory tissues. Mediat Inflamm 4(5):322–330

Dvorak P, Novak M, Kamaryt P et al (2012) Histological findings around electrodes in pacemaker and implantable cardioverter-defibrillator patients: comparison of steroid-eluting and non-steroid-eluting electrodes. Europace 14:117–123

Edlinger C, Paar V, Tuscher T et al (2019) Potential mechanisms of endothelialisation in individuals implanted with a leadless pacemaker systems: an experimental in vitro study. J Electrocardiol 55:72–77

Edlinger C, Krizanic F, Butter C, Bannehr M, Neuss M, Fejzic D, Hoppe UC, Lichtenauer M (2020) Economic assessment of traditional surgical valve replacement versus use of transfemoral intervention in degenerative aortic stenosis. Minerva Med 112(3):372–383

El Faquir N, Rocatello G, Rahhab Z, Bosmans J, De Backer O, Van Mieghem NM, Mortier P, De Jaegere PP (2020) Differences in clinical valve size selection and valve size selection for patient-specific computer simulation in transcatheter aortic valve replacement (TAVR): a retrospective multicenter analysis. Int J Card Imaging 36:123–129

Eliaz N (2019) Corrosion of metallic biomaterials: a review. Materials (Basel) 12(3):407

Entman ML, Youker K, Shoji T, Kukielka G, Shappell SB, Taylor AA, Smith CW (1992) Neutrophil induced oxidative injury of cardiac myocytes. A compartmented system requiring CD11b/CD18-ICAM-1 adherence. J Clin Invest 90:1335–1345

Esposito M, Kennergren C, Holmström N et al (2002) Morphologic and immunohistochemical observations of tissues surrounding retrieved transvenous pacemaker leads. J Biomed Mater Res 63:548–558

Fage SW, Muris J, Jakobsen SS et al (2016) Titanium: a review on exposure, release, penetration, allergy, epidemiology, and clinical reactivity. Contact Dermatitis 74:323–345

Fischer G, Stieh J, Uebing A et al (2003) Experience with transcatheter closure of secundum atrial septal defects using the Amplatzer septal occluder: a single Centre study in 236 consecutive patients. Heart 89:199–204

Flick MJ, Du X, Witte DP, Jiroušková M, Soloviev DA, Busuttil SJ, Plow EF, Degen JL (2004) Leukocyte engagement of fibrin (ogen) via the integrin receptor α M β 2/Mac-1 is critical for host inflammatory response in vivo. J Clin Invest 113:1596–1606

Forssmann-Falck R (1997) Werner Forssmann: a pioneer of cardiology. Am J Cardiol 79:651–660

Fountain RB, Holmes DR, Chandrasekaran K et al (2006) The PROTECT AF (WATCHMAN left atrial appendage system for embolic PROTECTion in patients with atrial fibrillation) trial. Am Heart J 151:956–961

Frangogiannis NG (2012) Regulation of the inflammatory response in cardiac repair. Circ Res 110:159–173

Frangogiannis NG (2014) The inflammatory response in myocardial injury, repair, and remodelling. Nat Rev Cardiol 11:255–265

Frangogiannis NG (2017) The extracellular matrix in myocardial injury, repair, and remodeling. J Clin Invest 127:1600–1612

Frangogiannis NG, Lindsey ML, Michael LH, Youker KA, Bressler RB, Mendoza LH, Spengler RN, Smith CW, Entman ML (1998) Resident cardiac mast cells degranulate and release preformed TNF-α, initiating the cytokine cascade in experimental canine myocardial ischemia/reperfusion. Circulation 98:699–710

Frangogiannis NG, Mendoza LH, Lindsey ML, Ballantyne CM, Michael LH, Smith CW, Entman ML (2000) IL-10 is induced in the reperfused myocardium and may modulate the reaction to injury. J Immunol 165:2798–2808

Geissmann F, Jung S, Littman DR (2003) Blood monocytes consist of two principal subsets with distinct migratory properties. Immunity 19:71–82

Gilles S, Zahler S, Welsch U, Sommerhoff CP, Becker BF (2003) Release of TNF-α during myocardial reperfusion depends on oxidative stress and is prevented by mast cell stabilizers. Cardiovasc Res 60:608–616

Gloire G, Legrand-Poels S, Piette J (2006) NF-κB activation by reactive oxygen species: fifteen years later. Biochem Pharmacol 72:1493–1505

Goli A, Shroff S, Osman MN, Lucke J (2012) A case of gold-coated pacemaker for pacemaker allergy. The Journal 945

Gong Y, Koh D-R (2010) Neutrophils promote inflammatory angiogenesis via release of preformed VEGF in an in vivo corneal model. Cell Tissue Res 339:437–448

González-Rosa JM, Martín V, Peralta M, Torres M, Mercader N (2011) Extensive scar formation and regression during heart regeneration after cryoinjury in zebrafish. Development 138:1663–1674

Granier M, Laugaudin G, Massin F et al (2018) Occurrence of incomplete endothelialization causing residual permeability after left atrial appendage closure. J Invasive Cardiol 30:245–250

Gretzer C, Emanuelsson L, Liljensten E, Thomsen P (2006) The inflammatory cell influx and cytokines changes during transition from acute inflammation to fibrous repair around implanted materials. J Biomater Sci Polym Ed 17:669–687

Greutmann M, Greutmann-Yantiri M, Kretschmar O, Senn O, Roffi M, Jenni R, Luescher TF, Eberli FR (2009) Percutaneous PFO closure with Amplatzer PFO occluder: predictors of residual shunts at 6 months follow-up. Congenit Heart Dis 4:252–257

Gristina AG (1994) Implant failure and the immuno-incompetent fibro-inflammatory zone. Clin Orthop Relat Res 298:106–118

Grosse Meininghaus D, Kruells-Muench J, Peltroche-Llacsahuanga H (2020) First-in-man implantation of a gold-coated biventricular defibrillator: difficult differential diagnosis of metal hypersensitivity reaction vs chronic device infection. HeartRhythm Case Rep 6:304–307

Hagen PT, Scholz DG, Edwards WD (1984) Incidence and size of patent foramen ovale during the first 10 decades of life: an autopsy study of 965 normal hearts. Mayo Clin Proc 59:17–20

Hanna RN, Carlin LM, Hubbeling HG, Nackiewicz D, Green AM, Punt JA, Geissmann F, Hedrick CC (2011) The transcription factor NR4A1 (Nur77) controls bone marrow differentiation and the survival of Ly6C− monocytes. Nat Immunol 12:778

Haynes DR, Boyle SJ, Rogers SD et al (1998) Variation in cytokines induced by particles from different prosthetic materials. Clin Orthop Relat Res:223–230

Henderson E, Nash DH, Dempster WM (2011) On the experimental testing of fine Nitinol wires for medical devices. J Mech Behav Biomed Mater 4:261–268

Hettinger J, Richards DM, Hansson J, Barra MM, Joschko A-C, Krijgsveld J, Feuerer M (2013) Origin of monocytes and macrophages in a committed progenitor. Nat Immunol 14:821–830

Hofmann U, Frantz S (2015) Role of lymphocytes in myocardial injury, healing, and remodeling after myocardial infarction. Circ Res 116:354–367

Hopewell J (1990) The skin: its structure and response to ionizing radiation. Int J Radiat Biol 57:751–773

Ingram DA, Mead LE, Tanaka H, Meade V, Fenoglio A, Mortell K, Pollok K, Ferkowicz MJ, Gilley D, Yoder MC (2004) Identification of a novel hierarchy of endothelial progenitor cells using human peripheral and umbilical cord blood. Blood 104:2752–2760

Ipach I, Schäfer R, Mittag F et al (2012) The development of whole blood titanium levels after instrumented spinal fusion - is there a correlation between the number of fused segments and titanium levels? BMC Musculoskelet Disord 13:159

Ito M, Tada N, Hata M (2017) Balloon repositioning of Transcatheter Aortic Valve after migration into the left ventricular outflow tract, followed by Valve-in-Valve procedure. Tex Heart Inst J 44:274–278

Jana S (2019) Endothelialization of cardiovascular devices. Acta Biomater 99:53–71

Jose J, Richardt G, Abdel-Wahab M (2015) Balloon- or self-expandable TAVI: clinical equipoise? Interv Cardiol 10:103–108

Kawaguchi M, Takahashi M, Hata T, Kashima Y, Usui F, Morimoto H, Izawa A, Takahashi Y, Masumoto J,

Koyama J (2011) Inflammasome activation of cardiac fibroblasts is essential for myocardial ischemia/reperfusion injury. Circulation 123:594–604

Keiler J, Schulze M, Sombetzki M et al (2017) Neointimal fibrotic lead encapsulation - clinical challenges and demands for implantable cardiac electronic devices. J Cardiol 70:7–17

Klune JR, Dhupar R, Cardinal J, Billiar TR, Tsung A (2008) HMGB1: endogenous danger signaling. Mol Med 14:476–484

Knepp MD, Rocchini AP, Lloyd TR et al (2010) Long-term follow up of secundum atrial septal defect closure with the amplatzer septal occluder. Congenit Heart Dis 5:32–37

Komar M, Przewlocki T, Olszowska M, Sobien B, Podolec P (2014) The benefit of atrial septal defect closure in elderly patients. Clin Interv Aging 9:1101

Kono H, Kimura Y, Latz E (2014) Inflammasome activation in response to dead cells and their metabolites. Curr Opin Immunol 30:91–98

Kono H, Rock KL (2008) How dying cells alert the immune system to danger. Nat Rev Immunol 8:279–289

Kramer DB, Kesselheim AS (2015) The Watchman saga–closure at last? N Engl J Med 372:994–995

Krizanic F, Sievert H, Pfeiffer D et al (2010) The Occlutech Figulla PFO and ASD occluder: a new nitinol wire mesh device for closure of atrial septal defects. J Invasive Cardiol 22:182–187

Kumar P, Sarkar A, Kar SK (2019) Assessment of ventricular function in patients of atrial septal defect by strain imaging before and after correction. Ann Card Anaesth 22:41

Kypta A, Blessberger H, Kammler J, Lichtenauer M, Lambert T, Silye R, Steinwender C (2016) First autopsy description of changes 1 year after implantation of a leadless cardiac pacemaker: unexpected ingrowth and severe chronic inflammation. Can J Cardiol 32:1578. e1571–1578. e1572

Kypta A, Blessberger H, Lichtenauer M, Lambert T, Kammler J, Steinwender C (2015) Gold-coated pacemaker implantation for a patient with type IV allergy to titanium. Indian Pacing Electrophysiol J 15:291–292

Lacey DC, De Kok B, Clanchy FI et al (2009) Low dose metal particles can induce monocyte/macrophage survival. J Orthop Res 27:1481–1486

Lai S-L, Marín-Juez R, Stainier DY (2019) Immune responses in cardiac repair and regeneration: a comparative point of view. Cell Mol Life Sci 76:1365–1380

Latz E, Xiao TS, Stutz A (2013) Activation and regulation of the inflammasomes. Nat Rev Immunol 13:397–411

Le Gloan L, Legendre A, Iserin L, Ladouceur M (2018) Pathophysiology and natural history of atrial septal defect. J Thorac Dis 10:S2854

Lee PH, Song JK, Kim JS et al (2018) Cryptogenic stroke and high-risk patent foramen Ovale: the DEFENSE-PFO trial. J Am Coll Cardiol 71:2335–2342

Leon MB, Smith CR, Mack M et al (2010) Transcatheter aortic-valve implantation for aortic stenosis in patients who cannot undergo surgery. N Engl J Med 363:1597–1607

Li D-W, Liu Z-Q, Wei J, Liu Y, Hu L-S (2012) Contribution of endothelial progenitor cells to neovascularization. Int J Mol Med 30:1000–1006

Luttikhuizen DT, Harmsen MC, Luyn MJV (2006) Cellular and molecular dynamics in the foreign body reaction. Tissue Eng 12:1955–1970

Madhkour R, Wahl A, Praz F, Meier B (2019) Amplatzer patent foramen ovale occluder: safety and efficacy. Expert Rev Med Devices 16:173–182

Maleckis K, Anttila E, Aylward P et al (2018) Nitinol stents in the Femoropopliteal artery: a mechanical perspective on material, design, and performance. Ann Biomed Eng 46:684–704

Martinez FO, Gordon S (2014) The M1 and M2 paradigm of macrophage activation: time for reassessment. F1000prime Rep 6

Mas JL, Derumeaux G, Chatellier G (2017) Trials of patent foramen Ovale closure. N Engl J Med 377:2599–2600

Masura J, Gavora P, Formanek A et al (1997) Transcatheter closure of secundum atrial septal defects using the new self-centering amplatzer septal occluder: initial human experience. Catheter Cardiovasc Diagn 42:388–393

Masura J, Gavora P, Podnar T (2005) Long-term outcome of transcatheter secundum-type atrial septal defect closure using Amplatzer septal occluders. J Am Coll Cardiol 45:505–507

Mcdonald B, Pittman K, Menezes GB, Hirota SA, Slaba I, Waterhouse CC, Beck PL, Muruve DA, Kubes P (2010) Intravascular danger signals guide neutrophils to sites of sterile inflammation. Science 330:362–366

Meerkin D, Butnaru A, Dratva D et al (2013) Early safety of the Amplatzer cardiac plug™ for left atrial appendage occlusion. Int J Cardiol 168:3920–3925

Meier B (2005) Closure of patent foramen ovale: technique, pitfalls, complications, and follow up. Heart 91:444–448

Mocatta TJ, Pilbrow AP, Cameron VA, Senthilmohan R, Frampton CM, Richards AM, Winterbourn CC (2007) Plasma concentrations of myeloperoxidase predict mortality after myocardial infarction. J Am Coll Cardiol 49:1993–2000

Mollova M, Bersell K, Walsh S, Savla J, Das LT, Park S-Y, Silberstein LE, Dos Remedios CG, Graham D, Colan S (2013) Cardiomyocyte proliferation contributes to heart growth in young humans. Proc Natl Acad Sci 110:1446–1451

Mond HG, Proclemer A (2011) The 11th world survey of cardiac pacing and implantable cardioverter-defibrillators: calendar year 2009–a world Society of Arrhythmia's project. Pacing Clin Electrophysiol 34:1013–1027

Moons P, Sluysmans T, De Wolf D et al (2009) Congenital heart disease in 111 225 births in Belgium: birth prevalence, treatment and survival in the 21st century. Acta Paediatr 98:472–477

Mund JA, Estes ML, Yoder MC, Ingram DA Jr, Case J (2012) Flow cytometric identification and functional characterization of immature and mature circulating endothelial cells. Arterioscler Thromb Vasc Biol 32:1045–1053

Muzio M, Polntarutti N, Bosisio D, Prahladan M, Mantovani A (2000) Toll like receptor family (TLT) and signalling pathway. Eur Cytokine Netw 11:489–490

Nahrendorf M, Pittet MJ, Swirski FK (2010) Monocytes: protagonists of infarct inflammation and repair after myocardial infarction. Circulation 121:2437–2445

Nahrendorf M, Swirski FK, Aikawa E, Stangenberg L, Wurdinger T, Figueiredo J-L, Libby P, Weissleder R, Pittet MJ (2007) The healing myocardium sequentially mobilizes two monocyte subsets with divergent and complementary functions. J Exp Med 204:3037–3047

Nasakina EO, Sudarchikova MA, Sergienko KV et al (2019) Ion release and surface characterization of nanostructured Nitinol during long-term testing. Nanomaterials (Basel):9

Nashat H, Montanaro C, Li W, Kempny A, Wort SJ, Dimopoulos K, Gatzoulis MA, Babu-Narayan SV (2018) Atrial septal defects and pulmonary arterial hypertension. J Thorac Dis 10:S2953

Nassif M, Abdelghani M, Bouma BJ, Straver B, Blom NA, Koch KT, Tijssen JG, Mulder BJ, De Winter RJ (2016) Historical developments of atrial septal defect closure devices: what we learn from the past. Expert Rev Med Devices 13:555–568

Neuser J, Akin M, Bavendiek U, Kempf T, Bauersachs J, Widder JD (2016) Mid-term results of interventional closure of patent foramen ovale with the Occlutech Figulla® Flex II Occluder. BMC Cardiovasc Disord 16:1–7

Nicholls M (2020) André F. Cournand for cardiac catheterization: mark Nicholls focusses on the role of Professor André F. Cournand in the development of cardiac catheterization and the award of the 1956 Nobel Prize. Oxford University Press

Nishimura M, Naito S (2005) Tissue-specific mRNA expression profiles of human toll-like receptors and related genes. Biol Pharm Bull 28:886–892

Omdahl P, Eggen MD, Bonner MD et al (2016) Right ventricular anatomy can accommodate multiple Micra Transcatheter pacemakers. Pacing Clin Electrophysiol 39:393–397

Östberg AK, Dahlgren U, Sul YT et al (2015) Inflammatory cytokine release is affected by surface morphology and chemistry of titanium implants. J Mater Sci Mater Med 26:155

Pase L, Layton JE, Wittmann C, Ellett F, Nowell CJ, Reyes-Aldasoro CC, Varma S, Rogers KL, Hall CJ, Keightley MC (2012) Neutrophil-delivered myeloperoxidase dampens the hydrogen peroxide burst after tissue wounding in zebrafish. Curr Biol 22:1818–1824

Pedra CA, Pedra SF, Costa RN, Ribeiro MS, Nascimento W, Campanhã LOS, Santana MVT, Jatene IB, Assef JE, Fontes VF (2016) Mid-term outcomes

after percutaneous closure of the secundum atrial septal defect with the Figulla-Occlutech Device. J Interv Cardiol 29:208–215

Peuster M, Fink C, Von Schnakenburg C (2003) Biocompatibility of corroding tungsten coils: in vitro assessment of degradation kinetics and cytotoxicity on human cells. Biomaterials 24:4057–4061

Podnar T, Martanovic P, Gavora P et al (2001) Morphological variations of secundum-type atrial septal defects: feasibility for percutaneous closure using Amplatzer septal occluders. Catheter Cardiovasc Interv 53:386–391

Proclemer A, Ghidina M, Gregori D et al (2010) Trend of the main clinical characteristics and pacing modality in patients treated by pacemaker: data from the Italian pacemaker registry for the quinquennium 2003-07. Europace 12:202–209

Radovsky AS, et al (1988) Paired comparisons of steroid-eluting and nonsteroid endocardial pacemaker leads in dogs: electrical performance and morphologic alterations. Pacing Clin Electrophysiol 11(7):1085–1094. https://doi.org/10.1111/j.1540-8159.1988.tb03955.x. PMID: 2457888

Reddy VY, Exner DV, Cantillon DJ et al (2015) Percutaneous implantation of an entirely Intracardiac leadless pacemaker. N Engl J Med 373:1125–1135

Rehman J, Li J, Orschell CM, March KL (2003) Peripheral blood "endothelial progenitor cells" are derived from monocyte/macrophages and secrete angiogenic growth factors. Circulation 107:1164–1169

Renker M, Kim WK (2020) Choice of transcatheter heart valve: should we select the device according to each patient's characteristics or should it be "one valve fits all"? Ann Transl Med 8:961

Reynolds D, Duray GZ, Omar R, Soejima K, Neuzil P, Zhang S, Narasimhan C, Steinwender C, Brugada J, Lloyd M, Roberts PR, Sagi V, Hummel J, Bongiorni MG, Knops RE, Ellis CR, Gornick CC, Bernabei MA, Laager V, Stromberg K, Williams ER, Hudnall JH, Ritter P (2016) Micra Transcatheter Pacing Study Group A leadless intracardiac transcatheter pacing system. N Engl J Med 374:533–541. https://doi.org/10.1056/NEJMoa1511643

Ries MW, Kampmann C, Rupprecht HJ et al (2003) Nickel release after implantation of the Amplatzer occluder. Am Heart J 145:737–741

Rotman OM, Bianchi M, Ghosh RP, Kovarovic B, Bluestein D (2018) Principles of TAVR valve design, modelling, and testing. Expert Rev Med Devices 15:771–791

Ryhänen J, Kallioinen M, Tuukkanen J et al (1998) In vivo biocompatibility evaluation of nickel-titanium shape memory metal alloy: muscle and perineural tissue responses and encapsule membrane thickness. J Biomed Mater Res 41:481–488

Sadiq M, Kazmi T, Rehman AU et al (2012) Device closure of atrial septal defect: medium-term outcome with special reference to complications. Cardiol Young 22:71–78

Sandberg O, et al (2012) Etanercept does not impair healing in rat models of tendon or metaphyseal bone injury. Acta Orthop 83(3):305–310. https://doi.org/10.3109/17453674.2012.693018. Epub 2012 May 23. PMID: 22616743; PMCID: PMC3369160

Sattler S, Rosenthal N (2016) The neonate versus adult mammalian immune system in cardiac repair and regeneration. Biochimica et Biophysica Acta (BBA)-Mol Cell Res 1863:1813–1821

Saver JL, Committee RTS (2017) Trials of patent foramen Ovale closure. N Engl J Med 377:2600

Saw J, Lopes JP, Reisman M et al (2016) Cardiac computed tomography angiography for left atrial appendage closure. Can J Cardiol 32:1033.e1031–1033.e1039

Saylor DM, Craven BA, Chandrasekar V et al (2018) Predicting patient exposure to nickel released from cardiovascular devices using multi-scale modeling. Acta Biomater 70:304–314

Scalise F, Auguadro C, Sorropago G et al (2016) Long-term contrast echocardiography and clinical follow-up after percutaneous closure of patent foramen Ovale using two different atrial septal Occluder devices. J Interv Cardiol 29:406–413

Schwartz RS, Holmes DR, Van Tassel RA et al (2010) Left atrial appendage obliteration: mechanisms of healing and intracardiac integration. JACC Cardiovasc Interv 3:870–877

Sellers SL, Blanke P, Leipsic JA (2019) Bioprosthetic heart valve degeneration and dysfunction: focus on mechanisms and multidisciplinary imaging considerations. Radiol Cardiothorac Imaging 1:e190004

Shayan M, Chun Y (2015) An overview of thin film nitinol endovascular devices. Acta Biomater 21:20–34. https://doi.org/10.1016/j.actbio.2015.03.025. Epub 2015 Mar 31

Shukla R, Bansal V, Chaudhary M et al (2005) Biocompatibility of gold nanoparticles and their endocytotic fate inside the cellular compartment: a microscopic overview. Langmuir 21:10644–10654

Sievert H, Babic UU, Hausdorf G et al (1998) Transcatheter closure of atrial septal defect and patent foramen ovale with ASDOS device (a multi-institutional European trial). Am J Cardiol 82:1405–1413

Sigler M, Bartmus D, Paul T (2005a) Histology of a surgically removed stenotic modified Blalock-Taussig shunt after previous endovascular stenting. Heart 91:1097

Sigler M, Paul T, Grabitz RG (2005b) Biocompatibility screening in cardiovascular implants. Z Kardiol 94:383–391

Sims GP, Rowe DC, Rietdijk ST, Herbst R, Coyle AJ (2009) HMGB1 and RAGE in inflammation and cancer. Annu Rev Immunol 28:367–388

Singh M, Saini HK (2003) Resident cardiac mast cells and ischemia-reperfusion injury. J Cardiovasc Pharmacol Ther 8:135–148

Smiley ST, King JA, Hancock WW (2001) Fibrinogen stimulates macrophage chemokine secretion through toll-like receptor 4. J Immunol 167:2887–2894

Soehnlein O, Lindbom L (2010) Phagocyte partnership during the onset and resolution of inflammation. Nat Rev Immunol 10:427–439

Søndergaard L, Kasner SE, Rhodes JF et al (2017) Patent foramen Ovale closure or antiplatelet therapy for cryptogenic stroke. N Engl J Med 377:1033–1042

Stoeckel D, Pelton A, Duerig T (2004) Self-expanding nitinol stents: material and design considerations. Eur Radiol 14:292–301

Stokes KB, Bird T, Gunderson B (1991) The mythology of threshold variations as a function of electrode surface area. Pacing Clin Electrophysiol 14:1748–1751

Sun JS, Lin FH, Tsuang YH et al (2000) Effect of anti-inflammatory medication on monocyte response to titanium particles. J Biomed Mater Res 52:509–516

Talman V, Ruskoaho H (2016) Cardiac fibrosis in myocardial infarction—from repair and remodeling to regeneration. Cell Tissue Res 365:563–581

Tang L, Jennings TA, Eaton JW (1998) Mast cells mediate acute inflammatory responses to implanted biomaterials. Proc Natl Acad Sci 95:8841–8846

Thannickal VJ, Fanburg BL (2000) Reactive oxygen species in cell signaling. Am J Phys Lung Cell Mol Phys 279:L1005–L1028

Timmers L, Sluijter JP, Van Keulen JK, Hoefer IE, Nederhoff MG, Goumans M-J, Doevendans PA, Van Echteld CJ, Joles JA, Quax PH (2008) Toll-like receptor 4 mediates maladaptive left ventricular remodeling and impairs cardiac function after myocardial infarction. Circ Res 102:257–264

Tjong FV, Stam OC, Van Der Wal AC et al (2015) Postmortem histopathological examination of a leadless pacemaker shows partial encapsulation after 19 months. Circ Arrhythm Electrophysiol 8:1293–1295

Tramontana M, Bianchi L, Hansel K et al (2020) Nickel allergy: epidemiology, Pathomechanism, clinical patterns, treatment and prevention programs. Endocr Metab Immune Disord Drug Targets 20:992–1002

Ungersboeck A, Geret V, Pohler O et al (1995) Tissue reaction to bone plates made of pure titanium: a prospective, quantitative clinical study. J Mater Sci Mater Med 6:223–229

Wan E, Yeap XY, Dehn S, Terry R, Novak M, Zhang S, Iwata S, Han X, Homma S, Drosatos K (2013) Enhanced efferocytosis of apoptotic cardiomyocytes through myeloid-epithelial-reproductive tyrosine kinase links acute inflammation resolution to cardiac repair after infarction. Circ Res 113:1004–1012

Wang R, Kawashima H, Mylotte D, Rosseel L, Gao C, Aben J-P, Abdelshafy M, Onuma Y, Yang J, Soliman O (2021) Quantitative angiographic assessment of aortic regurgitation after transcatheter implantation of the Venus A-valve: comparison with other self-expanding valves and impact of a learning curve in a single Chinese center. Glob Heart 16(1):20

Weirather J, Hofmann UD, Beyersdorf N, Ramos GC, Vogel B, Frey A, Ertl G, Kerkau T, Frantz S (2014) Foxp3+ CD4+ T cells improve healing after myocardial infarction by modulating monocyte/macrophage differentiation. Circ Res 115:55–67

Welt FG, Rogers C (2002) Inflammation and restenosis in the stent era. Arterioscler Thromb Vasc Biol 22:1769–1776

Witten ML, Sheppard PR, Witten BL (2012) Tungsten toxicity. Chem Biol Interact 196:87–88

Yao B, La LB, Chen YC, Chang LJ, Chan EK (2012) Defining a new role of GW182 in maintaining miRNA stability. EMBO Rep 13:1102–1108

Yona S, Kim K-W, Wolf Y, Mildner A, Varol D, Breker M, Strauss-Ayali D, Viukov S, Guilliams M, Misharin A (2013) Fate mapping reveals origins and dynamics of monocytes and tissue macrophages under homeostasis. Immunity 38:79–91

Youker K, Smith CW, Anderson DC, Miller D, Michael LH, Rossen RD, Entman ML (1992) Neutrophil adherence to isolated adult cardiac myocytes. Induction by cardiac lymph collected during ischemia and reperfusion. J Clin Invest 89:602–609

Zahn EM, Wilson N, Cutright W et al (2001) Development and testing of the Helex septal occluder, a new expanded polytetrafluoroethylene atrial septal defect occlusion system. Circulation 104:711–716

Zdolsek J, Eaton JW, Tang L (2007) Histamine release and fibrinogen adsorption mediate acute inflammatory responses to biomaterial implants in humans. J Transl Med 5:1–6

Zhang Q, Hitchins VM, Schrand AM et al (2011) Uptake of gold nanoparticles in murine macrophage cells without cytotoxicity or production of pro-inflammatory mediators. Nanotoxicology 5:284–295

Zouggari Y, Ait-Oufella H, Waeckel L, Vilar J, Loinard C, Cochain C, Recalde A, Duriez M, Levy BI, Lutgens E (2009) Regulatory T cells modulate postischemic neovascularization. Circulation 120:1415

Zymek P, Bujak M, Chatila K, Cieslak A, Thakker G, Entman ML, Frangogiannis NG (2006) The role of platelet-derived growth factor signaling in healing myocardial infarcts. J Am Coll Cardiol 48: 2315–2323

Adv Exp Med Biol - Cell Biology and Translational Medicine (2022) 17: 23–55
https://doi.org/10.1007/5584_2022_717
© Springer Nature Switzerland AG 2022
Published online: 23 June 2022

Articular Cartilage Regeneration in Veterinary Medicine

Metka Voga and Gregor Majdic

Abstract

Cartilage is an avascular tissue with a limited rate of oxygen and nutrient diffusion, resulting in its inability to heal spontaneously. Articular cartilage defects eventually lead to osteoarthritis (OA), the endpoint of progressive destruction of cartilage. In companion animals, OA is the most common joint disease, and many pain management and surgical attempts have been made to find an appropriate treatment. Pain management of OA is usually the first choice of OA therapy, which is often managed with nonsteroidal anti-inflammatory drugs (NSAIDs). To avoid known negative side effects of NSAIDs, other approaches are being considered, such as the use of anti-nerve growth factor monoclonal antibodies (anti-NGF mAB), hyaluronic acid (HA), platelet-rich plasma (PRP), and mesenchymal stem cells (MSCs). The latter is increasingly being recognized as effective in reducing or even eliminating pain and lameness associated with OA. However, the in vivo mechanisms of MSC action do not relate to their differentiation potential, but rather to their immunomodulatory functions.

Achieving actual regeneration of cartilage to prevent OA from developing or even revert already existing OA condition has not yet been achieved. Several techniques have been tried to overcome cartilage's inability to regenerate, from osteochondral transplantation, autologous chondrocyte implantation (ACI), and matrix-induced ACI (MACI). Combinatory use of MSCs unique features and biomaterials is also being investigated with the aim to as much as possible recapitulate the native microenvironment of the cartilage, yet so far none of the methods have produced reliable and truly effective results. Although OA, for now, remains an incurable disease, novel techniques are being developed, rendering hope for the future accomplishment of actual cartilage regeneration. The aim of this chapter is firstly to summarize known and developing pain management options for OA, secondly to present surgical attempts to regenerate articular cartilage, and finally to present the attempts to improve existing regenerative treatment options using mesenchymal stem cells, with the vision for the possible use of developing strategies in veterinary medicine.

M. Voga and G. Majdic (✉)
Institute for Preclinical Sciences, University of Ljubljana, Ljubljana, Slovenia
e-mail: gregor.majdic@vf.uni-lj.si

Keywords

Articular cartilage · Biomaterials · Mesenchymal stem cells · Osteoarthritis · Regeneration · Veterinary medicine

Abbreviations

ACI	Autologous chondrocyte implantation
Anti-NGF mAB	Anti-nerve growth factor monoclonal antibodies
BMP2	Bone morphogenetic protein-2
CD105	Cluster of differentiation 105
CD73	Cluster of differentiation 73
CD90	Cluster of differentiation 90
CD45	Cluster of differentiation 45
CD34	Cluster of differentiation 34
CD14	Cluster of differentiation 14
CD11b	Cluster of differentiation 11b
CD79a	Cluster of differentiation 79a
CD19	Cluster of differentiation 19
COMP	Cartilage oligomeric matrix protein
ECM	Extracellular matrix
EGF	Epidermal growth factor
FGF	Fibroblast growth factor
GAGs	Glycosaminoglycans
HA	Hyaluronic acid
HIF-1α	Hypoxia-inducible factor-1alpha
HLA	Human leukocyte antigen
IGF	Insulin-like growth factor
MACI	Matrix-induced autologous chondrocyte implantation
MAT-3	Matrilin-3 protein
MMP13	Matrix metalloproteinase 13
MMP	Modified Maquet procedure
MSCs	Mesenchymal stem cells/ medicinal signaling cells
NGF	Nerve growth factor
NSAIDs	Nonsteroidal anti-inflammatory drugs
OA	Osteoarthritis
OCD	Osteochondritis dissecans
PDGF	Platelet-derived growth factor
PRP	Platelet-rich plasma
PCL	Polycaprolactone
PEG	Polyethylene glycol
PGA	Polyglycolic acid
PLA	Polylactic acid
PLGA	Polylactic-co-glycolic acid
PTHrP	Parathyroid hormone-related protein
ROS	Reactive oxygen species
RUNX2	Runt-related transcription factor 2
SOX9	SRY-box transcription factor 9
TGF-β	Transforming growth factor beta
TPLO	Tibial plateau leveling osteotomy
TSP-1	Thrombospondin-1
TTA	Tibial tuberosity advancement
VEGF	Vascular endothelial growth factor
2D	Two-dimensional
3D	Three-dimensional

1 Cartilage and Its (In)Ability to Heal

Cartilage is a connective tissue of mesodermal origin (Armiento et al. 2019). In the fetus, cartilage acts as a bone template and provides a structure for endochondral ossification (Chiara and Ranieri 2009). In the adult organism, cartilage remains in several areas in the body, such as joints, nose, ear, trachea, and intervertebral disks, playing a role of a supportive structure, shock absorber, flexibility, and movement (Hoshi et al. 2018). Four types of cartilaginous tissues are distinguished based on the cellularity, morphology, and extracellular matrix (ECM) composition: hyaline cartilage, fibrocartilage, and elastic and hypertrophic cartilage (Armiento et al. 2019). The development of certain cartilage type is dependent on the mechanical impact on the tissue. The most common is hyaline cartilage, the embryonic form of cartilage, present at the connection between the ribs and sternum, in the trachea, and on the joint surface where it resists compressive load and provides frictionless movement (Nürnberger et al. 2006). Major constituents of cartilage include a small number of cells, chondrocytes, and a large proportion of their product, ECM, embedded in an abundant interstitial fluid which represents the majority of tissue weight and is essential for joint lubrication and wear resistance (Bora Jr. and Miller 1987). In the vertebrate skeletal system, articular cartilage is highly organized (Nürnberger et al. 2006). Complex organization of articular cartilage rises from differentiation of the cartilage into four layers (superficial, middle, deep, and calcified zone), ECM compartmentalization (collagen type I predominating in the uppermost part of the zone and collagen type II in the middle and deep zone), and orientation of collagen fibers (Nürnberger et al. 2006) (forming Benninghoff arcades, oriented mostly parallel to the articulating surface with average fibril rotating through the tissue until the orientation of collagen fibers in the middle and deep zones near the interface with bone is perpendicular to the joint surface) (Benninghoff 1925). Despite well-established cartilaginous tissue structure, there are considerable variations between the species. For example, small species such as mice have higher cellularity than larger animals (Stockwell 1971), whereas cartilage thickness is higher in smaller animals (Stockwell 1971; Frisbie et al. 2006).

2 Osteoarthritis in Companion Animals

In the adult organism, cartilage lacks blood and lymph vessels, nerves, and perichondrium. Chondrocytes are thus sustained by nutrients, gases, and cytokines delivered by the synovial fluid (Stockwell 1978). Cartilage metabolism is relatively slow. Low rate of tissue turnover, ascribed to cartilage avascularity and limited rate of oxygen and nutrient diffusion from synovial fluid, results in cartilage inability to heal spontaneously (Hayes Jr. et al. 2001). Intrinsic repair mechanisms, even in minor cartilage defects, are insufficient for the regeneration of cartilage ad integrum (Nürnberger et al. 2006). Natural repairing process of hyaline cartilage results in mechanically inferior fibrocartilage that in comparison to hyaline cartilage contains high levels of type I collagen and only a small portion of glycosaminoglycans (GAGs) and collagen type II, making it less resilient to wear, with higher-friction motion between bones (Armiento et al. 2019). Cartilage injuries may often appear asymptomatic but symptoms appear with progressive cartilage destruction (Mehana et al. 2019; Janakiramanan et al. 2006). The loss and dysfunction of articular cartilage eventually lead to

osteoarthritis (OA), a clinical and pathological endpoint of progressive cartilage destruction, affecting both humans and animals worldwide. OA is a slowly progressing degenerative joint disease characterized by whole joint structural changes including varying degrees of osteophyte formation, subchondral bone change, and synovitis, leading to pain and loss of joint function (Dieppe and Lohmander 2005; Enomoto et al. 2019). OA is the most common joint disease in companion animals, especially dogs and horses (Gencoglu et al. 2020) and also geriatric cats (Clarke et al. 2005). Risk factors for OA in dogs are associated with genetics, breed and conformational predispositions, body weight, age, and neuter status (Anderson et al. 2020). In horses, changes in composition and structure properties of cartilage result from cartilage damage due to trauma, impact injuries, abnormal joint loading, excessive wear, or aging process (Gencoglu et al. 2020). In cats, idiopathic OA mediated by congenital, traumatic, infectious, nutritional, and immune-mediated causes is prevailing (Enomoto et al. 2019). The prevalence of OA is higher in older animals, but can also occur in young animals (Gencoglu et al. 2020; Anderson et al. 2020). Although the exact etiology of OA has yet to be identified, the environmental stress followed by metabolic changes in chondrocytes may play a key role in cartilage degeneration (Zheng et al. 2021): Adverse microenvironmental conditions lead to a switch in chondrocyte metabolism from a resting regulatory state in which oxidative phosphorylation is a leading metabolic process to highly metabolically active glycolysis (Zheng et al. 2021). The consequential increase in biosynthesis of inflammatory and degrative mediators and exposure of chondrocytes to proinflammatory cytokines, hypoxia, and nutrient stress are promoting signaling pathways of catabolism. Enhanced catabolism is followed by mitochondrial dysfunction, resulting in excessive production of reactive oxygen species (ROS) and oxidative damage, a hallmark of OA (Zheng et al. 2021; Mobasheri et al. 2017). The important consequence of ROS is the activation of AMP-activated protein kinase (AMPK) and consequential upregulation of the expression of collagen type I, proinflammatory cytokines, and matrix metalloproteinases (MMP) (Zheng et al. 2021). In particular, MMP13 is known to break down collagen type II, a key structural component of cartilage ECM. Matrix degradation products further promote inflammation and prevent the cycle of degeneration to break (Bedingfield et al. 2020).

3 Pain Management of OA

3.1 Conservative Treatment

OA is currently an incurable disease (Enomoto et al. 2019) and pain management is usually the first step in cartilage therapy. In veterinary medicine, nonsteroidal anti-inflammatory drugs NSAIDs are often the first choice for the treatment of OA and can be used for long-term management of the inflammatory component of OA pain. In addition to NSAIDs, gabapentin, amantadine, and tramadol can be administered when treatment with NSAIDs is not an option. Conservative treatment of OA also relies on the use of weight management, nutritional joint support, and physical rehabilitation including laser therapy, magnetic field therapy, shock wave therapy, massage, and balneotherapy (Zylinska et al. 2018; Rychel 2010). Unfortunately, existing therapies are often associated with severe side effects, such as potential renal, gastrointestinal, or hepatic adverse reactions, and are also often not sufficiently effective (Rychel 2010).

Additional conservative treatment option for treating OA is arthrocentesis or articular puncture, performed to inject supplements such as GAGs or HA to improve the natural qualities of HA, present in the articular fluid, and to increase the mobility of the joint (Zylinska et al. 2018). IM injections of polysulfated GAGs to dogs with OA resulted in improved lameness scores in 12 out of 16 dogs. Reduced lameness was ascribed to GAGs inhibition of cartilage oligomeric matrix protein (COMP) degradation seen as a decrease in serum COMP concentration (Fujiki et al. 2007). However, these results were short-lived similar to the HA treatment. Single intraarticular injection

of HA alone in dogs with naturally occurring hip OA also had only a temporary amelioration of the symptoms as measured by Canine Brief Pain Inventory. However, intraarticular injection of HA combined with corticosteroids appeared superior in positive effects compared to HA alone (Alves et al. 2020). Although intraarticular injection of HA and corticosteroids might prove useful for patients that cannot tolerate NSAIDs (Franklin and Cook 2013), based on the retrospective studies in dogs, there was weak or no evidence to support the use of HA for OA (Sanderson et al. 2009; Aragon et al. 2007). Evidence for the efficacy of HA is relatively weak due to the lack of control groups, and the limited numbers of controlled clinical studies make it difficult to suggest the superior effect of HA over the use of NSAID (Aragon et al. 2007).

3.2 Novel Pain Management Treatment Options

3.2.1 Platelet-Rich Plasma

In comparison to intraarticular injection of HA combined with corticosteroids, patient-based assessment scores in lameness and pain were better with intraarticular injection of autologous conditioned platelet-rich plasma (PRP) (Franklin and Cook 2013). PRP is an autologous product, containing an increased concentration of growth factors and bioactive proteins that may enhance the healing process on a cellular level. Besides bioactive factors such as serotonin, histamine, dopamine, calcium, and adenosine that have fundamental effects on the biological aspect of wound healing, PRP contains cytokines and growth factors, including transforming growth factor-β (TGF-β), platelet-derived growth factor (PDGF), insulin-like growth factor (IGF), fibroblast growth factor (FGF), epidermal growth factor (EGF), and vascular endothelial growth factor (VEGF) that play an important role in cell chemotaxis, proliferation, differentiation, and angiogenesis and therefore represent a potential to enhance healing of tendon, ligament, muscle, and bone (Foster et al. 2009). The advantage of PRP is primarily that it is a simple, rapid, cost-effective, and safe way to obtain a clinical improvement of animals affected by OA (Catarino et al. 2020), although diverse methods and devices used to evaluate pain and lameness among different studies make the results difficult to compare (Vilar et al. 2018). Several studies have shown beneficial, albeit temporary, results of intraarticular injection of PRP in the treatment of canine OA. A single injection of PRP into OA joints of dogs was shown to have a positive effect estimated by the lameness grades (Catarino et al. 2020) or force platform gait analysis (Vilar et al. 2018; Venator et al. 2020), but these effects only lasted for 3 to 6 months. Prolonging management of pain was achieved by combining PRP treatment with physical therapy (Cuervo et al. 2020). In comparison to dogs, intraarticular administration of PRP in horses with naturally occurring OA indicates variable changes in kinetic gait parameters (Mirza et al. 2016). Due to differences in PRP concentrations used in different studies, optimization of number of enriched platelets, the volume applied, and concentration of growth factors used for clinical application is needed. Furthermore, characteristics of PRP products differ considerably in the amount of blood processed, method of PRP preparation, and the amount of PRP produced (Franklin et al. 2015). Despite mentioned promising results, there is a lack of data supporting the use of a particular PRP for a specific medical condition, and a consensus on the actual benefits of PRP has not yet been established.

3.2.2 Anti-nerve Growth Factor Monoclonal Antibodies Therapy

A potential alternative to pharmacological pain management in dogs and cats is analgesia using anti-nerve growth factor monoclonal antibodies (anti-NGF mAB) therapy. NGF is a soluble signaling protein, belonging to a family of neurotrophin molecules. During development, NGF has an essential role in the development of sensory and sympathetic neurons, whereas in the adult organism, NGF takes an important part in the sensitization of nociceptors after tissue injury (Mantyh et al. 2011). NGF is produced and released by peripheral tissues such as

chondrocytes (Enomoto et al. 2019) and white adipose tissue depots (Ryan et al. 2008). NGF serum level was shown to be associated with stress-related conditions, for example, during transportation (Kawamoto et al. 1996) or exercise load (Matsuda et al. 1991; Ando et al. 2016), and was thus recognized as an important factor to evaluate stress status in an animal (Ando et al. 2020). Besides psychological stress, NGF was correlated also with the mechanical stress associated with OA. Isola et al. (Isola et al. 2011) reported that the concentration of NGF in synovial fluid in dogs with OA was significantly higher in comparison to healthy dogs, suggesting the involvement of NGF in OA inflammation. Similarly, as in dogs, NGF concentration in horses was also higher in synovial fluid from acutely inflamed joints and joints with chronic OA in comparison to healthy joints (Kendall et al. 2021). Some recent clinical studies used anti-NGF mAB to alleviate OA pain in animal patients and are limited to a few studies conducted on dogs and cats. For the treatment of inflammatory pain in dogs, rat anti-NGF mAB were fully caninized (Gearing et al. 2013). Canine-specific anti-NGF mAB were used intravenously in pilot, masked, placebo-controlled clinical studies to alleviate pain in dogs with degenerative joint disease (Lascelles et al. 2015). With 25 dogs included in the study, a positive analgesic effect, similar to that expected with NSAIDs, was recognized based on significantly improved patient-specific outcomes of pain and mobility and significantly increased objectively measured activity. Positive effects of the treatment were observed over 4 weeks after a single treatment with anti-NGF mAB (Lascelles et al. 2015). Similar observations were made in another study conducted by Webster et al. (Webster et al. 2014) where OA-associated pain was alleviated in dogs up to 4 weeks after IV treatment with anti-NGF mAB. Similarly, as in dogs, species-specific anti-NGF mAB were developed for pain treatment in cats (Gearing et al. 2016). In a study with 34 cats, feline-specific anti-NGF mAB were used subcutaneously to treat degenerative joint disease-associated pain. A positive analgesic effect was observed for 6 weeks during the study with significantly increased objectively measured activity (Gruen et al. 2016). Current evidence suggests that anti-NGF mAB therapy of OA in dogs and cats and possibly in horses could be an alternative to NSAIDs and other pharmacological drugs. The efficiency of a single injection of anti-NGF mAB seems to last 4–6 weeks, but further studies are needed to better understand the level of analgesia and to determine possible adverse side effects and the long-term safety of NGF use.

3.2.3 Mesenchymal Stem Cells/Medicinal Signaling Cells

Longer-lasting pain management effects of treating OA were accomplished using adult multipotent mesenchymal stem cells (MSCs). Stem cells are undifferentiated cells with the capability of self-renewal and differentiation into different specialized cells (Morrison et al. 1997). Compared to other stem cell types such as embryonic stem cells and induced pluripotent stem cells, MSCs were recognized as the most promising type of stem cells for therapy because of the relatively simple harvest techniques, isolation, and the absence of greater ethical concerns associated with their use (Sasaki et al. 2018). For both laboratory-based scientific investigations and preclinical studies, a set of standards to define human MSCs was proposed by the Mesenchymal and Tissue Stem Cell Committee of the International Society for Cellular Therapy (ISCT) (Dominici et al. 2006). In essence, (1) MSC must be plastic-adherent when maintained in standard culture conditions using tissue culture flasks; (2) 95% of the MSC population must express CD105, CD73, and CD90 and lack the expression of CD45, CD34, CD14, or CD11b, CD79a or CD19, and HLA class II; and (3) MSCs must be able to differentiate into osteoblasts, adipocytes, and chondroblasts under standardized *in vitro* differentiating conditions. For the identification of animal MSCs, minimal criteria are yet to be defined. MSCs are found in numerous tissues, which, when endogenously activated, act to replace dead, injured, or diseased tissue cells (Caplan 1991). MSCs of common veterinary patients, e.g., dogs, horses, and cats,

have been isolated from several tissues including the bone marrow and adipose tissue (Sasaki et al. 2018; Arevalo-Turrubiarte et al. 2019; Webb et al. 2012), umbilical cord (Zhang et al. 2018; Denys et al. 2020), umbilical cord blood (Kang et al. 2012; Koch et al. 2007), muscle and periosteum (Radtke et al. 2013; Kisiel et al. 2012), gingiva and periodontal ligament (Mensing et al. 2011), peripheral blood (Sato et al. 2016; Longhini et al. 2019), endometrium (Rink et al. 2017), and placenta (Carrade et al. 2011). MSCs have also been described in several joint tissues such as synovium (Sasaki et al. 2018), synovial fluid, and synovial membranes (Arevalo-Turrubiarte et al. 2019; Prado et al. 2015) and inside the infrapatellar fat pad. One of the important aspects of the therapeutic potential of MSCs is their ability to migrate into the damaged tissue and secrete immunomodulatory and trophic bioactive factors (Caplan 2017). Therapeutic properties of MSCs, ascribed to their immunomodulatory functions, are exhibited by paracrine action, secretion of extracellular vesicles, immunomodulation mediated by apoptosis, and mitochondrial transfer (Voga et al. 2020). In veterinary medicine, the therapeutic potential of MSCs is being exploited for the treatment of various organ systems. Musculoskeletal diseases have especially been proven indicative for MSC therapy, as was shown in horses with tendon injuries (Pacini et al. 2007; Godwin et al. 2012; Dyson 2004; Smith et al. 2013; Muir et al. 2016), bone spavin (Nicpon et al. 2013), and meniscal damages (Ferris et al. 2014). Notably, as recently reviewed by our group (Voga et al. 2020), remarkable clinical outcomes of MSC treatment have also been shown in dogs (Mohoric et al. 2016; Black et al. 2007; Vilar et al. 2013; Shah et al. 2018; Harman et al. 2016; Maki 2020; Kriston-Pal et al. 2020) and horses (Magri et al. 2019; Marinas-Pardo et al. 2018) with osteoarthritic conditions, showing as significant longer-termed reduction or even elimination of pain and lameness. Based on the results of these studies, MSC treatment for OA appears safe with promising clinical outcomes, showing reduced lameness and pain associated with OA, decreasing the need for use of anti-inflammatory drugs with their known side effects. However, in comparison to clinical evaluation, the long-term follow-ups with radiographic and CT imaging are scarce and often do not report improvements following MSC therapy, as recently reviewed by Brondeel et al. (2021). Some reduction in progression of OA, demonstrated with radiographic images, was shown in an equine model of OA in fetlock joints (Bertoni et al. 2021), but there is a need for the long-term follow-up imaging performed on actual patients where the progression of the disease is often very different from the experimentally induced pathologies. Demonstrated ability of MSCs to slow down or even stop OA progression is indicative of their well-established immunomodulatory function. One of the important features of MSCs is their tendency to home to injured or inflammation sites when administered in vivo. However, in contrast to the initial belief that MSCs differentiate and replace damaged tissue, evidence from recent years suggest that MSCs in vivo rarely or never differentiate into the tissue at the site (Guimaraes-Camboa et al. 2017; Meirelles Lda et al. 2009) but secrete bioactive factors. The in vitro multipotency of MSCs thus cannot be directly related to their mechanisms of action in vivo. To avoid the confusion originating from the discrepancy between the name and therapeutic potential of MSCs, it was proposed by Caplan that the term "mesenchymal stem cells" should be changed into "medicinal signaling cells" (MSCs) (Caplan 2017). The actual regeneration of cartilage to prevent OA from developing or even revert already existing OA condition, therefore, remains the topic of research, which is in recent years focusing on exploiting the in vitro differentiation capabilities of MSCs as a basis for finding novel potential solutions to address this issue.

The following part of this chapter focuses on the regenerative surgical attempts to treat cartilage defects, starting with the initial MSC-free attempts to regenerate cartilage, followed by the presentation of the studies exploiting the in vitro differentiation potential of MSCs for cartilage regeneration.

4 Surgical Treatment of OA and Cartilage Defects

4.1 Conventional Treatment Options

Conventional surgical treatment of OA is indicated when conservative therapy fails or is inadequate in alleviating pain and maintaining the function of the joint (Cook and Payne 1997). In dogs, several surgical techniques for OA have been developed. Surgeries may offer treatment of the primary cause, such as cranial cruciate ligament rupture, where tibial plateau leveling osteotomy (TPLO) (Slocum and Slocum 1993), tibial tuberosity advancement (TTA) (Lafaver et al. 2007), or modified Maquet procedure (MMP) (Ness 2016) is indicated. In cases where providing pain relief and lessening the progression of future OA is needed, salvage procedures are performed, such as femoral head and neck excision (indicated in coxofemoral luxation; severe coxofemoral OA; comminuted or complicated fractures of the femoral head, neck, or acetabulum; avascular necrosis of the femoral head; or failed total hip replacement) (Harper 2017a), arthrodesis (indicated for intractable articular fractures, luxations, subluxations, or failed total joint replacement) (McCarthy et al. 2020), and total joint replacement (indicated for patients with debilitating OA secondary to trauma or joint dysplasia) (Harper 2017b).

4.2 Reparative Treatment Techniques

In contrast to salvage surgical interventions used to treat irrevocably damaged articular cartilage by removal or replacement, reparative bone marrow stimulation techniques are used to expose the subchondral bone to stimulate bone marrow and improve cartilage vascularization, enabling the diffusion of nutrients from the subchondral bone into the cartilage and stimulating bone marrow cells to reach the avascular cartilage lesion and initiate a healing response (Stupina et al. 2015). In humans, the method of bone marrow stimulation is one of the most recommended reparative surgical techniques to treat OA (Gill and Steadman 2004). It can be achieved via drilling, chondral abrasion, or microfractures. The latter are of special interest as it can be performed arthroscopically. Light scraping, but not complete removal of calcified cartilage, is indicated to facilitate attachment of the reparative tissue to exposed calcified cartilage (Breinan et al. 2000). In veterinary medicine, objective evidence documenting the efficiency of bone marrow stimulation is not available. In a canine model of OA, chondral abrasion resulted in a fibrocartilage (Altman et al. 1992). In another study using a dog model of OA, subchondral tunneling of subchondral bone together with the injection of autologous bone marrow into the canals resulted in improved cartilage vascularization and consequently improved chondrocyte metabolism and functionality of cartilage (Stupina et al. 2012). Neither method of bone marrow stimulation resulted in hyaline cartilage formation, but rather in reparation of cartilage with the formation of fibrocartilaginous tissue. Similar was shown in horses. While microfractures increased the tissue volume in the defects (Frisbie et al. 1999) and did not cause any negative effects, this technique did not seem to have clinical effects in horses with stifle lameness diagnosed with naturally occurring OA (Cohen et al. 2009). Bone marrow stimulation results in the formation of fibrocartilage, with poor structural and mechanical properties that do not provide long-term efficacy of reparative surgical treatment techniques (Zylinska et al. 2018). Moreover, poor long-term wear characteristics of fibrocartilage do not prevent the progression of OA (Lane et al. 2004). Since the prevention of degenerative joint changes over time is one of the ultimate goals in the treatment of cartilage lesions (Burks et al. 2006), the limited intrinsic ability of cartilage to heal is proposed to alter with the regenerative treatment options that are therefore at the forefront of the cartilage treatment research.

4.3 Regenerative Treatment Options

A common feature of OA is cartilage defects that may either be associated with pain and decreased function or may appear asymptomatically (Janakiramanan et al. 2006). Either way, without treatment, cartilage defects may lead to progressive joint disease (Mehana et al. 2019; Burks et al. 2006). Treatment of cartilage defects is thus directed toward the regeneration of the defective cartilage and prevention of progression of the disease. Cartilage regeneration methods include osteochondral grafting, autologous chondrocyte implantation (ACI), matrix-induced ACI (MACI), and combinatory use of MSCs and biomaterials, aiming to replace the damaged cells and extracellular matrix while preserving the microarchitecture and biomechanical functions of the cartilage (Zylinska et al. 2018).

4.3.1 Osteochondral Transplantation

Osteochondral grafting is an attractive option for cartilage reconstruction because live homologous tissue is used. In humans, osteochondral and meniscal allograft transplantation in the knee has been performed for over 40 years (Rucinski et al. 2019; Familiari et al. 2018; De Armond et al. 2021). In animals, the majority of the studies are performed on animal models. One of the indications for using osteochondral grafts as a means for cartilage reconstruction in dogs is osteochondritis dissecans (OCD). OCD is an inflammatory condition that occurs when the diseased cartilage separates from the underlying bone. The disease can increase the risk of developing OA and it is an important cause of lameness in dogs (Schreiner et al. 2020). It was previously reported that no differences were detected between the surgical and medical treatment of OCD in 19 dogs. Medical treatment resulted in an even more rapid return to normal weight-bearing. Despite some clinical improvement, in most dogs, lameness continued and the disease progressed (Bouck et al. 1995). Albeit demonstrated to be technically feasible in canine caudocentral humeral head, medial humeral, and medial femoral condyle, positive clinical outcomes of osteochondral autograft transfer in dogs with OCD were short-termed, with minimal donor site morbidity (Fitzpatrick et al. 2010; Fitzpatrick et al. 2009; Fitzpatrick et al. 2012). The osteochondral graft may not even render clinical changes, as was shown in a canine model of full-thickness cartilage defect, where phalangeal osteochondral graft did not result in significant functional difference compared to the nongrafted group of dogs 6, 12, or 20 weeks after surgery (Dew and Martin 1992). In comparison to OCD in dogs, osteochondral grafts in the case of subchondral bone cysts in horses that can also lead to osteochondrosis (Bodo et al. 2004) resulted in the reconstruction of the articular surface, subchondral decompression, and a renewed cartilage gliding surface. Promising clinical outcomes demand further investigation of the suitability of treatment of subchondral bone cysts with osteochondral grafts (Bodo et al. 2004).

Even though studies on animals are for the most part conducted on animal models and not the actual patients, up to 20% of procedures are unsatisfactory (Huang et al. 2004). The clinical success of the grafts is dependent on the viability of cartilage cells, the capacity of host bone to join graft cartilage, and the host's immunologic tolerance. Integration of donor allograft into recipient's bone can thus be incomplete and can cause failure (Pritzker et al. 1977). Although function and quality of life, based on owner perception, seem to improve after osteochondral grafting (Cook et al. 2008), donor site morbidity is considered a major ethical concern albeit donor sites from canine stifle are currently the only reliable available source of canine donor osteochondral autograft material (Fitzpatrick et al. 2009). Morbidity associated with autografted tissue for treating osteochondral defects could be avoided using fresh allograft tissues. In a canine model of knee cartilage defect, allografts were shown to be similar to autografts regarding bone incorporation, articular cartilage composition, and biomechanical properties (Glenn Jr et al. 2006). Despite being a promising

solution for mismatch of transplanted cartilage, allografts may be immunogenic; hence the cartilage becomes vulnerable to direct injury by cytotoxic antibodies or lymphocytes or to indirect injury by inflammatory mediators and enzymes induced by the immune response. However, the literature on the immunogenicity of allografts is contradictory. In some studies, the severe immune response was demonstrated upon allograft transplantation, as shown by an induced inflammatory response, thinned, dull, and roughened cartilage of allografts, with the severely fibrotic and hyperplastic synovial membrane of the joints in dog models (Stevenson et al. 1989). In other studies, no immune response was detected (Glenn Jr et al. 2006; McCarty et al. 2016), or immune response was dependent on whether or not allografts were previously frozen or were vascularized (Stevenson et al. 1996). Freezing was reported to cause harm to the cartilage and thus lower the success rate of osteochondral transplantation (Stevenson et al. 1989). As it was demonstrated in a canine model, viable chondrocytes in osteochondral allografts at the time of transplantation are primarily responsible for the maintenance of donor articular cartilage health in the long term, confirming that not only storage but also procurement, processing, transportation, and clinical implantation are of great importance for allograft clinical use (Cook et al. 2016).

Novel systems for preserving osteochondral allografts, such as MOPS (Missouri Osteochondral Allograft Preservation System) (Cook et al. 2014), and novel methods for enhancing graft integration are being developed. A lack of osteochondral graft integration is one of the important problems in transplanting osteochondral grafts that can cause a treatment failure, especially since there is often a mismatch of transplanted cartilage regarding the contour and thickness of the injured surface (Huang et al. 2004; Hurtig et al. 2001). Also, transplantation of osteochondral grafts involves manual precise preparation of the donor graft and recipient bed. The process is user-dependent, not standardized, and subject to human error. A possible solution for bypassing the issue of insufficient supply of available donor tissue with accurate anatomical features is a fabrication of osteochondral constructs with the use of 3D printing techniques, improving the accuracy of anatomical architecture and topology, suggesting clinical relevance for large area cartilage repair (De Armond et al. 2021; Roach et al. 2015). Additionally, enhancing graft integration was attempted by using saturating grafts with bone marrow aspirate concentrate (Schreiner et al. 2020; Stoker et al. 2018) or PRP (Stoker et al. 2018), with the assumption that growth factors, cytokines, and other proteins contained in bone marrow aspirate concentrate may enhance osteoinductive, chemotactic, and neovascular signals needed for better graft integration. For example, in an in vitro study, bone marrow aspirate concentration was shown to be superior to PRP in enhancing integration potential for canine osteochondral allografts (Stoker et al. 2018). A combination of novel graft preservation and implantation techniques may therefore result in more satisfying clinical outcomes, as was demonstrated in a study where osteochondral allograft transplantation technique using fresh unicompartmental bipolar osteochondral and meniscal osteochondral allografts and application of bone marrow aspirate concentrate were used to treat medial compartment gonarthrosis in a canine model. Clinical, radiographic, and arthroscopic assessment of the graft and joint demonstrated the maintenance of the integrity of transplants and integration into the host tissue, leading to superior outcomes without early OA progression compared to NSAID controls (Schreiner et al. 2020).

While animal models provide crucial information about disease mechanisms, the artificially induced disease cannot recreate the natural in vivo environment (Cope et al. 2019). Studies conducted on actual veterinary patients are scarce, and extensive research is still needed to prove the efficacy and usefulness of osteochondral graft transplantation on actual patients. However, advancement in allograft transplantation in animal models suggests that osteochondral grafting is worthy of further investigation also in actual veterinary patients.

4.3.2 Autologous Chondrocyte Implantation

The lack of significant cellular activity in chondral defects was indicative for the researchers that chondrocytes are needed for articular cartilage regeneration (Shortkroff et al. 1996). Autologous chondrocyte implantation (ACI) was thus developed as an alternative for treating defects of articular cartilage. In humans with full-thickness cartilage defects, the procedure was described in 1994 by Peterson et al. (Brittberg et al. 1994): Cartilage slices were obtained from an uninvolved area of the injured knee during arthroscopy. Chondrocytes were then isolated and cultured for 14 to 21 days in the laboratory and then injected into the injured area under a periosteal flap taken from the proximal medial tibia. ACI seems to be advantageous over bone marrow stimulation techniques in that the cartilage that is formed is predominantly hyaline-like, containing collagen type II (Brittberg et al. 1994; Min et al. 2007; Cherubino et al. 2003). It was demonstrated by Min et al. that cartilage regeneration after ACI is correlated with at least 4-week-long survival of transplanted chondrocytes (Min et al. 2007). Fluorescently labeled chondrocytes implanted in the goat model were shown to integrate into the surrounding tissue and become a structural part of repaired tissue, rich in collagen type II and proteoglycans (Dell'Accio et al. 2003). In the canine model, ACI was shown to be superior to bone marrow stimulation techniques based on morphology, histology, and serum marker levels, with smooth surface, less fissure, and good border integration (Nganvongpanit et al. 2009). Similar as in dogs, in three horse models of cartilage lesions of fetlock joints in the forelimb, hyaline-like cartilage was formed after ACI treatment (Barnewitz et al. 2003). In the majority of animal models, ACI is investigated in full-thickness cartilage lesions. Partial-thickness cartilage lesions represent a more hostile environment for regeneration due to avascularity, poor cellularity, and smoothness of calcified cartilage. However, in patellofemoral joints in equine models, partial-thickness defects with intact calcified cartilage were proven to be a good indication for treatment with ACI. ACI improved cartilage healing (although less obviously as in full cartilage defects), as seen with improved histological, immunohistological, and biochemical scores, including defect filling with collagen type II and attachment to the surrounding cartilage (Nixon et al. 2011).

Although ACI has produced promising results, it was indicated in previous studies that the degree to which hyaline-like cartilage fills a defect is insufficient to integrate with surrounding tissue (Breinan et al. 1997). Significant effects after ACI treatment in dog models seem to be short-termed and degenerative changes are not prevented (Nixon et al. 2011). In attempts to enhance the filling of cartilage defects with the functional tissue, biomaterials were developed to serve as carriers of cells.

4.3.3   Matrix-Induced ACI (MACI)

In the original ACI technique, the periosteal cover was used since it was thought to have the chondrogenic potential (O'Driscoll and Fitzsimmons 2001) and stimulate subchondral bone remodeling (Russlies et al. 2005). However, with ACI, there are damage associated with periosteal harvest (Ueno et al. 2001), damage associated with the suturing of articular cartilage (Hunziker and Stahli 2008), and hypertrophy observed after periosteal grafting (Ueno et al. 2001). The downside of this method is also a non-homogenous distribution of chondrocytes due to the use of cellular suspension, together with the risk of leaking out in case of inadequate sealing (Haddo et al. 2004). These limitations were improved by using the matrix-induced ACI (MACI), where alternative covers, such as porcine-derived type I/III collagen membrane, are used. The bilayered structure of a membrane is cell occlusive at the compact side, protecting cells from diffusion and mechanical impact, and the porous side consists of collagen fibers, allowing for cell invasion and attachment (Haddo et al. 2004). Autologous chondrocytes are seeded onto the membrane, enabling the membrane to be attached to the defect with the fibrin glue eliminating periosteal harvest, and procedure is faster and with less extensive

exposure, as surgical implantation could be achieved via arthroscopy or mini-arthrotomy (Cherubino et al. 2003). Besides facilitating the handling of the cells, scaffolds are also useful for immobilization and broader distribution of the cells (Nuernberger et al. 2011). The procedure is traditionally performed by arthrotomy (Cherubino et al. 2003), but arthroscopy was also shown to be possible, as was shown in some studies with equine models that underwent arthroscopic implantation of cell-polymer (Ibarra et al. 2006; Masri et al. 2007) or cell-collagen membrane constructs (Frisbie et al. 2008; Nixon et al. 2017). In several studies of equine joint defect models, treatment with MACI resulted in significantly improved cartilage compared to spontaneously healing empty controls, as shown by arthroscopy, gross healing, histology scores, and mechanical analysis (Nixon et al. 2017; Nixon et al. 2015; Griffin et al. 2015). Materials other than collagenous membranes were also used for MACI, for example, PGLA, used in eight horse models and were shown to efficiently contain a large number of chondrocytes without the risk of cell loss when implanted arthroscopically with the use of a fluid pump (Masri et al. 2007). Although ACI and MACI have produced promising results and MACI treatment indeed improved cartilage healing, characterization of MACI graft implant in animal models showed that formed tissue has inferior shear properties to native cartilage (Nixon et al. 2015; Griffin et al. 2015; Lee et al. 2003). The loss of chondrocyte capacity to produce hyaline cartilage might be associated with the cell dedifferentiation occurring during chondrocyte culturing (Rakic et al. 2017).

Although increasing the dose of articular chondrocytes was shown to improve articular cartilage repair in a sheep model (Guillen-Garcia et al. 2014), chondrocytes cultured in vitro are prone to spontaneous dedifferentiation, albeit less so when cultured in a 3D environment. It was shown by Sanz-Ramos et al. (2014) that chondrocytes cultured in a 3D collagen environment possessed a better chondrogenic capacity in vitro and in vivo than the cells expanded on a plastic surface (Sanz-Ramos et al. 2014).

Interestingly, the extent of dedifferentiation seems to vary between species. For example, sheep chondrocytes were shown to be able of spontaneous redifferentiation into hyaline-like cartilage, whereas human chondrocytes were able to redifferentiate only when stimulated by chondrogenic inducers (Giannoni et al. 2005). In the equine model, chondrocyte redifferentiation was shown to be possible under the influence of 3D collagenous microenvironment, hypoxia, and BMP2 (bone morphogenetic protein-2) and RNA interference (Rakic et al. 2017). In comparison to human and equine chondrocytes, dog chondrocytes showed no capacity to redifferentiate regardless of the inducers present (Giannoni et al. 2005). The interspecies differences in chondrocyte characteristics in culture indicate that species should be considered when extrapolating data from one species to another and that differences between species in terms of chondrocyte phenotype stability during expansion might also result in different clinical outcomes when used in ACI. In addition to interspecies differences, chondrogenic differentiation of chondrocytes was dependent also on the number of passages and aging (De Angelis et al. 2020; Acosta et al. 2006; Veilleux et al. 2004), as well as whether the cells were osteoarthritic or not (Acosta et al. 2006). While, interestingly, adult donors showed a more stable expression of some chondrogenic markers, chondrocytes from elderly animals dedifferentiated at earlier passages, associated with a reduced proliferative capacity (De Angelis et al. 2020). Chondrocyte dedifferentiation could therefore be controlled from different aspects of donor and culture factors.

Another hurdle in using ACI/MACI for the treatment of chondral defects is a need for a two-step surgery. In 2006 the evidence that ACI could be delivered without cell expansion was presented. It was proposed that mechanical fragmentation of cartilage was sufficient to mobilize embedded chondrocytes through the increased surface of tissue area. In goats, cartilage fragments were placed on resorbable scaffold hyaline-like tissue (Lu et al. 2006). The procedure was adopted also in horse models with autologous cartilage fragments on a polymer scaffold implanted in a defect within the equine femoral trochlea. Compared to two-step ACI treatment, one-step treatment with minced cartilage achieved an even higher score in arthroscopic, histologic, and immunohistochemistry evaluation and prompted a phase 1 clinical study in humans (Frisbie et al. 2009). In a study performed in dogs, it was demonstrated that 100-μm-sized cartilage particles yielded the highest number of cells and provided the most optimal cartilage regeneration, based on the autologous intrafacial implantation of the microcartilage together with the absorbable scaffold and the slow release system of the basic fibroblast growth factor (Nishiwaki et al. 2017). Another possibility to overcome the need for two-step surgery was proposed by Bekkers et al. who showed that a one-stage procedure could be achieved by combining chondrocytes or chondrons with bone marrow mononuclear cells or MSCs. In a goat model, such implantation outperformed microfracture (Bekkers et al. 2013a, b).

Despite promising results associated with ACI/MACI for treatment of chondral defects, there are still many challenges that have not yet been overcome, such as insufficient integration of implanted chondrocytes, insufficient capacity of chondrocytes to produce hyaline cartilage, dedifferentiation of cultured chondrocytes, the need for two-step surgery, and the harvesting procedure that may result in changes in the articular cartilage that potentially represent a risk of becoming clinically relevant (Lee et al. 2000). This is why in recent years other treatment options for cartilage defects are increasingly being investigated. MSCs as possible substitute cells for chondrocytes are the focus of the most recent research. MSCs seem promising candidates for replacing chondrocytes because of their immunomodulatory properties and their ability to differentiate into several specialized cells, including chondrocytes. At the same time, many novel biomaterials are at the forefront of cartilage regeneration research, aiming to (i) resemble native cartilage tissue to provide the most optimal environment for chondrogenic differentiation of MSCs and (ii) simultaneously develop clinically relevant biocompatible material for in vivo implantation.

5 Attempts to Improve Existing Regenerative Treatment Options with the Use of Mesenchymal Stem Cells

5.1 Chondrogenic Differentiation of MSCs

MSCs have in recent years received significant interest in veterinary and human medicine due to their immunomodulatory and multilineage differentiation properties. Under appropriate culture conditions, MSCs can be induced toward differentiation into different lineages such as adipocyte, osteocyte, and chondrocyte lineages (Dennis et al. 1999). Although there are some reports on spontaneous chondrogenic differentiation of MSC ascribed to either high cell density (Bosnakovski et al. 2004; Dudakovic et al. 2014), presence (Fortier et al. 1998) or absence (Cho et al. 2018) of fetal bovine serum (FBS) in cell culture media, early passages (De Bari et al. 2001), or tissue source (Naruse et al. 2004), chondrogenesis on a standard 2D polystyrene surface is commonly induced with specific culture conditions such as chondrogenic differentiation media, high cell density, and highly humid atmosphere. Chondrogenic differentiation of MSCs is commonly performed in two ways. One technique is a pellet culture – a scaffold-free three-dimensional (3D) culture with high cellular density, where cells are grown in polystyrene conical tubes to form a spherical aggregate at the bottom of a tube (Johnstone et al. 1998). Another method is a micromass culture system where cells are placed in the microwell cell culture plate as droplets of cells with high density that become coalesced to form micromasses of cartilaginous tissue (Mello and Tuan 1999). During early chondrogenesis progenitor cells condense and express collagen type I. By the 5th day, collagen type II is detected and type X collagen is detected by the 14th day. The presence of aggrecan and link protein in the cell aggregates demonstrate that aggregating proteoglycans of the cartilaginous tissue are synthesized by the newly differentiating cells (Yoo et al. 1998). Commonly recognized markers of chondrogenesis in MSCs are SOX9, collagen type II, aggrecan, GAG, and COMP (De Angelis et al. 2020). In chondrogenic differentiating media, growth factors and hormones, namely, TGF-β and dexamethasone (Li and Pei 2018; Mwale et al. 2006), are often used to induce chondrogenesis. TGF-β upregulates chondrogenesis by enhancing SOX9 expression and inhibiting osteoblast differentiation by repressing expression of RUNX2 (Pei et al. 2009), while dexamethasone potentiates the growth factor-induced chondrogenesis of MSCs in vitro, although its influence is not indispensable for chondrogenic differentiation of MSCs as it is dependable on tissue source and microenvironment of MSCs (Shintani and Hunziker 2011). Besides TGF-β, other growth factors, namely, IHH and BMP2 (Steinert et al. 2012; An et al. 2010), FGF (Handorf and Li 2011), and IGF (An et al. 2010; Patil et al. 2012), were also shown to be inducers of chondrogenesis of human MSCs. However, the molecular mechanisms of chondrogenesis are not yet fully understood.

5.2 Hypertrophy Associated with Chondrogenic Differentiation of MSCs

Due to their rapid expansion in culture, trilineage differentiation potential, and easier retrieval that is not associated with articular cartilage damage as opposed to chondrocytes, using MSCs over articular chondrocytes is thought to be advantageous, especially since chondrogenesis of MSCs can be achieved with relatively simple procedures on a standard polystyrene surface. However, the undesirable effect of differentiating MSCs toward chondrogenic lineage is the constitutive expression of hypertrophic markers in MSCs. Hypertrophic markers include collagen type X, MMP13, VEGF (Chen et al. 2019), and a novel biomarker, thrombospondin-1 (TSP-1), known by its antiangiogenic properties and recently described as an antihypertrophic protein (Cortes et al. 2021; Gelse et al. 2011). The chondrocyte hypertrophy stage can ultimately lead to apoptosis, vascular invasion, and ossification, similarly as in the growing cartilage (Bruderer et al. 2014; Mueller and Tuan 2008). Notably, hypertrophy-related changes can also be related to pathological conditions such as OA (Tchetina et al. 2005; Walker et al. 1995; Nakase et al. 2002). Importantly, it was shown that chondrogenically differentiated MSCs with expressed hypertrophy-associated genes result in mineralization, related to endochondral ossification when transplanted to ectopic sites in severe combined immunodeficient mice (Pelttari et al. 2006). The main hesitation associated with the clinical use of MSCs is therefore their inability to recapitulate stable articular chondrocyte phenotype. Indeed, the extent of the expression of hypertrophic factors might be dependent on the protocol for induction of chondrogenesis. Micromass culture was shown to be superior to pellet culture in that induced cartilaginous tissue was larger, more homogenous, and enriched in collagen type II, while the expression of hypertrophic markers was lower than in a pellet culture (Zhang et al. 2010). Yet, MSCs cultured under either of the two chondrogenic conditions are prone to hypertrophy and matrix calcification, unlike articular chondrocytes that under the same conditions maintain a non-hypertrophy phenotype (Pelttari et al. 2006). Hypertrophy correlated with both techniques is therefore undesirable as it may cause endochondral ossification in vivo.

Reduction of chondrocyte hypertrophy is extensively being investigated by using different techniques, such as co-culturing MSCs with chondrocytes; culturing MSCs in the hypoxic atmosphere; adding hormones, proteins, or other components to the culture media; silencing hypertrophic genes; or using biomaterials to imitate the natural cell environment. Some of these techniques offer promising results, although to date none have shown clinically relevant reduction, let alone complete prevention of hypertrophic differentiation.

5.3 Attempts at Reduction of MSC Hypertrophy

5.3.1 Co-culture

Chondrogenesis of MSCs greatly depends on the microenvironment, as soluble factors from surrounding tissue/cells or direct cell-cell contact can alter gene and protein expression profiles (Grassel and Ahmed 2007). The accurate regulation of key factors involved in chondrocyte hypertrophy might enable guidance of MSCs between chondral and endochondral pathways (Dreher et al. 2020). One of the ways to reduce hypertrophic differentiation of MSCs is thus co-culturing MSCs with chondrocytes, as it was previously shown that chondrocytes provide chondrogenic signals to MSCs via paracrine secretion of soluble factors including TGF-β1, IGF-1, and BMP2 (Liu et al. 2010). Inversely, chondrocytes were also shown to be affected by paracrine secretion of MSCs, as was shown by co-culturing human adipose or bone marrow-derived MSCs, leading to reduction of hypertrophy and dedifferentiation of chondrocytes, which was partially ascribed to HGF secretion by MSCs (Maumus et al. 2013). In rats, reduced hypertrophy by MSC and chondrocyte co-culture was demonstrated by increased expression of aggrecan and collagen type II together with a reduction of collagen type X and MMP13 formation (Ahmed et al. 2014). Similarly, hypertrophy reduction was shown in 3D in vitro environment with co-cultures of bovine MSCs and ACs (Meretoja et al. 2013). Effects of hypertrophy suppression were demonstrated in several other studies where MSCs were co-cultured with chondrocytes (Fischer et al. 2010; Ramezanifard et al. 2017; Amann et al. 2017). Since there is a lack of proper chondrogenic niche, it is a great challenge to stabilize ectopic chondrogenic differentiated MSC phenotype not only in vitro but also in vivo, e.g., in subcutaneous tissue. It was previously shown that the differentiation potential of MSCs is different in vitro when compared to implantation in vivo. Yang et al. (2009) demonstrated that the proliferation rate of bone marrow-derived rat MSCs cultured in vitro in a 3D environment was similar to self-renewal capacity during in vivo implantation (Yang et al. 2009), whereas trilineage differentiation potential was suppressed in vivo in comparison to in vitro conditions. However, it was shown by Liu et al. (2010) that chondrogenic niche within subcutaneous environment could be created by co-transplantation of MSCs and articular chondrocytes, as was shown with bone marrow-derived porcine MSCs and articular chondrocytes. Chondrogenic signals were provided by the secretion of soluble factors by chondrocytes, including TGF-β1, IGF-1, and BMP2, and not by cell-cell interactions (Liu et al. 2010). Interestingly, there are some reports about the inability of articular chondrocytes to prevent hypertrophy of MSCs in pellet cultures (Giovannini et al. 2010). Similarly, nasal chondrocytes were not able to prevent MSC hypertrophy and calcification in vivo unless parathyroid hormone-related protein (PTHrP) was added to the culture (Anderson-Baron et al. 2020).

5.3.2 PTHrP

PTHrP along with its receptors is generally accepted as an inhibitor of chondrocyte development during chondrogenesis of the growth plate (Kronenberg 2003) and is a commonly reported factor to reduce hypertrophy. Fischer et al. showed that when cultured in a chondrocyte-conditioned medium together with PTHrP, expression of collagen type X, the activity of alkaline phosphatase, and matrix calcification in human MSCs were reduced. Pulsed rather than constant application of PTHrP was shown to be even more effective in the reduction of endochondral differentiation (Fischer et al. 2014). PTHrP was shown to be effective in the reduction of endochondral ossification in several other studies investigating the effect of PTHrP on human MSCs (Mwale et al. 2010; Weiss et al. 2010; Mueller et al. 2013). However, although PTHrP was shown to reduce hypertrophy, it was also reported to simultaneously reduce GAG synthesis and thus have a negative effect on chondrogenesis in human MSCs (Browe et al.

2019). Therefore, further research is needed to better understand the role of PTHrP in the chondrogenesis of MSCs.

5.3.3 Matrilin-3

Besides PTHrP, a non-collagenous ECM protein matrilin-3 (MAT3) was reported to play a regulatory role in cartilage homeostasis. It was previously shown that mutation or deletion of human MAT3 is associated with the early onset of cartilage degenerative diseases (Stefansson et al. 2003; Borochowitz et al. 2004). Indicative chondroprotective properties of MAT3 were supported in a study conducted on human and mice chondrocytes, where it was shown that MAT3 was responsible for the upregulation of cartilage matrix components such as collagen type II and aggrecan. Moreover, it was shown to slow down cartilage degeneration by downregulation of matrix-degrading enzymes, namely, collagenase MMP13 and aggrecanase ADAMTS-4 and ADAMTS-5 (Jayasuriya et al. 2012). The role of MAT3 in slowing cartilage degeneration was shown also in vivo, where MAT3-primed MSCs suspension slowed the progression of cartilage degeneration in the medial meniscus OA mouse model (Muttigi et al. 2020). In addition to its chondroprotective role, MAT3 was also shown to significantly reduce hypertrophy in chondrocytes and MSCs. In hypertrophic chondrocytes, MAT3 acts as a BP-2 antagonist as it was shown to inhibit BMP/SMAD 1 activity leading to downregulation of collagen X expression and thus inhibition of premature chondrocyte hypertrophy (Yang et al. 2014). In hypertrophic human adipose-derived MSCs, MAT3 significantly reduced the expression of hypertrophic markers such as collagen type X, RUNX2, and ALP (Muttigi et al. 2020). In a study conducted by Liu et al. (2018) where the chondroprotective role of MAT3 was demonstrated in vivo as well as in vitro, the role of MAT3 was ascribed to its function in promoting the expression of HIF1-α. Hypoxia-inducible factor-1alpha (HIF-1α) was shown to be a key mediator in the cellular response to hypoxia (Kanichai et al. 2008) and vital in articular cartilage homeostasis (Liu et al. 2018).

5.3.4   Hypoxia

Since the articular cartilage microenvironment is relatively low in partial oxygen pressure (~ 1–5% O_2) (Gale et al. 2019; Brighton and Heppenstall 1971), a low-oxygen environment for cell chondrogenic differentiation culture conditions was proposed as opposed to standard incubator culture conditions ($\sim 21\%$ O_2). In fetal mice forelimb organ culture, HIF-1α was shown to regulate chondrocyte differentiation and function during endochondral ossification through triggering BMP2 activation and suppressing the activity of alkaline phosphatase and suppressing collagen type X expression (Hirao et al. 2006). When combined with BMP2, hypoxia and BMP2 synergistically promote the expansion of proliferating chondrocyte zone and inhibit chondrocyte hypertrophy and ossification (Zhou et al. 2015). In chondrocytes, hypoxia promoted chondrocyte rather than osteoblast commitment by suppressing collagen type X mediated by downregulation of RUNX2 activity (Hirao et al. 2006). Interestingly, in chondrocytes, hypoxic culture conditions were shown to induce the expression of PTHrP in a HIF-1alpha-dependent manner (Pelosi et al. 2013). Combining hypoxia and exogenous PTHrP may therefore result in an additive effect in maintaining high levels of GAGs while reducing ALP activity (Browe et al. 2019). Similar effects of hypoxia that were shown with chondrocytes were also shown with MSCs. Kanichai et al. demonstrated that a hypoxic cell environment together with chondrogenic culture conditions significantly enhances collagen II expression and proteoglycan deposition in rat MSCs (Kanichai et al. 2008). HIF-1α in human and murine MSCs, similarly as in chondrocytes, potentiated the expression of BMP2-induced chondrogenic markers and inhibited expression of RUNX2 and osteogenic markers in vitro (Zhou et al. 2015). As in chondrocytes, where hypoxia was shown to induce the expression of PTHrP, hypoxia was also shown to induce PTHrP and reduce MEF2C expression in human MSCs, demonstrating a pathway by which hypoxia attenuates hypertrophy (Browe et al. 2019). Based on the published results from human and murine stem cells, hypoxia seems to enhance chondrogenesis while suppressing hypertrophy. In addition, hypoxia was shown to enhance chondrogenesis also in canine and equine MSCs (Lee et al. 2016; Ranera et al. 2013). Interestingly, in another study investigating the effect of hypoxia on chondrogenesis of equine MSCs, hypoxia did not significantly increase the chondrogenesis of either synovium or bone marrow-derived MSCs, but it did downregulate the expression of hypertrophic marker collagen type X (Gale et al. 2019). Moreover, when studying hypertrophy of bovine MSCs and ACs cultured in a 3D microenvironment under different atmospheric conditions, hypertrophy was reduced in co-cultures of MSCs and ACs in both normoxic and hypoxic conditions, whereas culturing MSCs alone even increases hypertrophic differentiation in hypoxia compared to normoxic conditions (Meretoja et al. 2013). These studies indicate the possibility that there is a difference in susceptibility of MSC to hypoxic conditions between species. The effect of hypoxic culture conditions on suppressing hypertrophy in MSC chondrogenic differentiation might also be dependent on the tissue source of MSCs (Gale et al. 2019). Further studies are therefore needed to more accurately establish the role of hypoxia in MSC chondrogenesis.

Silencing genes associated with hypertrophy is another possible approach in stabilizing chondrogenic phenotype, as was demonstrated in a study conducted on equine bone marrow-derived MSCs, where it was shown that silencing the hypertrophic genes might prevent the persistence of collagen I expression and increase the collagen type II/collagen type I ratio. Introducing siRNA to cells targeting col1a1 resulted in 50% inhibition of col1 expression, suggesting the need for further exploration of the knockout strategy to limit hypertrophic differentiation of MSCs (Branly et al. 2018).

Besides abovementioned attempts to revert hypertrophy, there are also some reports of other possible ways to reduce chondrogenic differentiation-related hypertrophy. For example, it was previously shown that TGF-β and high doses of steroid hormones together with the absence of thyroid hormones inhibit the induction of hypertrophy (Mueller and Tuan 2008; Karl et al. 2014). Pei et al. showed that TGF–βinduced chondrogenesis was enhanced when synovium-derived MSCs were transfected with histone deacetylase 4, while type X collagen expression was simultaneously reduced (Pei et al. 2009). One of the reported agents to suppress the expression of hypertrophic genes is XAT (xanthotoxin), a furanocoumarin, also named methoxsalen, otherwise used in treating various skin diseases in humans such as vitiligo and psoriasis. It was previously shown to be able to prevent bone loss in ovariectomized mice through inhibition of RANKL-induced osteoclastogenesis (Dou et al. 2016). In the following study examining the effect of XAT on chondrocyte hypertrophic differentiation, it was shown that XAT inactivates the p38-MAPK/HDAC4 signaling pathway leading to reduced degradation of HDAC4 and inhibition of RUNX2 and thus participates in maintaining chondrocyte phenotype in regenerated cartilage (Cao et al. 2017). Hypertrophy of IPSC during chondrogenesis was also reduced using lithium-containing bioceramics with bioactive ionic components (Hu et al. 2020).

Studies investigating different options to revert hypertrophy provide promising results and offer the potential for new ways of maintaining chondrogenic differentiation by suppressing endochondral ossification. However, in most of these studies, MSCs were cultured in a standard 2D environment, which is fundamentally different from their natural environment, and none of the methods described above have provided satisfactory results, preventing the application of differentiated cells in clinical use for cartilage regeneration. To further address this issue, other approaches in the induction of chondrogenic differentiation of MSCs and cartilage regeneration are being investigated, with the focus on recapitulating MSCs native environment.

5.4 Biomaterials for Mimicking Native Cartilage Tissue

5.4.1 The Influence of the 3D Structure on MSCs

The importance of mimicking cellular natural microenvironment lies in spatially and temporally complex signaling that directs the cellular phenotype. The cell, together with the ECM, growth factors, hormones, and other molecules, is connected into an entity, which guides the functioning of individual organs and the whole organism (Tibbitt and Anseth 2009). The interaction of stem cells and their niches creates a dynamic system that is being imitated by in vitro niche models to move closer to the possibility of the therapeutic use of chondrogenic differentiated MSCs. 3D cell culture mimics mechanical and biochemical properties of the natural cellular environment and consequently provides a better insight into the physiological function of MSCs (Jensen and Teng 2020), which is especially important from the therapeutic aspect of using MSCs (Egger et al. 2019). Studies investigating the influence of the 3D environment on MSCs have shown that the 3D environment provides better conditions for expressing biological mechanisms, including cell number, vitality, morphology, proliferation, differentiation, response to environmental signals, intercellular communication, migration, angiogenesis stimulation, immune system avoidance, gene expression, and protein synthesis. 3D cell environment has thus been shown to be more suitable for cell culture than 2D (Antoni et al. 2015). In 3D cultures using carriers or biomaterials, four basic groups of materials are used – polymeric, ceramic, metallic, and composite materials (Kapusetti et al. 2019) – among which the most commonly used are hydrogels, polymeric materials, hydrophilic glass fibers, and organoids (Jensen and Teng 2020).

5.4.2 Influence of Biomaterial Properties on MSCS

The mechanical, surface, and chemical properties of the biomaterial are recognized as crucial in controlling cell fate (Martino et al. 2012). Stem cells are known to be sensitive to the mechanical properties of biomaterials and can recognize a solid substrate even when they are not in direct contact with it (Schaap-Oziemlak et al. 2014). Their adhesion to the substrate depends on the elasticity of the biomaterial, suggesting that even the smallest changes in the mechanical properties of the biomaterial can affect stem cell differentiation. Thus, the different elasticities of the biomaterial have different effects on cell adhesion, proliferation, and differentiation potential. For example, higher biomaterial strength leads to greater potential for osteogenic differentiation due to increased integrin activation, and softer biomaterials increase expression of II type collagen and lipoprotein lipase, markers for adipogenic and chondrogenic differentiation, respectively (Xu et al. 2013). In addition to the mechanical properties of the biomaterial, the surface properties also play an important role in the fate of MSCs. Stem cells do not bind directly to the surface of the biomaterial. In proteinaceous solution, e.g., in cell culture medium, stem cells bind indirectly to the surface of the biomaterial by binding to pre-bound proteins because of their slower movement compared to proteins (Tamada and Ikada 1993). The binding of cells to proteins depends on the distribution and conformation of the proteins, the latter of which depends on the wettability and chemical composition of the biomaterial (Schaap-Oziemlak et al. 2014). Therefore, the manipulation of proteins bound to the surface of the biomaterial is of particular importance in controlling cell adhesion (Schaap-Oziemlak et al. 2014). The results of several studies also indicate the influence of the chemical properties of the biomaterial surface on the direction of cell differentiation (Ren et al. 2009; Curran et al. 2006; Benoit et al. 2008). The surface treatment of biomaterials with different chemical groups, e.g., methyl (-CH3), amino (-NH2), thiol (-SH), hydroxyl (-OH), or carboxyl (-COOH) groups, can have different effects on cell fate and lead MSCs to adipogenic, osteogenic, or chondrogenic differentiation (Curran et al. 2006; Benoit et al. 2008). However, the direction of cell differentiation in a 2D or 3D environment may differ with the addition of the same chemical group (Schaap-Oziemlak et al. 2014). Therefore, the 2D or 3D environment may affect the fate of MSCs differently depending on the functional chemical group.

5.4.3 General Structure of Biomaterials for Cell Encapsulation

In addition to the mechanical, surface, and chemical properties, the scaffold structure itself also importantly affects stem cells. 3D biomaterials can be microporous, nanofibrous, or composed as hydrogels. Microporous structure supports the encapsulation of cells, but due to the pore size (100 μm) being larger than the average cell diameter (10 μm), they represent a curved 2D microenvironment. Nanofibrous structures containing fibrillar ECM proteins provide a better approximation of the natural cellular environment, but their mechanical properties are too weak to handle the stress required for mechanotransduction. Hydrogels do not have these limitations, making them a suitable biomaterial for the development of an ECM-like environment. The network structure of interconnected polymer chains allows for high water content and transport of oxygen, nutrients, waste, and other soluble molecules. Hydrogels can be composed from a range of natural or synthetic materials that exhibit a wide range of different mechanical and chemical properties (Tibbitt and Anseth 2009). Compared to synthetic hydrogels, natural hydrogels not only enable but also promote their cell activities. Natural hydrogels are usually composed of ECM proteins such as collagen, fibrin, hyaluronic acid, or components from other biological sources such as chitosan (Ribeiro et al. 2017), alginate (Sun and Tan 2013), and silk (Kundu et al. 2013).

5.4.4 Natural Biomaterials to Promote MSC Chondrogenesis

For cartilage regeneration, various scaffold materials have been developed. Most commonly used biomaterials for cartilage tissue regeneration are of natural origin, which are biocompatible, contain bioactive molecules such as RGD tripeptides that enable cell adhesion, but have in most cases poor mechanical properties and high degradation rate. Natural biomaterials are composed either of polymers, for example, agarose, alginate, chitosan, and hyaluronate, or of proteins, such as collagen, gelatin, fibrin, and silk (Ge et al. 2012). On the other hand, synthetic polymers such as polyglycolic acid (PGA), polylactic acid (PLA), poly(lactic-co-glycolic acid (PLGA), or poly(ethylene glycol) (PEG) lack the binding sites for adhesion molecules and have been shown to promote the undesirable endochondral ossification (Salonius et al. 2020), but usually provide with controllable degradation rate, high reproducibility, and easy manipulation to form specific shapes (Ahmed and Hincke 2010). Due to the advantages and disadvantages of either natural or synthetic materials, hybrid materials are also thought of as promising materials for providing microenvironment resembling cartilage tissue that is suitable for induction of stem cell chondrogenesis. Below, the commonly used biomaterials for induction of chondrogenesis are described.

Collagen

One of the most extensively used biomaterials in tissue engineering is collagen as it is a key component of cartilage ECM. It is also biocompatible and easy to manipulate with. Bioactive domains in its structure allow for good adhesion of cells. Type I/III collagen membrane has been frequently used in MACI therapy (Haddo et al. 2004). However, there are several disadvantages associated with the use of collagen as a scaffold. Firstly, the use of collagen is associated with the risk of immunogenicity (Kim et al. 2020a, b). Secondly, there is also a possibility of prion transmission (Raftery et al. 2016). Thirdly, collagen does not possess suitable mechanical strength to withstand the in vivo forces (Ahmed and Hincke 2010; Raftery et al. 2016), and lastly, culturing MSCs on collagen does not prevent hypertrophic differentiation of MSCs, as shown by human bone marrow-derived MSCs cultured either on commercial type I/III membrane or collagen/polylactide composite scaffolds, both resulting in a hypertrophic state of the cells (Salonius et al. 2020).

Regarding the immunogenicity of collagen, atelocollagen – telopeptides-free collagen – provides a biomaterial with no immunogenic activity. For treatment of chondral defects in human medicine, atelocollagen combined with microdrilling is used as an enhancement of traditional microfracture technique using the off-the-shelf product (Kim et al. 2020a). Atelocollagen, obtained by salt precipitation, was also tested for chondrogenesis of MSCs. Compared to type I collagen, type I atelocollagen enhanced chondrogenic markers' expression of human adipose-derived MSCs. Moreover, reduction of chondrogenic markers' expression RUNX2, osterix, and MMP13 was observed in cells cultured on atelocollagen, indicating better suitability of atelocollagen compared to collagen for in vitro cartilage engineering applications (Kim et al. 2020b). As a less immunogenic alternative to collagen, gelatin is also used. It is produced from processed bovine or porcine bones and skin and is usually used in combination with other materials to combine positive properties of both (Ahmed and Hincke 2010). For example, the gelatin-alginate scaffold was used to demonstrate that the proliferation rate of bone marrow-derived rat MSCs cultured in vitro on the scaffold was similar to self-renewal capacity during in vivo implantation (Yang et al. 2009).

To avoid the risk of prion transmission, other sources of collagen, besides mammal, are being investigated, such as salmon skin. However, it was shown that salmon skin-derived collagen is inferior to bovine-derived collagen in several terms such as porosity, pore size, architecture, compressive modulus, capacity for water uptake, and rat MSC proliferation and differentiation (Raftery et al. 2016).

In structural and load-bearing performance, collagen plays a pivotal role, while surrounding polysaccharides are needed for internal stress management and elastic reinforcement of collagen and absorption of fluids due to their hydrophilic nature. A protein-polysaccharide scaffold was therefore thought of as a promising material for induction of stem cell chondrogenesis. When used either alone or cross-linked with dextran or chitosan, the PEG-chitosan construct was determined as the most appropriate in inducing chondrogenesis as well as in reducing hypertrophy in human bone marrow-derived MSCs (Sartore et al. 2021). To improve the mechanical strength of the scaffold, chitosan is also increasingly studied and often used in combination with collagen. The addition of chitosan to collagen not only improved the mechanical strength of collagen but also increased compressive strength and swelling ratio and prolonged the degradation rate (Raftery et al. 2016).

Hyaluronic Acid

In addition to collagen, hyaluronic acid (HA) is one of the promising biomaterials in use for chondrogenic induction of stem cells. Hyaluronic acid is a natural component of the cartilage ECM. However, HA is highly degradable in vivo and cannot bind proteins with high affinity because of the lack of negatively charged sulfate groups. Sulfated HA was therefore fabricated to encapsulate human MSCs. The sulfated HA exhibited slower degradation, improved protein sequestration, and promoted chondrogenesis. Furthermore, it suppressed hypertrophy in vitro and in vivo in the OA rat model, due to improved growth factor retention (Feng et al. 2017). When HA was added as a supplementation to a collagen hydrogel, it was shown to stimulate chondrogenic differentiation of adipose-derived human MSCs in a dose-dependent manner. Among different concentrations from 0 to 5%, 1% HA showed the best overall results in terms of SOX and Coll type II expression. Furthermore, exchanging 25% of human articular chondrocytes with 75% of adipose-derived human MSCs didn't change the chondrogenic potential of MSCs, but reduced

hypertrophy and improved biomechanical properties (Amann et al. 2017).

Silk Fibroin

One of the promising biomaterials for use in tissue engineering is silk fibroin, derived from the silkworm *Bombyx mori*. It is biocompatible, has suitable mechanical properties, and is produced in bulk in the textile industry (Kundu et al. 2013). In comparison to other natural biomaterials used for tissue engineering, SF provides a remarkable combination of strength, toughness, and elasticity that are ascribed to its crystallinity, hydrogen bonding, and numerous small β-sheet crystals (Altman et al. 1992). Another advantage of SF is its ability to take the form of different shapes such as hydrogels, tubes, sponges, composites, fibers, microspheres, and films that could be used in tissue engineering (Rockwood et al. 2011). It was previously reported that silk fibroin can aid in MSC differentiation when combined with different components. It was previously shown that silk fibroin with incorporated L-ascorbic acid 2-phosphate significantly promoted collagen type I in mouse fibroblast L929 cells (Fan et al. 2012). It was shown to promote osteogenic differentiation and mineralization of human ADMSCs (Gandhimathi 2015), and in another study, it was shown that silk fibroin scaffold combined with PRP effectively induced chondrogenesis of human ADMSCs (Rosadi et al. 2019). Interestingly, it was shown by Barlian et al. that silk fibroin combined with silk spidroin promoted better chondrogenesis of human Wharton jelly's MSCs than silk fibroin alone and that cell culture medium supplemented with PRP promoted higher GAG accumulation in comparison with medium supplemented with ascorbic acid (Barlian et al. 2018). Contrary to mentioned studies where combining silk fibroin with other components was needed to induce chondrogenesis in MSCs, we have shown in our previous research that SF alone could also induce chondrogenesis in canine adipose-derived MSCs, possibly as a species-specific effect.

Decellularized Cartilage Matrix

Besides natural biomaterials such as collagen, hyaluronic acid, gelatin, chitosan, or silk fibroin, which have provided some promising results regarding chondrogenic differentiation of MSCs and reducing their hypertrophy phenotype, other ways for more accurate recapitulation of the cartilage microenvironment are being exploited. Among them, decellularized cartilage scaffolds have shown promise in providing the structural integrity of engineered tissues, better load-bearing ability, and functioning as a reservoir of signaling molecules, e.g., cytokines and growth factors, providing a specific microenvironment similar to native tissue. A hybrid natural ECM scaffold/artificial polymer polycaprolactone (PCL) was developed by combining ECM produced by bovine chondrocytes co-cultured with rabbit MSCs on electrospun microfibrous PCL. This hybrid scaffold was shown to have a positive effect on rabbit MSCs on aggrecan, collagen II, and collagen II/I expression compared to PCL controls (Levorson et al. 2014). Further, Yang et al. developed a cartilage ECM-derived acellular matrix by physically shattering human cartilage, followed by decellularization, freeze drying, and cross-linking techniques. They showed that ECM enabled attachment, proliferation, and chondrogenic differentiation of canine bone marrow-derived MSCs (Yang et al. 2008). ECM scaffold was also shown to be beneficial in reducing loss of chondrogenic phenotype as shown by using ECM scaffold derived from porcine chondrocytes seeded with rabbit MSCs in vivo compared with PGA scaffold (Choi et al. 2010).

In comparison to other mentioned biomaterials, decellularized cartilage ECM is advantageous in that it importantly recapitulates the native cartilage structure. However, achieving the complexity of articular cartilage structure regarding the mechanical stimulation to which the articular cartilage is constantly subjected and related orientation of collagen fibrils is especially challenging. The effect of mechanical loading and orientation of collagen fibrils on cartilage regeneration potential has been investigated in several studies.

5.4.5 Role of Mechanical Stimulation in Cartilage Regeneration

Since articular cartilage is subjected to constant movement and mechanical load, mechanical stimulation was proposed as a factor to affect ECM development. For example, it was shown in chicken micromass cultures that mechanical loading significantly augmented cartilage matrix production and upregulated expression of collagen type III, aggrecan, and hyaluronan synthases through enhanced expression of SOX9 and protein kinase A activity (Juhasz et al. 2014). Improvement of cartilage formation with reduction of hypertrophy was demonstrated to depend on several parameters, such as loading intensity, duration, and frequency of mechanical stimulation (Thorpe et al. 2012; Haugh et al. 2011; Zhang et al. 2015; O'Conor et al. 2013; Li et al. 2010; Bian et al. 2012). Optimal mechanical load, therefore, plays a crucial role during in vitro chondrogenesis of MSCs. Although mechanical forces importantly regulate MSC chondrogenic gene expression, sustained TGF-β exposure is usually also necessary for mechanically based chondrogenic improvement (Zhang et al. 2015; Huang et al. 2010; Goldman and Barabino 2016). Also, the dosage of growth factor was shown to importantly affect hypertrophy, in that only high levels of TGF-β stabilized chondrogenic phenotype (Zhang et al. 2015; Bian et al. 2012). There are, however, reports on mechanically induced proteoglycan synthesis in the absence of chondrogenic cytokines (Kisiday et al. 2009). In a study investigating the influence of mechanical load on porcine bone marrow-derived MSCs cultured on agarose or fibrin scaffolds, the mechanical load was even shown to override the influence of specific substrates, scaffolds, or hydrogels that have been shown to regulate MSC fate (Thorpe et al. 2012). In contrast to studies supporting the effectiveness of mechanical load in MSC chondrogenesis, it was shown that in the initiation stage of cartilage repair, the mechanical load may not necessarily positively affect the cell fate. In a study investigating the effect of chondrogenic priming of equine peripheral blood MSCs on adhesion and incorporation into cartilage explants, it was shown that mechanical loading reduced the adhesion of cells and altered integration of MSCs into isolated cartilage explants (Spaas et al. 2015). These results are consistent with other studies investigating the effect of biomaterial properties on cell chondrogenesis, showing that mechanical properties can influence cells in terms of their spreading, migration, and differentiation (Toh et al. 2012; Vainieri et al. 2020). This indicates that adjusting biomaterial properties to match mechanical properties, alongside composition and architecture of cartilage, may prevent the incorporation of cells into the cartilage and consequently alter initiation steps of tissue repair (Vainieri et al. 2020). In support of these data, it was also previously demonstrated that mechanical load was associated with bone formation. Mechanical load led to the expression of NGF in mice osteoblasts, followed by the activation of NGF-receptor-positive sensory neurons, resulting in osteogenic cues and bone mass formation (Tomlinson et al. 2017). The data indicate that removing the mechanical load could have a positive effect on MSC in enabling them to reestablish joint homeostasis. Due to the contradictory results from different studies investigating mechanical load on MSCs, further research of the biomechanics, especially early in the disease course, will be needed to provide the data on which MSC repair strategies are needed for optimal cartilage regeneration (McGonagle et al. 2017).

5.4.6 Importance of Biomaterial Architecture

The mechanical performance of articular cartilage directly correlates with the complexity of its structure. Scaffold geometry, recapitulating native orientation of collagen fibrils forming Benninghoff arcades (Benninghoff 1925), thus also seems to play an important role in regulating the cartilage-like activity of cells. For example, bone marrow-derived porcine MSCs expressed collagen type II and synthesized GAGs to a greater extent when cultured on aligned polycaprolactone (PCL) microfibers than on randomly oriented scaffold that was more supportive of an endochondral phenotype as indicated by

higher expression of bone morphogenetic protein-2 (BMP2) and type I collagen gene (Olvera et al. 2017). Similarly, mimicking aligned structures of ECM fibrils in cartilage tissue led to better chondrogenesis of human BM-MSC in a nanofibrous scaffold compared to a scaffold with randomly aligned nanofibers (Zamanlui et al. 2018). Furthermore, it was shown that chondrocytes respond differently to geometrically different scaffolds, for example, nanofibrous poly (L-lactide) scaffold more efficiently promotes the cartilage-like activity of bovine chondrocytes than microfibrous scaffolds (Li et al. 2006). A similar tendency of cells toward favoring nanoultrastructure of the scaffold was shown for MSCs. Culturing human MSCs on nanofibrous polycaprolactone resulted in an increased expression of aggrecan compared to MSCs cultured on a microfibrous scaffold (Schagemann et al. 2013). These studies indicate that nano-topographical geometry with aligned structures is favored by cell types such as chondrocytes and MSCs.

To further improve the imitation of the complex structure of cartilage tissue, Nurnberger et al. (2021) have fabricated decellularized articular cartilage scaffold treated for GAG removal and engraved with a CO_2 laser to create the well-defined structure of native cartilage. With the laser, lines and crossed lines were created allowing enough space for homogenous distribution and for the new matrix to be generated. Interestingly, it was shown that new collagen fibers perpendicularly aligned to the cartilage superficial zone, corresponding to the natural alignment of the collagen fibers, deeming superior over scaffolds that promote random matrix deposition (Nurnberger et al. 2021).

One of the novel techniques used for creating complex 3D scaffold structures is 3D bioprinting, as was shown by printing decellularized ECM cross-linked with gelatin methacrylate. Bioactive factors and cells were quantitatively and accurately placed within to form a bionic multifunctional scaffold to recognize, bind, and recruit endogenous stem cells to the site. Scaffold with implanted aptamers for specifically recognizing and recruiting adipose-derived stem cells, together with TGF-β for stem cell chondrogenesis, resulted in a great improvement of in vivo cartilage full-thickness defects in rabbit models (Yang et al. 2021). Similarly, as in rabbit models, pig models of cartilage defects were used for testing 3D-printed hybrid scaffolds made of gelatine and hydroxyapatite. Gelatine-hydroxyapatite scaffolds, compared to gelatine scaffolds or blank controls, were shown to be the best in reducing hypertrophic markers and repairing cartilage injuries (Huang et al. 2021). 3D bioprinting allows for the fabrication of complicated yet stable structures of tissue analogs and is thus considered a very promising technology, holding considerable potential for articular cartilage repair.

The architectural complexity of cartilage tissue and its constant subjection to mechanical forces demands an understanding of complex mechanisms required for induction of stable chondrogenic phenotype with minimizing the upregulation of hypertrophic genes. Challenges faced in scaffold fabrication are achieving a layered structure mimicking highly specific hierarchical ultrastructure arrangement of ECM of cartilage, mechanical environment for cells resembling native cartilage, and providing physical and biochemical cues to control the biological environment of cells. Mimicking native mechanotransduction pathways may thus be a promising way in creating the desired environment for controlled and stable chondrogenesis. Although cartilaginous tissue structure is well established, its simulation in vitro has proven very challenging, yet novel technologies and increasing acquisition of comprehensive knowledge in regenerative medicine and tissue engineering are encouraging for future cartilage treatment options in both veterinary and human medicine.

6 Summary

Cartilage's unique characteristics encourage scientist to develop methods to overcome its inability to heal. So far, medication-mediated treatment is often the first choice of therapy; however, the therapy is focused on relieving the symptoms but

cannot induce repair or regeneration and is often associated with severe side effects. Due to cartilage avascularity, bone marrow stimulation techniques were developed, which have shown some short-term beneficial effects but resulted in a formation of fibrocartilage, which is mechanically insufficient to bear loading stress. Further attempts at repairing cartilage were focused on using native tissue to produce osteochondral grafts. The main disadvantages of this method are the limited amount of donor cartilage availability, donor site morbidity, and the lack of osteochondral graft integration. To overcome the lack of significant cellular activity with osteochondral grafts, ACI was proposed. ACI seemed to be advantageous over other techniques in that the cartilage that formed was predominantly hyaline-like, containing collagen type II. However, there was an issue with the non-homogenous distribution of chondrocytes and the consequential need for periosteal coverage, resulting in damage associated with periosteal harvest. The latter was overcome with the use of MACI. Although MACI treatment improved cartilage healing, the tissue formed was still inferior to the native hyaline cartilage. Moreover, cultivating chondrocytes is associated with chondrocyte dedifferentiation and thus potentially variable treatment results. Although this was shown as possible to overcome with one-step surgery where minced cartilage instead of isolated chondrocytes were used, novel methods to substitute the use of chondrocytes are being developed. MSCs' immunomodulatory properties and multilineage differentiation ability make them attractive candidates as an alternative to chondrocytes. However, the generation of cartilage tissue from MSC is challenging as in vitro chondrogenic differentiation of MSC reflects endochondral ossification unable to maintain a stable hyaline stage. Hypertrophic development of MSCs leads to the bone formation on ectopic sites and is thus unsuitable for cartilage therapy in vivo. Other approaches in the induction of stable chondrogenic phenotype of MSCs are being investigated, with the focus on recapitulating MSCs native environment and providing MSCs the best options to express their biological function. Many novel biomaterials are thus at the forefront of cartilage regeneration research, from standard collagen-based matrices to novel decellularized ECM cell carriers. Recapitulating the exact architecture of cartilage tissue has proven challenging yet of great importance for cartilage tissue engineering. Despite advances made in biomaterial-based stem cells therapies, each scaffold material currently used in tissue engineering approaches is still limited in possessing all the requirements needed for cartilage regeneration. Moreover, the knowledge of stem cell mechanisms of action is still elusive. A more detailed comprehensive understanding of the MSC mechanisms of action and their responses to complex structural, architectural, and geometrical properties of biomaterials is therefore needed to find the most appropriate way of delivering stable cartilage tissue formation. Combining technologies and knowledge of different scientific fields is essential for engineering a biomaterial that would fundamentally contribute to cartilage regeneration. The collaboration of scientists from interdisciplinary fields is thus of key importance for the further development of advanced cartilage therapies. Looking forward, one can be hopeful that, based on the novel cutting-edge technologies being available and progressive knowledge acquisition, we are on the verge of future developmental breakthroughs in the field of cartilage regeneration.

References

Acosta CA et al (2006) Gene expression and proliferation analysis in young, aged, and osteoarthritic sheep chondrocytes effect of growth factor treatment. J Orthop Res 24(11):2087–2094

Ahmed TA, Hincke MT (2010) Strategies for articular cartilage lesion repair and functional restoration. Tissue Eng Part B Rev 16(3):305–329

Ahmed MR et al (2014) Combination of ADMSCs and chondrocytes reduces hypertrophy and improves the functional properties of osteoarthritic cartilage. Osteoarthr Cartil 22(11):1894–1901

Altman RD et al (1992) Preliminary observations of chondral abrasion in a canine model. Ann Rheum Dis 51(9):1056–1062

Alves JC et al (2020) A pilot study on the efficacy of a single intra-articular administration of triamcinolone acetonide, hyaluronan, and a combination of both for clinical management of osteoarthritis in police working dogs. Front Vet Sci 7:512523

Amann E et al (2017) Hyaluronic acid facilitates chondrogenesis and matrix deposition of human adipose derived mesenchymal stem cells and human chondrocytes co-cultures. Acta Biomater 52:130–144

An C et al (2010) IGF-1 and BMP-2 induces differentiation of adipose-derived mesenchymal stem cells into chondrocytes-like cells. Ann Biomed Eng 38(4):1647–1654

Anderson KL et al (2020) Risk factors for canine osteoarthritis and its predisposing arthropathies: a systematic review. Front Vet Sci 7:220

Anderson-Baron M et al (2020) Suppression of hypertrophy during in vitro chondrogenesis of cocultures of human mesenchymal stem cells and nasal chondrocytes correlates with lack of in vivo calcification and vascular invasion. Front Bioeng Biotechnol 8:572356

Ando I et al (2016) Changes in serum NGF levels after the exercise load in dogs: a pilot study. J Vet Med Sci 78(11):1709–1712

Ando I et al (2020) Evaluation of stress status using the stress map for guide dog candidates in the training stage using variations in the serum cortisol with nerve growth factor and magnesium ions. Vet Anim Sci 10:100129

Antoni D et al (2015) Three-dimensional cell culture: a breakthrough in vivo. Int J Mol Sci 16(3):5517–5527

Aragon CL, Hofmeister EH, Budsberg SC (2007) Systematic review of clinical trials of treatments for osteoarthritis in dogs. J Am Vet Med Assoc 230(4):514–521

Arevalo-Turrubiarte M et al (2019) Analysis of mesenchymal cells (MSCs) from bone marrow, synovial fluid and mesenteric, neck and tail adipose tissue sources from equines. Stem Cell Res 37:101442

Armiento AR, Alini M, Stoddart MJ (2019) Articular fibrocartilage – why does hyaline cartilage fail to repair? Adv Drug Deliv Rev 146:289–305

Barlian A et al (2018) Chondrogenic differentiation of adipose-derived mesenchymal stem cells induced by L-ascorbic acid and platelet rich plasma on silk fibroin scaffold. Peer J 6:e5809

Barnewitz D et al (2003) Tissue engineering: new treatment of cartilage alterations in degenerative joint diseases in horses – preliminary results of a long term study. Berl Munch Tierarztl Wochenschr 116(3-4):157–161

Bedingfield SK et al (2020) Matrix-targeted nanoparticles for MMP13 RNA interference blocks post-traumatic osteoarthritis. bioRxiv:2020.01.30.925321

Bekkers JE et al (2013a) Single-stage cell-based cartilage regeneration using a combination of chondrons and mesenchymal stromal cells: comparison with microfracture. Am J Sports Med 41(9):2158–2166

Bekkers JE et al (2013b) One-stage focal cartilage defect treatment with bone marrow mononuclear cells and chondrocytes leads to better macroscopic cartilage regeneration compared to microfracture in goats. Osteoarthr Cartil 21(7):950–956

Benninghoff A (1925) Form und Bau der Gelenkknorpel in ihren Beziehungen zur Funktion. Z Zellforsch Mikrosk Anat 2(5):783–862

Benoit DS et al (2008) Small functional groups for controlled differentiation of hydrogel-encapsulated human mesenchymal stem cells. Nat Mater 7(10):816–823

Bertoni L et al (2021) Evaluation of allogeneic bone-marrow-derived and umbilical cord blood-derived mesenchymal stem cells to prevent the development of osteoarthritis in an equine model. Int J Mol Sci 22(5)

Bian L et al (2012) Dynamic compressive loading enhances cartilage matrix synthesis and distribution and suppresses hypertrophy in hMSC-laden hyaluronic acid hydrogels. Tissue Eng Part A 18(7-8):715–724

Black LL et al (2007) Effect of adipose-derived mesenchymal stem and regenerative cells on lameness in dogs with chronic osteoarthritis of the coxofemoral joints: a randomized, double-blinded, multicenter, controlled trial. Vet Ther 8(4):272–284

Bodo G et al (2004) Autologous osteochondral grafting (mosaic arthroplasty) for treatment of subchondral cystic lesions in the equine stifle and fetlock joints. Vet Surg 33(6):588–596

Bora FW Jr, Miller G (1987) Joint physiology, cartilage metabolism, and the etiology of osteoarthritis. Hand Clin 3(3):325–336

Borochowitz ZU et al (2004) Spondylo-epi-metaphyseal dysplasia (SEMD) matrilin 3 type: homozygote matrilin 3 mutation in a novel form of SEMD. J Med Genet 41(5):366–372

Bosnakovski D et al (2004) Chondrogenic differentiation of bovine bone marrow mesenchymal stem cells in pellet cultural system. Exp Hematol 32(5):502–509

Bouck GR, Miller CW, Taves CL (1995) A comparison of surgical and medical treatment of fragmented coronoid process and osteochondritis dissecans of the canine elbow. Vet Comp Orthop Traumatol 08(04):177–183

Branly T et al (2018) Improvement of the chondrocyte-specific phenotype upon equine bone marrow mesenchymal stem cell differentiation: influence of culture time, transforming growth factors and type I collagen siRNAs on the differentiation index. Int J Mol Sci 19(2)

Breinan HA et al (1997) Effect of cultured autologous chondrocytes on repair of chondral defects in a canine model. J Bone Joint Surg Am 79(10):1439–1451

Breinan HA et al (2000) Healing of canine articular cartilage defects treated with microfracture, a type-II collagen matrix, or cultured autologous chondrocytes. J Orthop Res 18(5):781–789

Brighton CT, Heppenstall RB (1971) Oxygen tension in zones of the epiphyseal plate, the metaphysis and diaphysis. An in vitro and in vivo study in rats and rabbits. J Bone Joint Surg Am 53(4):719–728

Brittberg M et al (1994) Treatment of deep cartilage defects in the knee with autologous chondrocyte transplantation. N Engl J Med 331(14):889–895

Brondeel C et al (2021) Review: mesenchymal stem cell therapy in canine osteoarthritis research: "Experientia Docet" (Experience will teach us). Front Vet Sci 8:668881

Browe DC et al (2019) Hypoxia activates the PTHrP-MEF2C pathway to attenuate hypertrophy in mesenchymal stem cell derived cartilage. Sci Rep 9(1):13274

Bruderer M et al (2014) Role and regulation of RUNX2 in osteogenesis. Eur Cell Mater 28:269–286

Burks RT et al (2006) The use of a single osteochondral autograft plug in the treatment of a large osteochondral lesion in the femoral condyle: an experimental study in sheep. Am J Sports Med 34(2):247–255

Cao Z et al (2017) Hypertrophic differentiation of mesenchymal stem cells is suppressed by xanthotoxin via the p38MAPK/HDAC4 pathway. Mol Med Rep 16(3):2740–2746

Caplan AI (1991) Mesenchymal stem cells. J Orthop Res 9(5):641–650

Caplan AI (2017) Mesenchymal stem cells: time to change the name! Stem Cells Transl Med 6(6):1445–1451

Carrade DD et al (2011) Clinicopathologic findings following intra-articular injection of autologous and allogeneic placentally derived equine mesenchymal stem cells in horses. Cytotherapy 13(4):419–430

Catarino J et al (2020) Treatment of canine osteoarthritis with allogeneic platelet-rich plasma: review of five cases. Open Vet J 10(2):226–231

Chen S et al (2019) MicroRNA-218 promotes early chondrogenesis of mesenchymal stem cells and inhibits later chondrocyte maturation. BMC Biotechnol 19(1):6

Cherubino P et al (2003) Autologous chondrocyte implantation using a bilayer collagen membrane: a preliminary report. J Orthop Surg (Hong Kong) 11(1):10–15

Chiara G, Ranieri C (2009) Cartilage and bone extracellular matrix. Curr Pharm Des 15(12):1334–1348

Cho H, Lee A, Kim K (2018) The effect of serum types on chondrogenic differentiation of adipose-derived stem cells. Biomater Res 22:6

Choi KH et al (2010) The chondrogenic differentiation of mesenchymal stem cells on an extracellular matrix scaffold derived from porcine chondrocytes. Biomaterials 31(20):5355–5365

Clarke SP et al (2005) Prevalence of radiographic signs of degenerative joint disease in a hospital population of cats. Vet Rec 157(25):793–799

Cohen JM et al (2009) Long-term outcome in 44 horses with stifle lameness after arthroscopic exploration and debridement. Vet Surg 38(4):543–551

Cook JL, Payne JT (1997) Surgical treatment of osteoarthritis. Vet Clin North Am Small Anim Pract 27(4):931–944

Cook JL, Hudson CC, Kuroki K (2008) Autogenous osteochondral grafting for treatment of stifle osteochondrosis in dogs. Vet Surg 37(4):311–321

Cook JL et al (2014) A novel system improves preservation of osteochondral allografts. Clin Orthop Relat Res 472(11):3404–3414

Cook JL et al (2016) Importance of donor chondrocyte viability for osteochondral allografts. Am J Sports Med 44(5):1260–1268

Cope PJ et al (2019) Models of osteoarthritis: the good, the bad and the promising. Osteoarthr Cartil 27(2): 230–239

Cortes I et al (2021) A scaffold- and serum-free method to mimic human stable cartilage validated by secretome. Tissue Eng Part A 27(5-6):311–327

Cuervo B et al (2020) Objective comparison between platelet rich plasma alone and in combination with physical therapy in dogs with osteoarthritis caused by hip dysplasia. Animals (Basel) 10(2)

Curran JM, Chen R, Hunt JA (2006) The guidance of human mesenchymal stem cell differentiation in vitro by controlled modifications to the cell substrate. Biomaterials 27(27):4783–4793

De Angelis E et al (2020) Gene expression markers in horse articular chondrocytes: chondrogenic differentiation IN VITRO depends on the proliferative potential and ageing. Implication for tissue engineering of cartilage. Res Vet Sci 128:107–117

De Armond CC et al (2021) Three-dimensional-printed custom guides for bipolar coxofemoral osteochondral allograft in dogs. PLoS One 16(2):e0244208

De Bari C, Dell'Accio F, Luyten FP (2001) Human periosteum-derived cells maintain phenotypic stability and chondrogenic potential throughout expansion regardless of donor age. Arthritis Rheum 44(1):85–95

Dell'Accio F et al (2003) Expanded phenotypically stable chondrocytes persist in the repair tissue and contribute to cartilage matrix formation and structural integration in a goat model of autologous chondrocyte implantation. J Orthop Res 21(1):123–131

Dennis JE et al (1999) A quadripotential mesenchymal progenitor cell isolated from the marrow of an adult mouse. J Bone Miner Res 14(5):700–709

Denys M et al (2020) Biosafety evaluation of equine Umbilical Cord-derived Mesenchymal Stromal Cells (UC-MSCs) by systematic pathogen screening in peripheral maternal blood and paired UC-MSCs. Biopreserv Biobank

Dew TL, Martin RA (1992) Functional, radiographic, and histologic assessment of healing of autogenous osteochondral grafts and full-thickness cartilage defects in the talus of dogs. Am J Vet Res 53(11): 2141–2152

Dieppe PA, Lohmander LS (2005) Pathogenesis and management of pain in osteoarthritis. Lancet 365(9463): 965–973

Dominici M et al (2006) Minimal criteria for defining multipotent mesenchymal stromal cells. The international society for cellular therapy position statement. Cytotherapy 8(4):315–317

Dou C et al (2016) Xanthotoxin prevents bone loss in ovariectomized mice through the inhibition of RANKL-induced osteoclastogenesis. Osteoporos Int 27(7):2335–2344

Dreher SI et al (2020) Significance of MEF2C and RUNX3 regulation for endochondral differentiation of human mesenchymal progenitor cells. Front Cell Dev Biol 8:81

Dudakovic A et al (2014) High-resolution molecular validation of self-renewal and spontaneous differentiation in clinical-grade adipose-tissue derived human mesenchymal stem cells. J Cell Biochem 115(10): 1816–1828

Dyson SJ (2004) Medical management of superficial digital flexor tendonitis: a comparative study in 219 horses (1992–2000). Equine Vet J 36(5):415–419

Egger D et al (2019) From 3D to 3D: isolation of mesenchymal stem/stromal cells into a three-dimensional human platelet lysate matrix. Stem Cell Res Ther 10(1):248

Enomoto M et al (2019) Anti-nerve growth factor monoclonal antibodies for the control of pain in dogs and cats. Vet Rec 184(1):23

Familiari F et al (2018) Clinical outcomes and failure rates of osteochondral allograft transplantation in the knee: a systematic review. Am J Sports Med 46(14): 3541–3549

Fan L et al (2012) Vitamin C-reinforcing silk fibroin nanofibrous matrices for skin care application. RCS Adv 2:4110–4119

Feng Q et al (2017) Sulfated hyaluronic acid hydrogels with retarded degradation and enhanced growth factor retention promote hMSC chondrogenesis and articular cartilage integrity with reduced hypertrophy. Acta Biomater 53:329–342

Ferris DJ et al (2014) Clinical outcome after intra-articular administration of bone marrow derived mesenchymal stem cells in 33 horses with stifle injury. Vet Surg 43(3):255–265

Fischer J et al (2010) Human articular chondrocytes secrete parathyroid hormone-related protein and inhibit hypertrophy of mesenchymal stem cells in coculture during chondrogenesis. Arthritis Rheum 62(9): 2696–2706

Fischer J et al (2014) Intermittent PTHrP(1-34) exposure augments chondrogenesis and reduces hypertrophy of mesenchymal stromal cells. Stem Cells Dev 23(20): 2513–2523

Fitzpatrick N, Yeadon R, Smith TJ (2009) Early clinical experience with osteochondral autograft transfer for treatment of osteochondritis dissecans of the medial humeral condyle in dogs. Vet Surg 38(2): 246–260

Fitzpatrick N et al (2010) Osteochondral autograft transfer for treatment of osteochondritis dissecans of the caudocentral humeral head in dogs. Vet Surg 39(8): 925–935

Fitzpatrick N et al (2012) Osteochondral autograft transfer for the treatment of osteochondritis dissecans of the medial femoral condyle in dogs. Vet Comp Orthop Traumatol 25(2):135–143

Fortier LA et al (1998) Isolation and chondrocytic differentiation of equine bone marrow-derived mesenchymal stem cells. Am J Vet Res 59(9):1182–1187

Foster TE et al (2009) Platelet-rich plasma: from basic science to clinical applications. Am J Sports Med 37(11):2259–2272

Franklin SP, Cook JL (2013) Prospective trial of autologous conditioned plasma versus hyaluronan plus corticosteroid for elbow osteoarthritis in dogs. Can Vet J 54(9):881–884

Franklin SP, Garner BC, Cook JL (2015) Characteristics of canine platelet-rich plasma prepared with five commercially available systems. Am J Vet Res 76(9): 822–827

Frisbie DD et al (1999) Arthroscopic subchondral bone plate microfracture technique augments healing of large chondral defects in the radial carpal bone and medial femoral condyle of horses. Vet Surg 28(4): 242–255

Frisbie DD, Cross MW, McIlwraith CW (2006) A comparative study of articular cartilage thickness in the stifle of animal species used in human pre-clinical studies compared to articular cartilage thickness in the human knee. Vet Comp Orthop Traumatol 19(3): 142–146

Frisbie DD et al (2008) Evaluation of autologous chondrocyte transplantation via a collagen membrane in equine articular defects: results at 12 and 18 months. Osteoarthr Cartil 16(6):667–679

Frisbie DD et al (2009) In vivo evaluation of autologous cartilage fragment-loaded scaffolds implanted into equine articular defects and compared with autologous chondrocyte implantation. Am J Sports Med 37(Suppl 1):71S–80S

Fujiki M et al (2007) Effects of treatment with polysulfated glycosaminoglycan on serum cartilage oligomeric matrix protein and C-reactive protein concentrations, serum matrix metalloproteinase-2 and -9 activities, and lameness in dogs with osteoarthritis. Am J Vet Res 68(8):827–833

Gale AL et al (2019) The effect of hypoxia on chondrogenesis of equine synovial membrane-derived and bone marrow-derived mesenchymal stem cells. BMC Vet Res 15(1):201

Gandhimathi C (2015) Controlled release of dexamethasone in PCL/silk fibroin/ascorbic acid nanoparticles for the initiation of adipose derived stem cells into osteogenesis. J Drug Metab Toxicol 6(1):2

Ge Z et al (2012) Functional biomaterials for cartilage regeneration. J Biomed Mater Res A 100(9): 2526–2536

Gearing DP et al (2013) A fully caninised anti-NGF monoclonal antibody for pain relief in dogs. BMC Vet Res 9: 226

Gearing DP et al (2016) In vitro and in vivo characterization of a fully felinized therapeutic anti-nerve growth factor monoclonal antibody for the treatment of pain in cats. J Vet Intern Med 30(4):1129–1137

Gelse K et al (2011) Thrombospondin-1 prevents excessive ossification in cartilage repair tissue induced by osteogenic protein-1. Tissue Eng Part A 17(15-16): 2101–2112

Gencoglu H et al (2020) Undenatured type II Collagen (UC-II) in Joint health and disease: a review on the current knowledge of companion animals. Animals (Basel) 10(4)

Giannoni P et al (2005) Species variability in the differentiation potential of in vitro-expanded articular chondrocytes restricts predictive studies on cartilage repair using animal models. Tissue Eng 11(1-2):237–248

Gill TJ, Steadman JR (2004) Bone marrow stimulation techniques: microfracture, drilling, and abrasion. In: Cole BJ, Malek MM (eds) Articular cartilage lesions: a practical guide to assessment and treatment. Springer, New York, pp 63–72

Giovannini S et al (2010) Micromass co-culture of human articular chondrocytes and human bone marrow mesenchymal stem cells to investigate stable neocartilage tissue formation in vitro. Eur Cell Mater 20:245–259

Glenn RE Jr et al (2006) Comparison of fresh osteochondral autografts and allografts: a canine model. Am J Sports Med 34(7):1084–1093

Godwin EE et al (2012) Implantation of bone marrow-derived mesenchymal stem cells demonstrates improved outcome in horses with overstrain injury of the superficial digital flexor tendon. Equine Vet J 44(1):25–32

Goldman SM, Barabino GA (2016) Hydrodynamic loading in concomitance with exogenous cytokine stimulation modulates differentiation of bovine mesenchymal stem cells towards osteochondral lineages. BMC Biotechnol 16:10

Grassel S, Ahmed N (2007) Influence of cellular microenvironment and paracrine signals on chondrogenic differentiation. Front Biosci 12:4946–4956

Griffin DJ et al (2015) Mechanical characterization of matrix-induced autologous chondrocyte implantation (MACI(R)) grafts in an equine model at 53 weeks. J Biomech 48(10):1944–1949

Gruen ME et al (2016) A feline-specific anti-nerve growth factor antibody improves mobility in cats with degenerative joint disease-associated pain: a pilot proof of concept study. J Vet Intern Med 30(4):1138–1148

Guillen-Garcia P et al (2014) Increasing the dose of autologous chondrocytes improves articular cartilage repair: histological and molecular study in the sheep animal model. Cartilage 5(2):114–122

Guimaraes-Camboa N et al (2017) Pericytes of multiple organs do not behave as mesenchymal stem cells in vivo. Cell Stem Cell 20(3):345–359 e5

Haddo O et al (2004) The use of chondrogide membrane in autologous chondrocyte implantation. Knee 11(1): 51–55

Handorf AM, Li WJ (2011) Fibroblast growth factor-2 primes human mesenchymal stem cells for enhanced chondrogenesis. PLoS One 6(7):e22887

Harman R et al (2016) A prospective, randomized, masked, and placebo-controlled efficacy study of intraarticular allogeneic adipose stem cells for the treatment of osteoarthritis in dogs. Front Vet Sci 3:81

Harper TAM (2017a) Femoral head and neck excision. Vet Clin North Am Small Anim Pract 47(4):885–897

Harper TAM (2017b) INNOPLANT total hip replacement system. Vet Clin North Am Small Anim Pract 47(4): 935–944

Haugh MG et al (2011) Temporal and spatial changes in cartilage-matrix-specific gene expression in mesenchymal stem cells in response to dynamic compression. Tissue Eng Part A 17(23-24):3085–3093

Hayes DW Jr, Brower RL, John KJ (2001) Articular cartilage. Anatomy, injury, and repair. Clin Podiatr Med Surg 18(1):35–53

Hirao M et al (2006) Oxygen tension regulates chondrocyte differentiation and function during endochondral ossification. J Biol Chem 281(41):31079–31092

Hoshi K et al (2018) Biological aspects of tissue-engineered cartilage. Histochem Cell Biol 149(4): 375–381

Hu Y et al (2020) A lithium-containing biomaterial promotes chondrogenic differentiation of induced pluripotent stem cells with reducing hypertrophy. Stem Cell Res Ther 11(1):77

Huang FS et al (2004) Effects of small incongruities in a sheep model of osteochondral autografting. Am J Sports Med 32(8):1842–1848

Huang AH et al (2010) Long-term dynamic loading improves the mechanical properties of chondrogenic mesenchymal stem cell-laden hydrogel. Eur Cell Mater 19:72–85

Huang J et al (2021) 3D printed gelatin/hydroxyapatite scaffolds for stem cell chondrogenic differentiation and articular cartilage repair. Biomater Sci 9(7): 2620–2630

Hunziker EB, Stahli A (2008) Surgical suturing of articular cartilage induces osteoarthritis-like changes. Osteoarthr Cartil 16(9):1067–1073

Hurtig M et al (2001) Arthroscopic mosaic arthroplasty in the equine third carpal bone. Vet Surg 30(3):228–239

Ibarra C et al (2006) Tissue engineered arthroscopic repair of experimental cartilage lesions in horses (SS-47). Arthroscop J Arthroscop Relat Surg 22-(6, Supplement):e24

Isola M et al (2011) Nerve growth factor concentrations in the synovial fluid from healthy dogs and dogs with secondary osteoarthritis. Vet Comp Orthop Traumatol 24(4):279–284

Janakiramanan N et al (2006) Osteoarthritis cartilage defects: does size matter? Curr Rheumatol Rev 2(4): 311–317

Jayasuriya CT et al (2012) Matrilin-3 induction of IL-1 receptor antagonist is required for up-regulating collagen II and aggrecan and down-regulating ADAMTS-5 gene expression. Arthritis Res Ther 14(5):R197

Jensen C, Teng Y (2020) Is it time to start transitioning from 2D to 3D cell culture? Front Mol Biosci 7:33

Johnstone B et al (1998) In vitro chondrogenesis of bone marrow-derived mesenchymal progenitor cells. Exp Cell Res 238(1):265–272

Juhasz T et al (2014) Mechanical loading stimulates chondrogenesis via the PKA/CREB-Sox9 and PP2A pathways in chicken micromass cultures. Cell Signal 26(3):468–482

Kang BJ et al (2012) Comparing the osteogenic potential of canine mesenchymal stem cells derived from adipose tissues, bone marrow, umbilical cord blood, and Wharton's jelly for treating bone defects. J Vet Sci 13(3):299–310

Kanichai M et al (2008) Hypoxia promotes chondrogenesis in rat mesenchymal stem cells: a role for AKT and hypoxia-inducible factor (HIF)-1alpha. J Cell Physiol 216(3):708–715

Kapusetti G, More N, Choppadandi M (2019) Introduction to ideal characteristics and advanced biomedical applications of biomaterials. In: Paul S (ed) Biomedical engineering and its applications in healthcare. Springer, Singapore, pp 171–204

Karl A et al (2014) Thyroid hormone-induced hypertrophy in mesenchymal stem cell chondrogenesis is mediated by bone morphogenetic protein-4. Tissue Eng Part A 20(1-2):178–188

Kawamoto K et al (1996) Nerve growth factor activity detected in equine peripheral blood of horses with fever after truck transportation. J Equin Sci 7(2):43–46

Kendall A et al (2021) Nerve growth factor in the equine joint. Vet J 267:105579

Kim SJ et al (2020a) Articular cartilage repair using autologous collagen-induced chondrogenesis (ACIC): a pragmatic and cost-effective enhancement of a traditional technique. Knee Surg Sports Traumatol Arthrosc 28(8):2598–2603

Kim SA et al (2020b) Atelocollagen promotes chondrogenic differentiation of human adipose-derived mesenchymal stem cells. Sci Rep 10(1):10678

Kisiday JD et al (2009) Dynamic compression stimulates proteoglycan synthesis by mesenchymal stem cells in the absence of chondrogenic cytokines. Tissue Eng Part A 15(10):2817–2824

Kisiel AH et al (2012) Isolation, characterization, and in vitro proliferation of canine mesenchymal stem cells derived from bone marrow, adipose tissue, muscle, and periosteum. Am J Vet Res 73(8):1305–1317

Koch TG et al (2007) Isolation of mesenchymal stem cells from equine umbilical cord blood. BMC Biotechnol 7: 26

Kriston-Pal E et al (2020) A regenerative approach to canine osteoarthritis using allogeneic, adipose-derived mesenchymal stem cells. Safety results of a long-term follow-up. Front Vet Sci 7:510

Kronenberg HM (2003) Developmental regulation of the growth plate. Nature 423(6937):332–336

Kundu B et al (2013) Silk fibroin biomaterials for tissue regenerations. Adv Drug Deliv Rev 65(4):457–470

Lafaver S et al (2007) Tibial tuberosity advancement for stabilization of the canine cranial cruciate ligament-deficient stifle joint: surgical technique, early results, and complications in 101 dogs. Vet Surg 36(6): 573–586

Lane JG et al (2004) Follow-up of osteochondral plug transfers in a goat model: a 6-month study. Am J Sports Med 32(6):1440–1450

Lascelles BD et al (2015) A canine-specific anti-nerve growth factor antibody alleviates pain and improves mobility and function in dogs with degenerative joint disease-associated pain. BMC Vet Res 11:101

Lee CR et al (2000) Effects of harvest and selected cartilage repair procedures on the physical and biochemical properties of articular cartilage in the canine knee. J Orthop Res 18(5):790–799

Lee CR et al (2003) Effects of a cultured autologous chondrocyte-seeded type II collagen scaffold on the healing of a chondral defect in a canine model. J Orthop Res 21(2):272–281

Lee J et al (2016) Chondrogenic potential and anti-senescence effect of hypoxia on canine adipose mesenchymal stem cells. Vet Res Commun 40(1):1–10

Levorson EJ et al (2014) Cell-derived polymer/extracellular matrix composite scaffolds for cartilage regeneration, Part 2: construct devitalization and determination of chondroinductive capacity. Tissue Eng Part C Methods 20(4):358–372

Li J, Pei M (2018) A protocol to prepare decellularized stem cell matrix for rejuvenation of cell expansion and cartilage regeneration. Methods Mol Biol 1577:147–154

Li WJ, Jiang YJ, Tuan RS (2006) Chondrocyte phenotype in engineered fibrous matrix is regulated by fiber size. Tissue Eng 12(7):1775–1785

Li Z et al (2010) Chondrogenesis of human bone marrow mesenchymal stem cells in fibrin-polyurethane composites is modulated by frequency and amplitude of dynamic compression and shear stress. Tissue Eng Part A 16(2):575–584

Liu X et al (2010) In vivo ectopic chondrogenesis of BMSCs directed by mature chondrocytes. Biomaterials 31(36):9406–9414

Liu Q et al (2018) Suppressing mesenchymal stem cell hypertrophy and endochondral ossification in 3D cartilage regeneration with nanofibrous poly(l-lactic acid) scaffold and matrilin-3. Acta Biomater 76:29–38

Longhini ALF et al (2019) Peripheral blood-derived mesenchymal stem cells demonstrate immunomodulatory potential for therapeutic use in horses. PLoS One 14(3):e0212642

Lu Y et al (2006) Minced cartilage without cell culture serves as an effective intraoperative cell source for cartilage repair. J Orthop Res 24(6):1261–1270

Magri C et al (2019) Comparison of efficacy and safety of single versus repeated intra-articular injection of allogeneic neonatal mesenchymal stem cells for treatment of osteoarthritis of the metacarpophalangeal/metatarsophalangeal joint in horses: A clinical pilot study. PLoS One 14(8):e0221317

Maki CB et al (2020) Intra-articular administration of allogeneic adipose derived MSCs reduces pain and lameness in dogs with hip osteoarthritis: a double blinded, randomized, placebo controlled pilot study. Front Vet Sci 7:570

Mantyh PW et al (2011) Antagonism of nerve growth factor-TrkA signaling and the relief of pain. Anesthesiology 115(1):189–204

Marinas-Pardo L et al (2018) Allogeneic adipose-derived mesenchymal stem cells (Horse Allo 20) for the treatment of osteoarthritis-associated lameness in horses: characterization, safety, and efficacy of intra-articular treatment. Stem Cells Dev 27(17):1147–1160

Martino S et al (2012) Stem cell-biomaterial interactions for regenerative medicine. Biotechnol Adv 30(1): 338–351

Masri M et al (2007) Matrix-encapsulation cell-seeding technique to prevent cell detachment during arthroscopic implantation of matrix-induced autologous chondrocytes. Arthroscopy 23(8):877–883

Matsuda H et al (1991) Nerve growth factor-like activity detected in equine peripheral blood after running exercise. Zentralbl Veterinarmed A 38(7):557–559

Maumus M et al (2013) Adipose mesenchymal stem cells protect chondrocytes from degeneration associated with osteoarthritis. Stem Cell Res 11(2):834–844

McCarty EC et al (2016) Fresh osteochondral allograft versus autograft: twelve-month results in isolated canine knee defects. Am J Sports Med 44(9): 2354–2365

McCarthy J et al (2020) Elbow arthrodesis using a medially positioned plate in 6 dogs. Vet Comp Orthop Traumatol 33(1):51–58

McGonagle D, Baboolal TG, Jones E (2017) Native joint-resident mesenchymal stem cells for cartilage repair in osteoarthritis. Nat Rev Rheumatol 13(12):719–730

Mehana EE, Khafaga AF, El-Blehi SS (2019) The role of matrix metalloproteinases in osteoarthritis pathogenesis: an updated review. Life Sci 234:116786

Meirelles Lda S et al (2009) Mechanisms involved in the therapeutic properties of mesenchymal stem cells. Cytokine Growth Factor Rev 20(5-6):419–427

Mello MA, Tuan RS (1999) High density micromass cultures of embryonic limb bud mesenchymal cells: an in vitro model of endochondral skeletal development. In Vitro Cell Dev Biol Anim 35(5):262–269

Mensing N et al (2011) Isolation and characterization of multipotent mesenchymal stromal cells from the gingiva and the periodontal ligament of the horse. BMC Vet Res 7:42

Meretoja VV et al (2013) The effect of hypoxia on the chondrogenic differentiation of co-cultured articular chondrocytes and mesenchymal stem cells in scaffolds. Biomaterials 34(17):4266–4273

Min BH et al (2007) The fate of implanted autologous chondrocytes in regenerated articular cartilage. Proc Inst Mech Eng H 221(5):461–465

Mirza MH et al (2016) Gait changes vary among horses with naturally occurring osteoarthritis following intra-articular administration of autologous platelet-rich plasma. Front Vet Sci 3:29

Mobasheri A et al (2017) The role of metabolism in the pathogenesis of osteoarthritis. Nat Rev Rheumatol 13(5):302–311

Mohoric L et al (2016) Blinded placebo study of bilateral osteoarthritis treatment using adipose derived mesenchymal stem cells. Slov Vet Res 53(3):167–174

Morrison SJ et al (1997) Identification of a lineage of multipotent hematopoietic progenitors. Development 124(10):1929–1939

Mueller MB, Tuan RS (2008) Functional characterization of hypertrophy in chondrogenesis of human mesenchymal stem cells. Arthritis Rheum 58(5):1377–1388

Mueller MB et al (2013) Effect of parathyroid hormone-related protein in an in vitro hypertrophy model for mesenchymal stem cell chondrogenesis. Int Orthop 37(5):945–951

Muir P et al (2016) Autologous bone marrow-derived mesenchymal stem cells modulate molecular markers of inflammation in dogs with cruciate ligament rupture. PLoS One 11(8):e0159095

Muttigi MS et al (2020) Matrilin-3-primed adipose-derived mesenchymal stromal cell spheroids prevent mesenchymal stromal-cell-derived chondrocyte hypertrophy. Int J Mol Sci 21(23)

Mwale F et al (2006) Limitations of using aggrecan and type X collagen as markers of chondrogenesis in mesenchymal stem cell differentiation. J Orthop Res 24(8):1791–1798

Mwale F et al (2010) Effect of parathyroid hormone on type X and type II collagen expression in mesenchymal stem cells from osteoarthritic patients. Tissue Eng Part A 16(11):3449–3455

Nakase T et al (2002) Distribution of genes for parathyroid hormone (PTH)-related peptide, Indian hedgehog, PTH receptor and patched in the process of experimental spondylosis in mice. J Neurosurg 97(1 Suppl):82–87

Naruse K et al (2004) Spontaneous differentiation of mesenchymal stem cells obtained from fetal rat circulation. Bone 35(4):850–858

Ness MG (2016) The Modified Maquet Procedure (MMP) in dogs: technical development and initial clinical experience. J Am Anim Hosp Assoc 52(4):242–250

Nganvongpanit K et al (2009) Prospective evaluation of serum biomarker levels and cartilage repair by autologous chondrocyte transplantation and subchondral drilling in a canine model. Arthritis Res Ther 11(3):R78

Nicpon J, Marycz K, Grzesiak J (2013) Therapeutic effect of adipose-derived mesenchymal stem cell injection in horses suffering from bone spavin. Pol J Vet Sci 16(4):753–754

Nishiwaki H et al (2017) A novel method to induce cartilage regeneration with cubic microcartilage. Cells Tissues Organs 204(5-6):251–260

Nixon AJ et al (2011) Autologous chondrocyte implantation drives early chondrogenesis and organized repair in extensive full- and partial-thickness cartilage defects in an equine model. J Orthop Res 29(7):1121–1130

Nixon AJ et al (2015) A chondrocyte infiltrated collagen type I/III membrane (MACI(R) implant) improves cartilage healing in the equine patellofemoral joint model. Osteoarthr Cartil 23(4):648–660

Nixon AJ et al (2017) Matrix-induced Autologous Chondrocyte Implantation (MACI) using a cell-seeded collagen membrane improves cartilage healing in the equine model. J Bone Joint Surg Am 99(23):1987–1998

Nürnberger S et al (2006) Ultrastructural insights into the world of cartilage: electron microscopy of articular cartilage. Osteosynth Trauma Care 14(03):168–180

Nuernberger S et al (2011) The influence of scaffold architecture on chondrocyte distribution and behavior in matrix-associated chondrocyte transplantation grafts. Biomaterials 32(4):1032–1040

Nurnberger S et al (2021) Repopulation of decellularised articular cartilage by laser-based matrix engraving. EBioMedicine 64:103196

O'Driscoll SW, Fitzsimmons JS (2001) The role of periosteum in cartilage repair. Clin Orthop Relat Res 391:S190–S207

O'Conor CJ, Case N, Guilak F (2013) Mechanical regulation of chondrogenesis. Stem Cell Res Ther 4(4):61

Olvera D et al (2017) Modulating microfibrillar alignment and growth factor stimulation to regulate mesenchymal stem cell differentiation. Acta Biomater 64:148–160

Pacini S et al (2007) Suspension of bone marrow-derived undifferentiated mesenchymal stromal cells for repair of superficial digital flexor tendon in race horses. Tissue Eng 13(12):2949–2955

Patil AS, Sable RB, Kothari RM (2012) Role of insulin-like growth factors (IGFs), their receptors and genetic regulation in the chondrogenesis and growth of the mandibular condylar cartilage. J Cell Physiol 227(5):1796–1804

Pei M et al (2009) Histone deacetylase 4 promotes TGF-beta1-induced synovium-derived stem cell chondrogenesis but inhibits chondrogenically differentiated stem cell hypertrophy. Differentiation 78(5):260–268

Pelosi M et al (2013) Parathyroid hormone-related protein is induced by hypoxia and promotes expression of the differentiated phenotype of human articular chondrocytes. Clin Sci (Lond) 125(10):461–470

Pelttari K et al (2006) Premature induction of hypertrophy during in vitro chondrogenesis of human mesenchymal stem cells correlates with calcification and vascular invasion after ectopic transplantation in SCID mice. Arthritis Rheum 54(10):3254–3266

Prado AA et al (2015) Characterization of mesenchymal stem cells derived from the equine synovial fluid and membrane. BMC Vet Res 11:281

Pritzker KP et al (1977) Articular cartilage transplantation. Hum Pathol 8(6):635–651

Radtke CL et al (2013) Characterization and osteogenic potential of equine muscle tissue- and periosteal tissue-derived mesenchymal stem cells in comparison with bone marrow- and adipose tissue-derived mesenchymal stem cells. Am J Vet Res 74(5):790–800

Raftery RM et al (2016) Multifunctional biomaterials from the sea: Assessing the effects of chitosan incorporation into collagen scaffolds on mechanical and biological functionality. Acta Biomater 43:160–169

Rakic R et al (2017) RNA interference and BMP-2 stimulation allows equine chondrocytes redifferentiation in 3D-hypoxia cell culture model: application for matrix-induced autologous chondrocyte implantation. Int J Mol Sci 18(9)

Ramezanifard R, Kabiri M, Hanaee Ahvaz H (2017) Effects of platelet rich plasma and chondrocyte co-culture on MSC chondrogenesis, hypertrophy and pathological responses. EXCLI J 16:1031–1045

Ranera B et al (2013) Expansion under hypoxic conditions enhances the chondrogenic potential of equine bone marrow-derived mesenchymal stem cells. Vet J 195(2): 248–251

Ren YJ et al (2009) In vitro behavior of neural stem cells in response to different chemical functional groups. Biomaterials 30(6):1036–1044

Ribeiro JCV et al (2017) Versatility of chitosan-based biomaterials and their use as scaffolds for tissue regeneration. Sci World J 2017:8639898

Rink BE et al (2017) Isolation and characterization of equine endometrial mesenchymal stromal cells. Stem Cell Res Ther 8(1):166

Roach BL et al (2015) Fabrication of tissue engineered osteochondral grafts for restoring the articular surface of diarthrodial joints. Methods 84:103–108

Rockwood DN et al (2011) Materials fabrication from Bombyx mori silk fibroin. Nat Protoc 6(10): 1612–1631

Rosadi I et al (2019) In vitro study of cartilage tissue engineering using human adipose-derived stem cells induced by platelet-rich plasma and cultured on silk fibroin scaffold. Stem Cell Res Ther 10(1):369

Rucinski K et al (2019) Effects of compliance with procedure-specific postoperative rehabilitation protocols on initial outcomes after osteochondral and meniscal allograft transplantation in the knee. Orthop J Sports Med 7(11):2325967119884291

Russlies M et al (2005) Periosteum stimulates subchondral bone densification in autologous chondrocyte

transplantation in a sheep model. Cell Tissue Res 319(1):133–142

Ryan VH et al (2008) NGF gene expression and secretion by canine adipocytes in primary culture: upregulation by the inflammatory mediators LPS and TNFalpha. Horm Metab Res 40(12):861–868

Rychel JK (2010) Diagnosis, and treatment of osteoarthritis. Top Companion Anim Med 25(1):20–25

Salonius E et al (2020) Chondrogenic differentiation of human bone marrow-derived mesenchymal stromal cells in a three-dimensional environment. J Cell Physiol 235(4):3497–3507

Sanderson RO et al (2009) Systematic review of the management of canine osteoarthritis. Vet Rec 164(14): 418–424

Sanz-Ramos P et al (2014) Improved chondrogenic capacity of collagen hydrogel-expanded chondrocytes: in vitro and in vivo analyses. J Bone Joint Surg Am 96(13):1109–1117

Sartore L et al (2021) Polysaccharides on gelatin-based hydrogels differently affect chondrogenic differentiation of human mesenchymal stromal cells. Mater Sci Eng C Mater Biol Appl 126:112175

Sasaki A et al (2018) Canine mesenchymal stem cells from synovium have a higher chondrogenic potential than those from infrapatellar fat pad, adipose tissue, and bone marrow. PLoS One 13(8):e0202922

Sato K et al (2016) Isolation and characterisation of peripheral blood-derived feline mesenchymal stem cells. Vet J 216:183–188

Schaap-Oziemlak AM et al (2014) Biomaterial–stem cell interactions and their impact on stem cell response. RSC Adv 4(95):53307–53320

Schagemann JC et al (2013) Chondrogenic differentiation of bone marrow-derived mesenchymal stromal cells via biomimetic and bioactive poly-epsilon-caprolactone scaffolds. J Biomed Mater Res A 101(6):1620–1628

Schreiner AJ et al (2020) Unicompartmental bipolar osteochondral and meniscal allograft transplantation is effective for treatment of medial compartment gonarthrosis in a canine model. J Orthop Res

Shah K et al (2018) Outcome of allogeneic adult stem cell therapy in dogs suffering from osteoarthritis and other joint defects. Stem Cells Int 2018:7309201

Shintani N, Hunziker EB (2011) Differential effects of dexamethasone on the chondrogenesis of mesenchymal stromal cells: influence of microenvironment, tissue origin and growth factor. Eur Cell Mater 22:302–319. discussion 319-20

Shortkroff S et al (1996) Healing of chondral and osteochondral defects in a canine model: the role of cultured chondrocytes in regeneration of articular cartilage. Biomaterials 17(2):147–154

Slocum B, Slocum TD (1993) Tibial plateau leveling osteotomy for repair of cranial cruciate ligament rupture in the canine. Vet Clin North Am Small Anim Pract 23(4):777–795

Smith RK et al (2013) Beneficial effects of autologous bone marrow-derived mesenchymal stem cells in naturally occurring tendinopathy. PLoS One 8(9): e75697

Spaas JH et al (2015) Chondrogenic priming at reduced cell density enhances cartilage adhesion of equine allogeneic MSCs – a loading sensitive phenomenon in an organ culture study with 180 explants. Cell Physiol Biochem 37(2):651–665

Stefansson SE et al (2003) Genomewide scan for hand osteoarthritis: a novel mutation in matrilin-3. Am J Hum Genet 72(6):1448–1459

Steinert AF et al (2012) Indian hedgehog gene transfer is a chondrogenic inducer of human mesenchymal stem cells. Arthritis Res Ther 14(4):R168

Stevenson S et al (1989) The fate of articular cartilage after transplantation of fresh and cryopreserved tissue-antigen-matched and mismatched osteochondral allografts in dogs. J Bone Joint Surg Am 71(9): 1297–1307

Stevenson S, Shaffer JW, Goldberg VM (1996) The humoral response to vascular and nonvascular allografts of bone. Clin Orthop Relat Res 326:86–95

Stockwell RA (1971) The interrelationship of cell density and cartilage thickness in mammalian articular cartilage. J Anat 109(Pt 3):411–421

Stockwell RA (1978) Chondrocytes. J Clin Pathol Suppl (R Coll Pathol) 12:7–13

Stoker AM et al (2018) Bone marrow aspirate concentrate versus platelet rich plasma to enhance osseous integration potential for osteochondral allografts. J Knee Surg 31(4):314–320

Stupina TA, Stepanov MA, Teplen'kii MP (2015) Role of subchondral bone in the restoration of articular cartilage. Bull Exp Biol Med 158(6):820–823

Stupina TA, Makushin VD, Stepanov MA (2012) Experimental morphological study of the effects of subchondral tunnelization and bone marrow stimulation on articular cartilage regeneration. Bull Exp Biol Med 153(2):289–293

Sun J, Tan H (2013) Alginate-based biomaterials for regenerative medicine applications. Materials (Basel) 6(4):1285–1309

Tamada Y, Ikada Y (1993) Effect of preadsorbed proteins on cell adhesion to polymer surfaces. J Colloid Interface Sci 155(2):334–339

Tchetina EV, Squires G, Poole AR (2005) Increased type II collagen degradation and very early focal cartilage degeneration is associated with upregulation of chondrocyte differentiation related genes in early human articular cartilage lesions. J Rheumatol 32(5):876–886

Thorpe SD et al (2012) European Society of Biomechanics S.M. Perren Award 2012: the external mechanical environment can override the influence of local substrate in determining stem cell fate. J Biomech 45(15): 2483–2492

Tibbitt MW, Anseth KS (2009) Hydrogels as extracellular matrix mimics for 3D cell culture. Biotechnol Bioeng 103(4):655–663

Toh WS et al (2012) Modulation of mesenchymal stem cell chondrogenesis in a tunable hyaluronic acid hydrogel microenvironment. Biomaterials 33(15): 3835–3845

Tomlinson RE et al (2017) NGF-TrkA signaling in sensory nerves is required for skeletal adaptation to mechanical loads in mice. Proc Natl Acad Sci U S A 114(18):E3632–E3641

Ueno T et al (2001) Cellular origin of endochondral ossification from grafted periosteum. Anat Rec 264(4): 348–357

Vainieri ML et al (2020) Evaluation of biomimetic hyaluronic-based hydrogels with enhanced endogenous cell recruitment and cartilage matrix formation. Acta Biomater 101:293–303

Veilleux NH, Yannas IV, Spector M (2004) Effect of passage number and collagen type on the proliferative, biosynthetic, and contractile activity of adult canine articular chondrocytes in type I and II collagen-glycosaminoglycan matrices in vitro. Tissue Eng 10(1-2):119–127

Venator KP et al (2020) Assessment of a single intra-articular stifle injection of pure platelet rich plasma on symmetry indices in dogs with unilateral or bilateral stifle osteoarthritis from long-term medically managed cranial cruciate ligament disease. Vet Med (Auckl) 11: 31–38

Vilar JM et al (2013) Controlled, blinded force platform analysis of the effect of intraarticular injection of autologous adipose-derived mesenchymal stem cells associated to PRGF-Endoret in osteoarthritic dogs. BMC Vet Res 9:131

Vilar JM et al (2018) Effect of leukocyte-reduced platelet-rich plasma on osteoarthritis caused by cranial cruciate ligament rupture: a canine gait analysis model. PLoS One 13(3):e0194752

Voga M et al (2020) Stem cells in veterinary medicine-current state and treatment options. Front Vet Sci 7:278

Walker GD et al (1995) Expression of type-X collagen in osteoarthritis. J Orthop Res 13(1):4–12

Webb TL, Quimby JM, Dow SW (2012) In vitro comparison of feline bone marrow-derived and adipose tissue-derived mesenchymal stem cells. J Feline Med Surg 14(2):165–168

Webster RP, Anderson GI, Gearing DP (2014) Canine brief pain inventory scores for dogs with osteoarthritis before and after administration of a monoclonal antibody against nerve growth factor. Am J Vet Res 75(6): 532–535

Weiss S et al (2010) Impact of growth factors and PTHrP on early and late chondrogenic differentiation of human mesenchymal stem cells. J Cell Physiol 223(1):84–93

Xu X et al (2013) Cultivation and spontaneous differentiation of rat bone marrow-derived mesenchymal stem cells on polymeric surfaces. Clin Hemorheol Microcirc 55(1):143–156

Yang C et al (2009) The differential in vitro and in vivo responses of bone marrow stromal cells on novel porous gelatin-alginate scaffolds. J Tissue Eng Regen Med 3(8):601–614

Yang Q et al (2008) A cartilage ECM-derived 3-D porous acellular matrix scaffold for in vivo cartilage tissue engineering with PKH26-labeled chondrogenic bone marrow-derived mesenchymal stem cells. Biomaterials 29(15):2378–2387

Yang X et al (2014) Matrilin-3 inhibits chondrocyte hypertrophy as a bone morphogenetic protein-2 antagonist. J Biol Chem 289(50):34768–34779

Yang Z et al (2021) 3D-bioprinted difunctional scaffold for in situ cartilage regeneration based on aptamer-directed cell recruitment and growth factor-enhanced cell chondrogenesis. ACS Appl Mater Interfaces 13(20):23369–23383

Yoo JU et al (1998) The chondrogenic potential of human bone-marrow-derived mesenchymal progenitor cells. J Bone Joint Surg Am 80(12):1745–1757

Zamanlui S et al (2018) Enhanced chondrogenic differentiation of human bone marrow mesenchymal stem cells on PCL/PLGA electrospun with different alignments and compositions. Int J Polym Mater Polym Biomater 67(1):50–60

Zhang L et al (2010) Chondrogenic differentiation of human mesenchymal stem cells: a comparison between micromass and pellet culture systems. Biotechnol Lett 32(9):1339–1346

Zhang T et al (2015) Cross-talk between TGF-beta/SMAD and integrin signaling pathways in regulating hypertrophy of mesenchymal stem cell chondrogenesis under deferral dynamic compression. Biomaterials 38:72–85

Zhang BY et al (2018) Evaluation of the curative effect of umbilical cord mesenchymal stem cell therapy for knee arthritis in dogs using imaging technology. Stem Cells Int 2018:1983025

Zheng L et al (2021) The role of metabolism in chondrocyte dysfunction and the progression of osteoarthritis. Ageing Res Rev 66:101249

Zhou N et al (2015) HIF-1alpha as a regulator of BMP2-induced chondrogenic differentiation, osteogenic differentiation, and endochondral ossification in stem cells. Cell Physiol Biochem 36(1):44–60

Zylinska B et al (2018) Treatment of articular cartilage defects: focus on tissue engineering. In Vivo 32(6):1289–1300

Adv Exp Med Biol - Cell Biology and Translational Medicine (2022) 17: 57–72
https://doi.org/10.1007/5584_2022_719
© Springer Nature Switzerland AG 2022
Published online: 2 August 2022

Adult Stem Cell Therapy as Regenerative Medicine for End-Stage Liver Disease

Caecilia H. C. Sukowati and Claudio Tiribelli

Abstract

The increased incidence of end-stage liver disease (ESLD) causes a major burden on the global health system and population health. Liver transplantation (LT) is one of the most effective treatments for ESLD patients, but its practice is extensively hampered by the scarcity of liver donors, the limited number of transplantation centers, the complexity of the procedure, and postoperative complication. In parallel, vast growing advances in cellular biology and biotechnology have opened new alternatives in clinics, including the transplantation of adult stem cells for chronic diseases such as ESLD. Numerous types of stem cells, such as mesenchymal stem cells, hematopoietic stem cells, endothelial progenitor cells, and other cells, obtained from bone marrow, umbilical cord, adipose tissue, or peripheral blood had been isolated and given to ESLD patients all over the world. Many clinical data had demonstrated promising results, indicating its potential. However, conclusive protocol and agreement on adult stem cell definition and transplantation method are still lacking, and thus further research must still be conducted.

Keywords

Adult stem cells · Cell transplantation · End-stage liver disease · Regenerative medicine

Abbreviations

BM	Bone marrow
EPC	Endothelial progenitor cells
ESC	Embryonic stem cells
ESLD	End-stage liver disease
HBV	Hepatitis B virus
HCC	Hepatitis C virus
HCC	Hepatocellular carcinoma
HLO	Human liver organoid
HSC	Hematopoietic stem cells
iPSC	Induced pluripotent stem cells
LPC	Liver progenitor cells
LT	Liver transplantation
MELD	Model for End-Stage Liver Disease
MSC	Mesenchymal stromal/stem cells
PHH	Primary human hepatocytes

C. H. C. Sukowati (✉)
Fondazione Italiana Fegato ONLUS, AREA Science Park, Trieste, Italy

Eijkman Research Center for Molecular Biology, National Research and Innovation Agency of Indonesia (BRIN), Jakarta, Indonesia
e-mail: caecilia.sukowati@fegato.it; caecilia.sukowati@brin.go.id

C. Tiribelli
Eijkman Research Center for Molecular Biology, National Research and Innovation Agency of Indonesia (BRIN), Jakarta, Indonesia

1   End-Stage Liver Disease

Liver disease is one of the major health problems in the world. It accounts for approximately two million deaths per year worldwide, one million due to complications of cirrhosis and one million to viral hepatitis (hepatitis B virus (HBV) and hepatitis C virus (HCV)) and hepatocellular carcinoma (HCC). Chronic liver disease is usually caused by prolonged excess alcohol consumption, metabolic disorders, and viral hepatitis infection (Asrani et al. 2019).

The number of end-stage liver disease (ESLD) cases is increasing resulting in a greater burden on the healthcare system (Fricker and Serper 2019). ESLD, often interchangeably called liver failure or decompensated cirrhosis, is the final stage of chronic liver disease and is associated with a high degree of mortality. The annual rates of liver disease progression to decompensated stage range from 4% for HCV to 6–10% for alcoholic cirrhosis and 10% for HBV (Asrani et al. 2019).

Liver cirrhosis is characterized by a silent, asymptomatic course that may be undetectable for years. This is usually referred to compensated cirrhosis. When the portal pressure is increased and liver function is significantly reduced, the clinical phenotype is observed. Decompensation is marked by the development of overt clinical signs, the most frequent of which are ascites, bleeding, encephalopathy, and jaundice (European Association for the Study of the Liver. 2018; Haep et al. 2021).

Liver transplantation (LT) is one (if not the only one) of the most effective treatments for any patients with ESLD. LT would extend life expectancy of the patients regardless of the natural history of underlying liver disease where LT is expected to improve the quality of life. However, in practice, LT is hampered by the shortage of donor organs, the limited number of liver transplantation facilities, and the high cost (Harries et al. 2019). Recently, the possibility of living donor liver transplantation (LDLT) can be another option. However, LDLT needs immense and complicated technical operations. And still, the donor shortage remains a concern (Au and Chan 2019; Choudhary et al. 2022).

Following LT, further, the liver recipient might suffer postoperative complications, transplant rejection, and long-term immunosuppression side effects (Feng and Bucuvalas 2017). Further, de novo malignancies are often detected in liver transplant patients undergoing daily immunosuppression regimens, one of the leading causes of late death. The incidence of de novo malignancies among transplant patients is predicted up to four times higher than in the healthy population (Herrero 2012; Manzia et al. 2019).

Since 2002, the Model for End-Stage Liver Disease (MELD) has been used to rank liver transplant candidates for ESLD (Kamath et al. 2001; Wiesner et al. 2003). It is considered an effective strategy for prioritizing candidates with a higher transplant survival benefit over those with lower survival benefit (Luo et al. 2018). This scoring system predicts liver disease severity based on serum creatinine, serum total bilirubin, and INR. It was previously shown to be useful in predicting mortality in patients with compensated and decompensated cirrhosis (Wiesner et al. 2003).

In brief, MELD score ranks patient to number 6 to >40 using the formula ($0.967*\log_e$(creatinine (mg/dL)) + $0.378 \times \log_e$(bilirubin (mg/dL)) + $1.120 \times \log_e$(INR) + 0.6431) $\times$ 10) and is suitable as a disease severity index to determine organ allocation priorities (Kamath et al. 2001). Regardless of various revisions and updates (MELD 3.0, MELD-Na, etc.) (Nagai et al. 2018; Kim et al. 2021), the change in MELD score is used as an indicator to measure the benefits of therapy following LT or other treatment regimens.

2   Liver Development and Regeneration

Liver is not only the largest internal organ in the body; it is also capable to replenish its mass by self-regeneration capacity. From a liver phenotypic point of view, it reflects the broad metabolic functions of hepatocytes as well as the liver's unique vascular anatomy, having an inflow blood supply from both an arterial (hepatic artery) and venous (portal vein) sources (Haep et al. 2021).

Following liver injury, hepatocytes can proliferate to reinstate their morphological and physiological function. In the 1930s, liver regenerative ability in a murine model of partial hepatectomy (PH) had been evidenced. Following PH of around 70% of its total mass, the liver was recovered in about 1 week (Higgins 1931). Using thymidine tracking in the DNA, the restoration of liver mass and function was further demonstrated (Bucher and Swaffield 1964).

In the case of sustained damage such as fibrosis and impaired hepatocytes regeneration, the liver needs to activate its resident stem cells compartment. The canals of Hering and bile ductules in the human liver contain liver progenitor cells (LPC) that can differentiate toward the biliary and hepatocytic lineage (Theise et al. 1999; Libbrecht and Roskams 2002). The source of the LPC is still unclear. It has been variously demonstrated that adult mature hepatocytes can be reprogrammed into proliferative bipotent progenitor cells in response to chronic liver injury (Tarlow et al. 2014; Hu et al. 2018). A population of EpCAM+ cells has been identified within the canals of Hering and the bile ductules, serving as facultative bipotent progenitors capable of differentiating into hepatocytes and cholangiocytes (Safarikia et al. 2020).

During liver disease, the degeneration from healthy-functioning livers involves a dynamic process of hepatocyte damage leading to the reduction of hepatic function. As already known widely, the activation of stellate cells and the production of extracellular matrix (ECM) are the keystone of liver fibrosis. In the case of cirrhosis and ESLD, hepatocyte proliferation or liver regeneration is finished (Haep et al. 2021). Liver failure is also majorly influenced by the exposure to an inflammatory setting, a loss of cell-cell contact caused by cell death and ECM deposition, and changes in energy metabolism and transcriptional deprogramming of hepatocytes (e.g., HNF4α, HNF1, FOXA, HNF6, and C/EBP). Further, clinical manifestations in patients with ESLD are directly related to specific alternated metabolic pathways in failing hepatocytes (Haep et al. 2021).

ESLD is not only due to the lack of healthy hepatocytes but also to the disturbance of tissue architecture and the continuous deposition of inflammatory cells (Lorenzini et al. 2008). Thus, when ESLD occurs, it is hard for the liver to establish its capacity to regenerate.

3 Stem Cell Therapy

Cell therapy has been thought of as the source of liver regeneration (Fig. 1). For therapy applications, donor cells must act as fully functional differentiated cells, such as the expression of liver-specific markers and secretion of albumin and alpha-fetoprotein. Thus, careful protocol and cell characterization should be verified before the transplantation.

Freshly isolated primary human hepatocytes (PHH) are currently the benchmark cell type for cell therapy, but they are not readily available, dedifferentiate quickly, and rapidly die in culture (Hannoun et al. 2016). Several groups had reported methods to cryopreserve the PHH (Godoy et al. 2013; Sison-Young et al. 2017). Despite various optimization protocols, cryopreservation still has damaging effects on the viability and metabolic function (Hannoun et al. 2016). Further, a rather large number of cells (10–15% of liver mass) are needed to provide enough function (Fitzpatrick et al. 2009). So far, various studies had demonstrated the clinical application of hepatocyte transplantation in liver diseases (Lee et al. 2018). In chronic liver disease, however, there are some hassles with engraftment since the liver architecture is disrupted. It is one of the causes of the common failure in hepatocyte transplantation to date (Fitzpatrick et al. 2009).

Stem cells have the astonishing proliferative capacity, self-renewal ability, and differentiation properties. Due to their plasticity, stem cells have been proposed as a source for cell therapy. Embryonic stem cells (ESCs) are the most pluripotent cells that can become all cell types in the body. They are derived from the embryo, typically from the inner cell mass in the blastocyst.

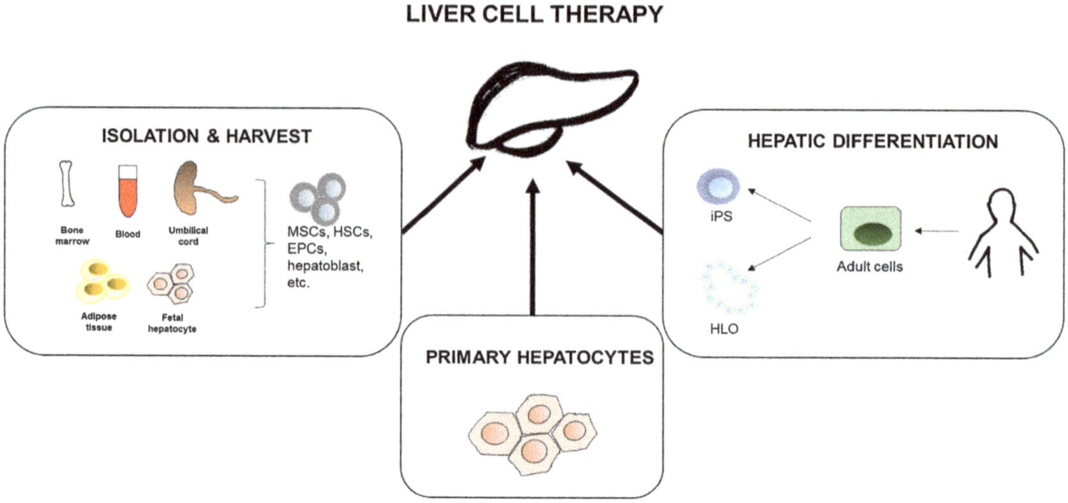

LIVER CELL THERAPY

Fig. 1 Organ sources of cellular therapy for the liver

Due to its pluripotency, the human ESCs would be potent tools in regenerative medicine such as Parkinson's disease, spinal cord injury, myocardial infarction, and many more (Mountford 2008). ESCs have been demonstrated to have potential in cell therapy for liver disease. ESCs transplantation had been demonstrated to reduce liver fibrosis and to engraft the liver in rodents (Heo et al. 2006; Sharma et al. 2008; Moriya et al. 2008; Haideri et al. 2017). It is important to notice, however, that ESCs implantation may be tumorigenic where teratoma can occur (Fujikawa et al. 2005; Blum and Benvenisty 2008; Hentze et al. 2009; Stachelscheid et al. 2013). In the human study, it also has a significant ethical dilemma because it involves the destruction of an embryo to obtain the ESCs.

Adult stem cells (or somatic stem cells) can be found in a small number of undifferentiated cells in a specific area of tissue or organ in the body. Even though they are not as multipotent as the ESCs, the adult stem cells can easily be obtained and differentiated into various cells. More importantly, these cells can be ideal sources for autologous stem cell transplantation to replenish tissue damage in the same patient. The bone marrow (BM) compartment is the major source of committed progenitor (stem) cells that can develop into mesenchymal lineages and hematopoietic cells (Masson et al. 2004).

In the beginning, it was assumed that adult stem cells could differentiate only into their maturation lineages. For instance, bone marrow stem cells could only differentiate into blood cells. However, more studies demonstrated that adult stem cells are multipotent and they can differentiate into various cells. For example, bone marrow-derived stem cells could regenerate de novo myocardium (Orlic et al. 2001); skeletal (Gussoni et al. 1999), adipocytic, chondrocytic, or osteocytic lineages (Pittenger et al. 1999); microglial and perivascular cells in the brain (Corti et al. 2002; Hess et al. 2004); as well as the liver cells (Petersen et al. 1999). The injection of these cells ameliorated the outcome of diseases.

To date, there have been numerous clinical studies on adult stem cell therapy for the treatment of ESLD registered in the public database (https://clinicaltrials.gov/) even though many of these studies' results are still unavailable. As the primary outcome, usually, these studies measure the improvement of the MELD score and liver function as the success of the treatment.

3.1 Mesenchymal Stem Cells

The mesenchymal stromal/stem cells (MSC) is the most common stem cells used in clinical therapy, in addition to being the most controversial. The term mesenchymal stem cells was firstly named in the late 1980s by Dr. Caplan for a cell type derived from bone marrow. These cells could be isolated and expanded in culture while maintaining their in vitro capacity to be induced to form a variety of mesodermal phenotypes and tissues (Caplan 1991). In regard to their multi-differentiation capacity and high self-renewal ability, MSC are a good option for promoting tissue regeneration and inhibiting fibrosis and, at the same time, lessening tissue inflammatory response (Xiang et al. 2022).

In the last three decades, however, the exponential growth of scientific articles had used this nomenclature across numerous isolated cells. In some cases, these cells are various tissue-specific cell types with the use of different cell-surface markers (Sipp et al. 2018), leading to confusion in the scientific community and clinical practice. A previous study demonstrated that "MSC" isolated from different anatomical sources (bone marrow, skeletal muscle, periosteum, and perinatal cord blood) actually differed widely in their transcriptomic signature and in vivo differentiation potential (Sacchetti et al. 2016).

Back in 2005, a working group of the International Society for Cellular Therapy (ISCT) acknowledged the MSC inconsistencies and ambiguities, and they recommended a new designation: multipotent mesenchymal stromal cells (Horwitz et al. 2005). The ISCT also proposed minimal criteria to define human MSC. First, MSC must be plastic-adherent when maintained in standard culture conditions. Second, MSC must express CD105, CD73, and CD90 and lack expression of CD45, CD34, CD14 or CD11b, CD79α or CD19, and HLA-DR surface molecules. Third, MSC must differentiate into osteoblasts, adipocytes, and chondroblasts in vitro (Dominici et al. 2006). Another term of medicinal signaling cells (also abbreviated as MSC) was proposed by Dr. Caplan to more accurately reflect the fact the tissue origin or disease and secrete bioactive factors that are immunomodulatory and trophic (regenerative) medicine (Caplan 2010; Caplan 2017). Regardless of the nomenclature, the capacity of the MSC in the repair of liver tissues had been widely studied with various results.

Several sources of MSC had been used in various clinical trials for ESLD, with the most common sources being umbilical cord (UC) and bone marrow (BM; autologous or allogenic). Some also take advantage of adipose tissue-derived MSC, naively or following cell differentiation (Nhung et al. 2015). As for adipose tissues, sources are broad, and cells can be collected from the subcutaneous tissue, viscera, omentum, inguinal fat pads, peritoneal fat, and other sources (Hu et al. 2019).

As expected, the results of these studies are variable. The injection of autologous BM with CD44+ phenotype had resulted in short benefit in treated patients, regardless of the delivery method (hepatic or peripheral transfusion) (Kharaziha et al. 2009; Peng et al. 2011; Amin et al. 2013; Salama et al. 2014). In these studies, MSC-injected patients had improvement in their liver function and MELD and CP scores compared to control. A meta-analysis of five studies showed that bone marrow infusion in the treatment of decompensated cirrhosis improved liver function without serious side effects at least for the first year (Pan et al. 2014). However, at least in one of the studies, the long-term outcomes were not markedly improved with no significant difference in the incidence of hepatocellular carcinoma (HCC) or mortality between the two groups (Peng et al. 2011).

A recent report from a Japanese clinical trial (UMIN Clinical Trials Registry UMIN000022601) using freshly isolated autologous adipose tissue-derived stem cells in seven patients also showed promising results. Stem cell transplantation improved serum albumin in six out of seven patients and prothrombin activity in five out of seven patients. No trial-related adverse events, which were serious or nonserious, were observed (Sakai et al. 2020; Sakai et al. 2021).

For donor transplantation, the infusion of allogenic MSC from donors was also considered a safe procedure. In patients with liver failure, donor MSC significantly increased the survival rate by improving liver function (reduction of ascites volume, increase of albumin, decrease of bilirubin, improvement of CP and MELD score) and decreasing the incidence of severe infections (Zhang et al. 2012; Lin et al. 2017; Schacher et al. 2021). In a longer study, upon allogeneic MSC infusion (obtained from donor BM, cord blood, and umbilical cord), MELD score improved at 6 months, 1 year, and 2 years of follow-up. No serious adverse events were observed during or after infusions of MSC in patients with decompensated cirrhosis as compared to control patients (Zhang et al. 2012; Liang et al. 2017). UC-derived MSC transfusion also increased liver function and survival rate in ACLF patients, either by intravenous infusion or hepatic arterial transfusion (Shi et al. 2012; Li et al. 2016).

However, in contrast, several studies showed no benefit of MSC transplantation. A previous study indicated the unsafety of the procedure, and even mortality, following cell transplantation. In a randomized, placebo-controlled trial, from 15 autologous MSC-injected patients, there were 3 deaths registered, while the rest of the patients did not show any improvement in liver function and CP or MELD score (Mohamadnejad et al. 2013). Another study had shown that in this study, even though it was considered safe and feasible, consecutive liver biopsy examinations suggested that MSC infusion via peripheral vessel could not reach the liver in a sufficient amount; thus there were no improvements in MELD scores and serum albumin (Kantarcıoğlu et al. 2015).

3.2 Hematopoietic Stem Cells

Hematopoietic stem cells (HSC) are the most accessible source of stem cells in the body. They give rise to all lineages of blood cell differentiation. In the beginning, it was thought that CD34 (CD34+ cells) is the HSC marker in mammals; however, then it was noticed that human CD34- also had self-renewing capability and acted as primitive HSC that could give rise to CD34+ cells (Zanjani et al. 1998; Wang et al. 2003; Sumide et al. 2018).

The differentiation capacity of HSC, especially to hepatic lineage, is still limited if not controversial. Previously, it was shown that HSC could become liver cells when co-cultured with injured liver separated by a barrier (Jang et al. 2004). In mouse model studies, the transplantation of human cord blood cells CD34+ was able to repopulate the liver (even though with a very low percentage) showing the contribution of HSC (Masson et al. 2004). However, this potentiality was challenged over time. A study showed that HSC expressed mRNAs of hepatic cell markers, but could not efficiently convert into hepatocytes in vitro even in the presence of cytokines or co-cultured hepatocytes (or tissue) (Lian et al. 2006). As mentioned by Thorgeirsson and Grisham, it seemed that the hematopoietic cells are only a minor contributor to hepatocyte formation under either physiological or pathological conditions. These cells, however, may provide cytokines and growth factors that promote hepatocyte functions by paracrine mechanisms (Thorgeirsson and Grisham 2006).

In the clinical study, the application of HSC transplantation in the ESLD had been another option, even though it is not as frequent as the MSC, in line with this limitation described above. One of the first studies comprised a rather small number (phase 1); autologous CD34 was injected into five patients with liver insufficiency. Patients were previously given subcutaneously granulocyte colony-stimulating factor (G-CSF) for 5 days to increase the number of harvested CD34+ cells from the circulation. Following portal vein or hepatic artery injection of these cells, four patients showed improvement in serum albumin (Gordon et al. 2006).

In another study which used the same method, 90 ESLD patients received G-CSF followed by autologous CD34+ and CD133+ HSC infusion in the portal vein. Up to 6 months of follow-up, around 50% had near normalization of liver enzymes and improvement in synthetic function, and 14% showed stable states, compared to control group (Salama et al. 2010). From the same group in another study, stem cell transplantation was done via portal vein infusion of 50% of HSC (CD34+/CD133+), and the other 50% were differentiated to MSC and infused systemically in a peripheral vein in the presence of growth factors. This procedure had a low incidence of complications and it improved CP and MELD score and degree of ascites of the patients. When the infusion was done in two sessions, the sustained response was continued throughout the follow-up period of 12 months (Zekri et al. 2015).

Another group had shown that the infusion of cell population with CD133+ marker (stem/progenitor cell (SPC)) in ESLD patients was feasible and safe and improved liver function transiently. The recollection of SPC after G-CSF treatment was associated with increased levels of selected cytokines potentially facilitating SPC function (Catani et al. 2017).

Hematopoietic cell isolation and injection from BM also had been performed. In this study, autologous mononuclear (CD34/CD45+) from BM was infused via the peripheral vein in nine patients. Following the procedures, no major adverse effects were noticed. Infused patients had significantly improved CP scores at 1 and 6 months together with improvement in liver biopsy (Terai et al. 2006).

3.3 Endothelial Progenitor Cells

The endothelial progenitor cells (EPC) were discovered around two decades ago (Asahara et al. 1997). These cells were purified by magnetic bead selection with the surface markers antigens CD34+ and Flk1+; in vitro, these cells differentiated into endothelial cells (Asahara et al. 1997). As in MSC the nomenclature of EPC is still under discussion, where another term "endothelial colony-forming cells (ECFC)" is also used (Prasain et al. 2012; Keighron et al. 2018). This disagreement in consensus needs a more precise characterization of these cells based on a pre-defined cellular phenotype and function (Medina et al. 2017).

By using a nonhuman primate model, the localization of injected autologous EPC/endothelial cells (EC) can be traced. At 14 days postinjection via the portal vein, these cells were found scattered in the intercellular spaces of hepatocytes at the hepatic tissues, indicating successful migration and reconstitution in the liver structure as the

functional EPC/EC (Qin et al. 2018). Another study examined the benefit of BM-EPC in a rat model of liver fibrosis/cirrhosis induced by carbon tetrachloride (Sakamoto et al. 2013; Lan et al. 2018). While EPC transplantation gave a beneficial result, combined transplantation of BM-EPC and BM-derived hepatocyte stem cells exhibited maximal treatment effect (Lan et al. 2018).

The transplantation of EPC in decompensated liver cirrhosis patients had been reported. In this phase 1–2 pilot clinical trial, autologous cells were harvested from the bone marrow of patients subjected to differentiation to EPC ex vivo. Following hepatic arterial administration in 11 patients, no treatment-related severe adverse events were observed. At 90 days posttransplantation, there was a significant improvement in MELD, and five of nine patients alive showed a decreased hepatic venous pressure gradient (D'Avola et al. 2017).

3.4 Fetal Human Hepatocytes

Fetal liver is becoming an available source of cells for the treatment of liver diseases. Group of Cardinale et al. defined fetal liver as the liver developed from 10 weeks of gestation, the timing when the hematopoietic progenitor cells migrate from the aorta-mesonephros-gonad region to colonize the liver (Giancotti et al. 2022). It contains hepatic stem/progenitor cells within the ductal plates and multipotent stem/progenitor cells within large intrahepatic bile ducts and extrahepatic bile ducts (Semeraro et al. 2013).

Still, limited information is available for the clinical application of fetal liver for ESLD. An Indian clinical study of fetal liver transplantation in 25 end-stage liver cirrhosis patients showed clinical improvement observed in terms of all clinical and biochemical parameters together with a decrease of MELD in 6 months' follow-up in all patients. These cells were obtained from fetal livers of spontaneous abortions from 16 to 20 weeks of gestation and showed positivity of EpCAM+ (Khan et al. 2010). A comparable result was obtained from a study in Italy. Following fetal liver transplantation in an ESLD patient, the MELD score decreased from 15 to 11 at 3-month and 10 at 18-month follow-up with no signs of encephalopathy. These cells expressed highly significant amounts of proliferation markers compared to adult hepatocytes (Gridelli et al. 2012).

3.5 Hepatic Lineage Differentiation

Several studies had taken another additional step for the application of the MSC. Taking advantage of the multipotency ability, MSCs obtained either from BM, UC, or adipose tissues can be subjected to a hepatic lineage differentiation in vitro before the infusion into the patient/recipient. For example, adipose-derived MSC can be differentiated into hepatocytes in 14 days' culture condition with hepatogenic medium containing dexamethasone, insulin, hepatocyte growth factor (HGF), and epidermal growth factor (EGF), followed by activation of the extracellular signal-regulated kinase (ERK)/mitogen-activated protein kinase (MAPK) signaling pathway (Liang et al. 2009).

One of the first clinical studies using this approach in ESLD was reported in 2011. In this study, upon the isolation and the phenotyping of the autologous BM-MSC, MSC was stimulated into hepatic cells using in the presence of HGF for 7 days. The hepatic-committed lineage was then evaluated by morphological, immunophenotyping, and albumin production. Cells were then injected via the intrasplenic or intrahepatic route. The result showed that MSC-infused patients had significant improvement in ascites and serum albumin, CP, and MELD score over the control group. No difference was observed between intrahepatic and intrasplenic groups (Amer et al. 2011).

Another study used a two-step MSC differentiation into the hepatic lineage, using HGF and FGF, continued by oncostatin and dexamethasone. In this phase 2 trial, however, cells were injected intravenously. MSC-received patients showed partial improvement in liver function tests and MELD score at 3 and 6 months post-infusion. However, there was no significant difference regarding clinical and laboratory findings for MSCs transplantation of either undifferentiated or differentiated cells (El-Ansary et al. 2012).

4 Cell Reprogramming

In the last decades, advances in molecular and cellular biology technologies open exponential opportunities in the manipulation of cellular fate. One of the greatest breakthroughs of the century is the discovery that mature cells can be reprogrammed to become immature, even pluripotent cells, leading to a greatly appreciated shared Nobel Prize in Physiology or Medicine 2012 awarded to Sir John B Gurdon and Shinya Yamanaka (https://www.nobelprize.org/prizes/medicine/2012/summary/).

Back in the 1960s, John Gurdon was successful in transplanting nuclei from fully differentiated cells from the intestine of a tadpole into the cell nucleus of a frog's egg cell. The egg developed into a fully functional cloned tadpole. The transplanted nucleus promoted the formation of a differentiated intestinal cell and at the same time contained the genetic information necessary for the formation of all other types of differentiated somatic cell in a normal feeding tadpole (Gurdon 1962). This nuclear transfer technique was then widely publicized several decades later with the cloning of Dolly sheep, published in 1997 by Wilmut et al. (1997).

In 2006, by using four defined transcription factors Oct3/4, Sox2, c-Myc, and Klf4 (OSKM factors), Takahashi and Yamanaka showed that mouse fibroblasts could be reprogrammed into an embryonic stage, namely, the induced pluripotent stem cells (iPSC). These iPSC cells exhibited ESCs morphology and growth properties and ESCs marker genes. Furthermore, subcutaneous transplantation of iPSC cells into nude mice resulted in variety of tissues from all three germ layers (Takahashi and Yamanaka 2006). In the following year, this technique was then proven in a human cell. Human iPSC cells were similar to human ESC in morphology, proliferation, surface antigens, gene expression, epigenetic status of pluripotent cell-specific genes, and telomerase activity (Takahashi et al. 2007). Because of its ESC-like pluripotency, iPSC is a valuable tool in the basic research on the mechanisms of tissue formation, cell therapy, and patient-specific cell development.

4.1 Induced Pluripotent Stem Cells (iPSC)

First data on the iPSC differentiation to functional hepatocytes was reported in 2009 by Song et al. (Song et al. 2009). They used iPSC cell lines 3U1 and 3U2 subjected to hepatic differentiation protocol composed of four stages: endoderm induction (activin A), hepatic specification (FGF4, BMP2), hepatoblast expansion (HGF, KGF), and hepatic maturation (oncostatin M, dexamethasone, N2, B27, nonessential amino acids, and β-mercaptoethanol). The differentiated cells exhibited mature hepatocyte functions including albumin secretion, glycogen synthesis, urea production, and inducible cytochrome P450 activity (Song et al. 2009). This process takes around 21 days.

A more rapid protocol was then demonstrated. In about 12 days, iPSC could be directed into mature hepatocytes by using the protocol of endodermal induction (activin A, Wnt3a, HGF), hepatic lineage commitment (in the presence of nonessential amino acids, β-mercaptoethanol, DMSO), and hepatic (oncostatin M, dexamethasone, ITS) (Chen et al. 2012). The cells had similar gene expression profile to mature hepatocytes. Besides its functionality as mature hepatocytes including cytochrome P450 enzyme activity, secreted urea, uptake of low-density lipoprotein (LDL), and glycogen storage, these induced hepatocyte-like cells rescued lethal fulminant hepatic failure in a NOD-SCID mouse model (Chen et al. 2012).

The induction of iPSC into bipotent hepatic progenitor cells (HPC) gave rise to both mature hepatocytes and cholangiocytes (Yanagida et al. 2013). The induced-HPC from iPSC resulted in CD13^{high}CD133+ cells, positive markers of hepatoblast. Spheroid formation of the HPC could be induced into hepatocytes

(dexamethasone, OSM) and cholangiocytes (EGF, HGF, R-spondin 1, Wnt-3a, A-83-01, and Y-27632) (Yanagida et al. 2013). The clinical application of iPSC was performed in several diseases such as degenerative and cardiovascular disease with various results (Martins et al. 2014; Bracha et al. 2017; Tsujimoto and Osafune 2021). However, for liver diseases, its application mostly is still conducted in a preclinical setting.

4.2 Human Liver Organoids (HLO)

Organoid biology is one of the fastest-growing interests in recent organ development and regeneration study. The capacity of isolated cells to self-assemble to form an entire organism was already reported in the early 1900s. When siliceous sponges are kept in confinement under proper conditions, they degenerate and gave rise to small masses of undifferentiated tissue which in turn grow and differentiate into perfect sponges (Wilson 1907).

Human liver organoids (HLO) derived from either adult stem/progenitors or pluripotent stem cells emulate the structure and cellular diversity of the human liver in vivo (Chang et al. 2021; Reza et al. 2021). Under a strict cell culture condition and the presence of correct growth factors (e.g., matrigel, TNFα), organoids can resemble a functional liver. A recent report even showed that from a single hepatocyte, organoids can be established and grown for multiple months while keeping its key morphological, functional, and gene expression features (Hu et al. 2018). However, when compared to the fetal culture, HLO derived from hepatocytes appeared to be more limited in their expansion times yet yielded organoids of very similar composition (Hu et al. 2018).

In the clinical application, HLO technology is not yet available, even though preclinical data in the animal model showed promising result. In a PH model in rat, the transplantation of HLO through portal vein is safer and more effective compared to monolayer cell transplantation, showing 70% replacement of the damaged liver (Tsuchida et al. 2019). Further, HLO in combination with co-culture with other cell lines and advanced bioengineering tools (sheet layers, microfluidics, 3D scaffold) will increase the differentiation efficiency and enhance the functional maturity.

5   General Perspective

Stem cell therapy is a promising alternative for the treatment of ESLD, especially when the availability of donor liver for LT is scarce. Thriving development of technology in stem cell isolation and maintenance, characterization, and in vitro differentiation to hepatic cells is growing fast, thus allowing an improved method in clinical application.

In ESLD, however, at least until now, stem cell therapy application is still rather far from ideal. The biology of stem cells is still needed to be explored. Clinicians and basic scientists must know whether the transplanted cells are multipotent and self-renewable or the cells' phenotype (Fig. 2), both in donor cells and in the recipient patient, including the protocol of administration, patient's status, safety, and efficacy. Further, vast differences in the source of the cells, type of the cells, transplantation protocol, and criteria of recipients render technical hitches.

The administration of stem cell injection (quantity and mode of delivery) may vary between laboratories based on each protocol and experience. Several studies were conducted to definite numbers of stem cells for the injection, while others calculate the body weight of the recipient. Similarly, several studies preferred intrahepatic administration while others via intrasplenic or peripheral vein. Therefore, so far, there is no definite indication or international consensus regarding the protocol of adult stem cells in ESLD patients.

Apart from a scientific perspective, the clinical application of cell therapy is related also to the vast speed of the internet spread. Advances in information technology significantly increase the global transfer of knowledge, including in the search for stem cell therapy in one click. As can be seen in cell therapy for regenerative medicine, the so-called stem cell tourism (Berger et al. 2016; Sipp 2017) is also a problem in hepatology and gastroenterology (Hermerén 2014). This problem requires prompt action for the regulation of cell therapy, from scientists, clinicians, professional associations, and government or authorities. Stem cell therapy for ESLD had shown some promising results, but more research and the definition of a better protocol are still significantly needed.

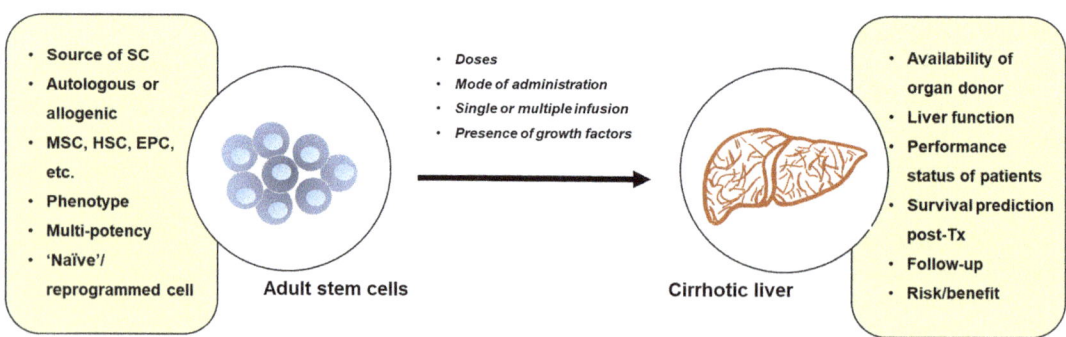

Fig. 2 Important factors for adult stem cell therapy for ESLD. Stem cell therapy would need to consider aspects both in the donor cells (source, types, phenotypes, potency) and in the recipient (liver status, patients' performance, risk/benefit), together with the mode of administration (site, presence of growth factor, doses) and correct protocol

References

Amer M-EM, El-Sayed SZ, El-Kheir WA, Gabr H, Gomaa AA, El-Noomani N, Hegazy M (2011) Clinical and laboratory evaluation of patients with end-stage liver cell failure injected with bone marrow-derived hepatocyte-like cells. Eur J Gastroenterol Hepatol 23(10): 936–941. https://doi.org/10.1097/MEG. 0b013e3283488b00

Amin MA, Sabry D, Rashed LA, Aref WM, el-Ghobary MA, Farhan MS, Fouad HA, Youssef YA-A (2013) Short-term evaluation of autologous transplantation of bone marrow-derived mesenchymal stem cells in patients with cirrhosis: Egyptian study. Clin Transpl 27(4):607–612. https://doi.org/10.1111/ctr. 12179

Asahara T, Murohara T, Sullivan A, Silver M, van der Zee R, Li T, Witzenbichler B, Schatteman G, Isner JM (1997) Isolation of putative progenitor endothelial cells for angiogenesis. Science 275(5302):964–967. https:// doi.org/10.1126/science.275.5302.964

Asrani SK, Devarbhavi H, Eaton J, Kamath PS (2019) Burden of liver diseases in the world. J Hepatol 70(1):151–171. https://doi.org/10.1016/j.jhep.2018. 09.014

Au KP, Chan ACY (2019) Is living donor liver transplantation justified in high model for end-stage liver disease candidates (35+)? Curr Opin Organ Transplant 24(5): 637–643. https://doi.org/10.1097/MOT. 0000000000000689

Berger I, Ahmad A, Bansal A, Kapoor T, Sipp D, Rasko JEJ (2016) Global distribution of businesses marketing stem cell-based interventions. Cell Stem Cell 19(2):158–162. https://doi.org/10.1016/j.stem.2016.07.015

Blum B, Benvenisty N (2008) The tumorigenicity of human embryonic stem cells. Adv Cancer Res 100: 133–158. https://doi.org/10.1016/S0065-230X(08) 00005-5

Bracha P, Moore NA, Ciulla TA (2017) Induced pluripotent stem cell-based therapy for age-related macular degeneration. Expert Opin Biol Ther 17(9):1113–1126. https://doi.org/10.1080/14712598.2017.1346079

Bucher NLR, Swaffield MN (1964) The rate of incorporation of labeled thymidine into the deoxyribonucleic acid of regenerating rat liver in relation to the amount of liver excised. Cancer Res 24(9):1611–1625

Caplan AI (1991) Mesenchymal stem cells. J Orthop Res 9(5):641–650. https://doi.org/10.1002/jor.1100090504

Caplan AI (2010) What's in a name? Tissue Eng Part A 16(8):2415–2417. https://doi.org/10.1089/ten.TEA. 2010.0216

Caplan AI (2017) Mesenchymal stem cells: time to change the name! Stem Cells Transl Med 6(6):1445–1451. https://doi.org/10.1002/sctm.17-0051

Catani L, Sollazzo D, Bianchi E, Ciciarello M, Antoniani C, Foscoli L, Caraceni P, Giannone FA, Baldassarre M, Giordano R, Montemurro T, Montelatici E, D'Errico A, Andreone P, Giudice V, Curti A, Manfredini R, Lemoli

RM (2017) Molecular and functional characterization of CD133+ stem/progenitor cells infused in patients with end-stage liver disease reveals their interplay with stromal liver cells. Cytotherapy 19(12):1447–1461. https:// doi.org/10.1016/j.jcyt.2017.08.001

Chang M, Bogacheva MS, Lou Y-R (2021) Challenges for the applications of human pluripotent stem cell-derived liver organoids. Front Cell Dev Biol 9:748576. https:// doi.org/10.3389/fcell.2021.748576

Chen Y-F, Tseng C-Y, Wang H-W, Kuo H-C, Yang VW, Lee OK (2012) Rapid generation of mature hepatocyte-like cells from human induced pluripotent stem cells by an efficient three-step protocol. Hepatology 55(4): 1193–1203. https://doi.org/10.1002/hep.24790

Choudhary NS, Saraf N, Dhampalwar S, Saigal S, Gautam D, Rastogi A, Bhangui P, Srinivasan T, Rastogi V, Mehrotra S, Soin AS (2022) Poor outcomes after recidivism in living donor liver transplantation for alcohol-related liver disease. J Clin Exp Hepatol 12(1): 37–42. https://doi.org/10.1016/j.jceh.2021.04.005

Corti S, Locatelli F, Donadoni C, Strazzer S, Salani S, Del Bo R, Caccialanza M, Bresolin N, Scarlato G, Comi GP (2002) Neuroectodermal and microglial differentiation of bone marrow cells in the mouse spinal cord and sensory ganglia. J Neurosci Res 70(6):721–733. https://doi.org/10.1002/jnr.10455

D'Avola D, Fernández-Ruiz V, Carmona-Torre F, Méndez M, Pérez-Calvo J, Prósper F, Andreu E, Herrero JI, Iñarrairaegui M, Fuertes C, Bilbao JI, Sangro B, Prieto J, Quiroga J (2017) Phase 1–2 pilot clinical trial in patients with decompensated liver cirrhosis treated with bone marrow-derived endothelial progenitor cells. Transl Res 188:80–91.e2. https://doi. org/10.1016/j.trsl.2016.02.009

Dominici M, Le Blanc K, Mueller I, Slaper-Cortenbach I, Marini F, Krause D, Deans R, Keating A, Prockop D, Horwitz E (2006) Minimal criteria for defining multipotent mesenchymal stromal cells. The international society for cellular therapy position statement. Cytotherapy 8(4):315–317. https://doi.org/10.1080/14653240600855905

El-Ansary M, Abdel-Aziz I, Mogawer S, Abdel-Hamid S, Hammam O, Teaema S, Wahdan M (2012) Phase II trial: undifferentiated versus differentiated autologous mesenchymal stem cells transplantation in Egyptian patients with HCV induced liver cirrhosis. Stem Cell Rev Rep 8(3):972–981. https://doi.org/10.1007/s12015-011-9322-y

European Association for the Study of the Liver (2018) EASL clinical practice guidelines for the management of patients with decompensated cirrhosis. J Hepatol 69(2):406–460. https://doi.org/10.1016/j.jhep.2018. 03.024

Feng S, Bucuvalas J (2017) Tolerance after liver transplantation: where are we? Liver Transplant 23(12): 1601–1614. https://doi.org/10.1002/lt.24845

Fitzpatrick E, Mitry RR, Dhawan A (2009) Human hepatocyte transplantation: state of the art. J Intern Med

266(4):339–357. https://doi.org/10.1111/j.1365-2796.2009.02152.x

Fricker ZP, Serper M (2019) Current knowledge, barriers to implementation, and future directions in palliative Care for end-Stage Liver Disease. Liver Transplant 25(5):787–796. https://doi.org/10.1002/lt.25434

Fujikawa T, Oh S-H, Pi L, Hatch HM, Shupe T, Petersen BE (2005) Teratoma formation leads to failure of treatment for type I diabetes using embryonic stem cell-derived insulin-producing cells. Am J Pathol 166(6):1781–1791. https://doi.org/10.1016/S0002-9440(10)62488-1

Giancotti A, D'Ambrosio V, Corno S, Pajno C, Carpino G, Amato G, Vena F, Mondo A, Spiniello L, Monti M, Muzii L, Bosco D, Gaudio E, Alvaro D, Cardinale V (2022) Current protocols and clinical efficacy of human fetal liver cell therapy in patients with liver disease: a literature review. Cytotherapy 24(4):376–384. https://doi.org/10.1016/j.jcyt.2021.10.012

Godoy P, Hewitt NJ, Albrecht U, Andersen ME, Ansari N, Bhattacharya S, Bode JG, Bolleyn J, Borner C, Böttger J, Braeuning A, Budinsky RA, Burkhardt B, Cameron NR, Camussi G, Cho C-S, Choi Y-J, Craig Rowlands J, Dahmen U, Damm G, Dirsch O, Donato MT, Dong J, Dooley S, Drasdo D, Eakins R, Ferreira KS, Fonsato V, Fraczek J, Gebhardt R, Gibson A, Glanemann M, Goldring CEP, Gómez-Lechón MJ, Groothuis GMM, Gustavsson L, Guyot C, Hallifax D, Hammad S, Hayward A, Häussinger D, Hellerbrand C, Hewitt P, Hoehme S, Holzhütter H-G, Houston JB, Hrach J, Ito K, Jaeschke H, Keitel V, Kelm JM, Kevin Park B, Kordes C, Kullak-Ublick GA, LeCluyse EL, Lu P, Luebke-Wheeler J, Lutz A, Maltman DJ, Matz-Soja M, McMullen P, Merfort I, Messner S, Meyer C, Mwinyi J, Naisbitt DJ, Nussler AK, Olinga P, Pampaloni F, Pi J, Pluta L, Przyborski SA, Ramachandran A, Rogiers V, Rowe C, Schelcher C, Schmich K, Schwarz M, Singh B, Stelzer EHK, Stieger B, Stöber R, Sugiyama Y, Tetta C, Thasler WE, Vanhaecke T, Vinken M, Weiss TS, Widera A, Woods CG, Xu JJ, Yarborough KM, Hengstler JG (2013) Recent advances in 2D and 3D in vitro systems using primary hepatocytes, alternative hepatocyte sources and non-parenchymal liver cells and their use in investigating mechanisms of hepatotoxicity, cell signaling and ADME. Arch Toxicol 87(8):1315–1530. https://doi.org/10.1007/s00204-013-1078-5

Gordon MY, Levicar N, Pai M, Bachellier P, Dimarakis I, Al-Allaf F, M'Hamdi H, Thalji T, Welsh JP, Marley SB, Davies J, Dazzi F, Marelli-Berg F, Tait P, Playford R, Jiao L, Jensen S, Nicholls JP, Ayav A, Nohandani M, Farzaneh F, Gaken J, Dodge R, Alison M, Apperley JF, Lechler R, Habib NA (2006) Characterization and clinical application of human CD34+ stem/progenitor cell populations mobilized into the blood by granulocyte colony-stimulating factor. Stem Cells 24(7):1822–1830. https://doi.org/10.1634/stemcells.2005-0629

Gridelli B, Vizzini G, Pietrosi G, Luca A, Spada M, Gruttadauria S, Cintorino D, Amico G, Chinnici C, Miki T, Schmelzer E, Conaldi PG, Triolo F, Gerlach JC (2012) Efficient human fetal liver cell isolation protocol based on vascular perfusion for liver cell-based therapy and case report on cell transplantation. Liver Transplant 18(2):226–237. https://doi.org/10.1002/lt.22322

Gurdon JB (1962) The developmental capacity of nuclei taken from intestinal epithelium cells of feeding tadpoles. J Embryol Exp Morphol 10:622–640

Gussoni E, Soneoka Y, Strickland CD, Buzney EA, Khan MK, Flint AF, Kunkel LM, Mulligan RC (1999) Dystrophin expression in the mdx mouse restored by stem cell transplantation. Nature 401(6751):390–394. https://doi.org/10.1038/43919

Haep N, Florentino RM, Squires JE, Bell A, Soto-Gutierrez A (2021) The inside-out of end-stage liver disease: hepatocytes are the keystone. Semin Liver Dis 41(2):213–224. https://doi.org/10.1055/s-0041-1725023

Haideri SS, McKinnon AC, Taylor AH, Kirkwood P, Starkey Lewis PJ, O'Duibhir E, Vernay B, Forbes S, Forrester LM (2017) Injection of embryonic stem cell derived macrophages ameliorates fibrosis in a murine model of liver injury. NPJ Regen Med 2:14. https://doi.org/10.1038/s41536-017-0017-0

Hannoun Z, Steichen C, Dianat N, Weber A, Dubart-Kupperschmitt A (2016) The potential of induced pluripotent stem cell derived hepatocytes. J Hepatol 65(1):182–199. https://doi.org/10.1016/j.jhep.2016.02.025

Harries L, Gwiasda J, Qu Z, Schrem H, Krauth C, Amelung VE (2019) Potential savings in the treatment pathway of liver transplantation: an inter-sectorial analysis of cost-rising factors. Eur J Health Econ 20(2):281–301. https://doi.org/10.1007/s10198-018-0994-y

Hentze H, Soong PL, Wang ST, Phillips BW, Putti TC, Dunn NR (2009) Teratoma formation by human embryonic stem cells: evaluation of essential parameters for future safety studies. Stem Cell Res 2(3):198–210. https://doi.org/10.1016/j.scr.2009.02.002

Heo J, Factor VM, Uren T, Takahama Y, Lee J-S, Major M, Feinstone SM, Thorgeirsson SS (2006) Hepatic precursors derived from murine embryonic stem cells contribute to regeneration of injured liver. Hepatology 44(6):1478–1486. https://doi.org/10.1002/hep.21441

Hermerén G (2014) Human stem-cell research in gastroenterology: experimental treatment, tourism and biobanking. Best Pract Res Clin Gastroenterol 28(2):257–268. https://doi.org/10.1016/j.bpg.2014.02.002

Herrero JI (2012) Screening of de novo tumors after liver transplantation. J Gastroenterol Hepatol 27(6):1011–1016. https://doi.org/10.1111/j.1440-1746.2011.06981.x

Hess DC, Abe T, Hill WD, Studdard AM, Carothers J, Masuya M, Fleming PA, Drake CJ, Ogawa M (2004) Hematopoietic origin of microglial and perivascular cells in brain. Exp Neurol 186(2):134–144. https://doi.org/10.1016/j.expneurol.2003.11.005

Higgins G (1931) Experimental pathology of the liver. I. Restoration of the liver of the white rat following partial surgical removal. Arch Pathol 12:186–202

Horwitz EM, Le Blanc K, Dominici M, Mueller I, Slaper-Cortenbach I, Marini FC, Deans RJ, Krause DS, Keating A, International Society for Cellular Therapy (2005) Clarification of the nomenclature for MSC: the International Society for Cellular Therapy position statement. Cytotherapy 7(5):393–395. https://doi.org/10.1080/14653240500319234

Hu H, Gehart H, Artegiani B, LÖpez-Iglesias C, Dekkers F, Basak O, van Es J, Chuva de Sousa Lopes SM, Begthel H, Korving J, van den Born M, Zou C, Quirk C, Chiriboga L, Rice CM, Ma S, Rios A, Peters PJ, de Jong YP, Clevers H (2018) Long-term expansion of functional mouse and human hepatocytes as 3D organoids. Cell 175(6):1591–1606.e19. https://doi.org/10.1016/j.cell.2018.11.013

Hu C, Zhao L, Li L (2019) Current understanding of adipose-derived mesenchymal stem cell-based therapies in liver diseases. Stem Cell Res Ther 10(1):199. https://doi.org/10.1186/s13287-019-1310-1

Jang Y-Y, Collector MI, Baylin SB, Diehl AM, Sharkis SJ (2004) Hematopoietic stem cells convert into liver cells within days without fusion. Nat Cell Biol 6(6):532–539. https://doi.org/10.1038/ncb1132

Kamath PS, Wiesner RH, Malinchoc M, Kremers W, Therneau TM, Kosberg CL, D'Amico G, Dickson ER, Kim WR (2001) A model to predict survival in patients with end-stage liver disease. Hepatology 33(2):464–470. https://doi.org/10.1053/jhep.2001.22172

Kantarcıoğlu M, Demirci H, Avcu F, Karslıoğlu Y, Babayiğit MA, Karaman B, Öztürk K, Gürel H, Akdoğan Kayhan M, Kaçar S, Kubar A, Öksüzoğlu G, Ural AU, Bağcı S (2015) Efficacy of autologous mesenchymal stem cell transplantation in patients with liver cirrhosis. Turk J Gastroenterol 26(3):244–250. https://doi.org/10.5152/tjg.2015.0074

Keighron C, Lyons CJ, Creane M, O'Brien T, Liew A (2018) Recent advances in endothelial progenitor cells toward their use in clinical translation. Front Med 5:354. https://doi.org/10.3389/fmed.2018.00354

Khan AA, Shaik MV, Parveen N, Rajendraprasad A, Aleem MA, Habeeb MA, Srinivas G, Raj TA, Tiwari SK, Kumaresan K, Venkateswarlu J, Pande G, Habibullah CM (2010) Human fetal liver-derived stem cell transplantation as supportive modality in the management of end-stage decompensated liver cirrhosis. Cell Transplant 19(4):409–418. https://doi.org/10.3727/096368910X498241

Kharaziha P, Hellström PM, Noorinayer B, Farzaneh F, Aghajani K, Jafari F, Telkabadi M, Atashi A, Honardoost M, Zali MR, Soleimani M (2009) Improvement of liver function in liver cirrhosis patients after autologous mesenchymal stem cell injection: a phase I–II clinical trial. Eur J Gastroenterol Hepatol 21(10):1199–1205. https://doi.org/10.1097/MEG.0b013e32832a1f6c

Kim WR, Mannalithara A, Heimbach JK, Kamath PS, Asrani SK, Biggins SW, Wood NL, Gentry SE, Kwong AJ (2021) MELD 3.0: the model for end-stage liver disease updated for the modern era. Gastroenterology 161(6):1887–1895.e4. https://doi.org/10.1053/j.gastro.2021.08.050

Lan L, Liu R, Qin L-Y, Cheng P, Liu B-W, Zhang B-Y, Ding S-Z, Li X-L (2018) Transplantation of bone marrow-derived endothelial progenitor cells and hepatocyte stem cells from liver fibrosis rats ameliorates liver fibrosis. World J Gastroenterol 24(2):237–247. https://doi.org/10.3748/wjg.v24.i2.237

Lee CA, Sinha S, Fitzpatrick E, Dhawan A (2018) Hepatocyte transplantation and advancements in alternative cell sources for liver-based regenerative medicine. J Mol Med 96(6):469–481. https://doi.org/10.1007/s00109-018-1638-5

Li Y-H, Xu Y, Wu H-M, Yang J, Yang L-H, Yue-Meng W (2016) Umbilical cord-derived mesenchymal stem cell transplantation in hepatitis B virus related acute-on-chronic liver failure treated with plasma exchange and Entecavir: a 24-month prospective study. Stem Cell Rev Rep 12(6):645–653. https://doi.org/10.1007/s12015-016-9683-3

Lian G, Wang C, Teng C, Zhang C, Du L, Zhong Q, Miao C, Ding M, Deng H (2006) Failure of hepatocyte marker-expressing hematopoietic progenitor cells to efficiently convert into hepatocytes in vitro. Exp Hematol 34(3):348–358. https://doi.org/10.1016/j.exphem.2005.12.004

Liang L, Ma T, Chen W, Hu J, Bai X, Li J, Liang T (2009) Therapeutic potential and related signal pathway of adipose-derived stem cell transplantation for rat liver injury. Hepatol Res 39(8):822–832. https://doi.org/10.1111/j.1872-034X.2009.00506.x

Liang J, Zhang H, Zhao C, Wang D, Ma X, Zhao S, Wang S, Niu L, Sun L (2017) Effects of allogeneic mesenchymal stem cell transplantation in the treatment of liver cirrhosis caused by autoimmune diseases. Int J Rheum Dis 20(9):1219–1226. https://doi.org/10.1111/1756-185X.13015

Libbrecht L, Roskams T (2002) Hepatic progenitor cells in human liver diseases. Semin Cell Dev Biol 13(6):389–396. https://doi.org/10.1016/s1084952102001258

Lin B-L, Chen J-F, Qiu W-H, Wang K-W, Xie D-Y, Chen X-Y, Liu Q-L, Peng L, Li J-G, Mei Y-Y, Weng W-Z, Peng Y-W, Cao H-J, Xie J-Q, Xie S-B, Xiang AP, Gao Z-L (2017) Allogeneic bone marrow-derived mesenchymal stromal cells for hepatitis B virus-related acute-on-chronic liver failure: a randomized controlled trial. Hepatology 66(1):209–219. https://doi.org/10.1002/hep.29189

Lorenzini S, Gitto S, Grandini E, Andreone P, Bernardi M (2008) Stem cells for end stage liver disease: how far have we got? World J Gastroenterol 14(29):4593–4599. https://doi.org/10.3748/wjg.14.4593

Luo X, Leanza J, Massie AB, Garonzik-Wang JM, Haugen CE, Gentry SE, Ottmann SE, Segev DL (2018) MELD as a metric for survival benefit of liver transplantation. Am J Transplant 18(5):1231–1237. https://doi.org/10.1111/ajt.14660

Manzia TM, Angelico R, Gazia C, Lenci I, Milana M, Ademoyero OT, Pedini D, Toti L, Spada M, Tisone G, Baiocchi L (2019) De novo malignancies after liver transplantation: the effect of immunosuppression-personal data and review of literature. World J Gastroenterol 25(35):5356–5375. https://doi.org/10.3748/wjg.v25.i35.5356

Martins AM, Vunjak-Novakovic G, Reis RL (2014) The current status of iPS cells in cardiac research and their potential for tissue engineering and regenerative medicine. Stem Cell Rev Rep 10(2):177–190. https://doi.org/10.1007/s12015-013-9487-7

Masson S, Harrison DJ, Plevris JN, Newsome PN (2004) Potential of hematopoietic stem cell therapy in hepatology: a critical review. Stem Cells 22(6):897–907. https://doi.org/10.1634/stemcells.22-6-897

Medina RJ, Barber CL, Sabatier F, Dignat-George F, Melero-Martin JM, Khosrotehrani K, Ohneda O, Randi AM, Chan JKY, Yamaguchi T, Van Hinsbergh VWM, Yoder MC, Stitt AW (2017) Endothelial progenitors: a consensus statement on nomenclature. Stem Cells Transl Med 6(5):1316–1320. https://doi.org/10.1002/sctm.16-0360

Mohamadnejad M, Alimoghaddam K, Bagheri M, Ashrafi M, Abdollahzadeh L, Akhlaghpoor S, Bashtar M, Ghavamzadeh A, Malekzadeh R (2013) Randomized placebo-controlled trial of mesenchymal stem cell transplantation in decompensated cirrhosis. Liver Int 33(10):1490–1496. https://doi.org/10.1111/liv.12228

Moriya K, Yoshikawa M, Ouji Y, Saito K, Nishiofuku M, Matsuda R, Ishizaka S, Fukui H (2008) Embryonic stem cells reduce liver fibrosis in CCl4-treated mice. Int J Exp Pathol 89(6):401–409. https://doi.org/10.1111/j.1365-2613.2008.00607.x

Mountford JC (2008) Human embryonic stem cells: origins, characteristics and potential for regenerative therapy. Transfus Med 18(1):1–12. https://doi.org/10.1111/j.1365-3148.2007.00807.x

Nagai S, Chau LC, Schilke RE, Safwan M, Rizzari M, Collins K, Yoshida A, Abouljoud MS, Moonka D (2018) Effects of allocating livers for transplantation based on model for end-stage liver disease–sodium scores on patient outcomes. Gastroenterology 155(5):1451–1462.e3. https://doi.org/10.1053/j.gastro.2018.07.025

Nhung TH, Nam NH, Nguyen NTK, Nghia H, Van Thanh N, Ngoc PK, Van Pham P (2015) A comparison of the chemical and liver extract-induced hepatic differentiation of adipose derived stem cells. Vitro Cell Dev Biol Anim 51(10):1085–1092. https://doi.org/10.1007/s11626-015-9939-2

Orlic D, Kajstura J, Chimenti S, Jakoniuk I, Anderson SM, Li B, Pickel J, McKay R, Nadal-Ginard B, Bodine DM, Leri A, Anversa P (2001) Bone marrow cells regenerate infarcted myocardium. Nature 410(6829):701–705. https://doi.org/10.1038/35070587

Pan X-N, Zheng L-Q, Lai X-H (2014) Bone marrow-derived mesenchymal stem cell therapy for decompensated liver cirrhosis: a meta-analysis. World J Gastroenterol 20(38):14051–14057. https://doi.org/10.3748/wjg.v20.i38.14051

Peng L, Xie D, Lin B-L, Liu J, Zhu H, Xie C, Zheng Y, Gao Z (2011) Autologous bone marrow mesenchymal stem cell transplantation in liver failure patients caused by hepatitis B: short-term and long-term outcomes. Hepatology 54(3):820–828. https://doi.org/10.1002/hep.24434

Petersen BE, Bowen WC, Patrene KD, Mars WM, Sullivan AK, Murase N, Boggs SS, Greenberger JS, Goff JP (1999) Bone marrow as a potential source of hepatic oval cells. Science 284(5417):1168–1170. https://doi.org/10.1126/science.284.5417.1168

Pittenger MF, Mackay AM, Beck SC, Jaiswal RK, Douglas R, Mosca JD, Moorman MA, Simonetti DW, Craig S, Marshak DR (1999) Multilineage potential of adult human mesenchymal stem cells. Science 284(5411):143–147. https://doi.org/10.1126/science.284.5411.143

Prasain N, Meador JL, Yoder MC (2012) Phenotypic and functional characterization of endothelial Colony forming cells derived from human umbilical cord blood. JoVE J Vis Exp 62:e3872. https://doi.org/10.3791/3872

Qin M, Guan X, Zhang Y, Shen B, Liu F, Zhang Q, Ma Y, Jiang Y (2018) Evaluation of ex vivo produced endothelial progenitor cells for autologous transplantation in primates. Stem Cell Res Ther 9(1):14. https://doi.org/10.1186/s13287-018-0769-5

Reza HA, Ryo O, Takebe T (2021) Organoid transplant approaches for the liver. Transpl Int. https://doi.org/10.1111/tri.14128

Sacchetti B, Funari A, Remoli C, Giannicola G, Kogler G, Liedtke S, Cossu G, Serafini M, Sampaolesi M, Tagliafico E, Tenedini E, Saggio I, Robey PG, Riminucci M, Bianco P (2016) No identical "mesenchymal stem cells" at different times and sites: human committed progenitors of distinct origin and differentiation potential are incorporated as adventitial cells in microvessels. Stem Cell Rep 6(6):897–913. https://doi.org/10.1016/j.stemcr.2016.05.011

Safarikia S, Carpino G, Overi D, Cardinale V, Venere R, Franchitto A, Onori P, Alvaro D, Gaudio E (2020) Distinct EpCAM-positive stem cell niches are engaged in chronic and neoplastic liver diseases. Front Med 7:479. https://doi.org/10.3389/fmed.2020.00479

Sakai Y, Fukunishi S, Takamura M, Inoue O, Takashima S, Usui S, Seki A, Nasti A, Ho TTB, Kawaguchi K, Asai A, Tsuchimoto Y, Yamashita T, Yamashita T, Mizukoshi E, Honda M, Imai Y, Yoshimura K, Murayama T, Wada T, Harada K, Higuchi K, Kaneko S (2020) Regenerative therapy for liver cirrhosis based on intrahepatic arterial infusion of autologous subcutaneous adipose tissue-derived regenerative (stem) cells: protocol for a confirmatory multicenter uncontrolled clinical trial. JMIR Res Protoc 9(3):e17904. https://doi.org/10.2196/17904

Sakai Y, Fukunishi S, Takamura M, Kawaguchi K, Inoue O, Usui S, Takashima S, Seki A, Asai A, Tsuchimoto Y, Nasti A, Bich Ho TT, Imai Y, Yoshimura K, Murayama T, Yamashita T, Arai K, Yamashita T, Mizukoshi E, Honda M, Wada T, Harada K, Higuchi K, Kaneko S (2021) Clinical trial of autologous adipose tissue-derived regenerative (stem) cells therapy for exploration of its safety and efficacy. Regen Ther 18:97–101. https://doi.org/10.1016/j.reth.2021.04.003

Sakamoto M, Nakamura T, Torimura T, Iwamoto H, Masuda H, Koga H, Abe M, Hashimoto O, Ueno T, Sata M (2013) Transplantation of endothelial progenitor cells ameliorates vascular dysfunction and portal hypertension in carbon tetrachloride-induced rat liver cirrhotic model. J Gastroenterol Hepatol 28(1):168–178. https://doi.org/10.1111/j.1440-1746.2012.07238.x

Salama H, Zekri A-RN, Bahnassy AA, Medhat E, Halim HA, Ahmed OS, Mohamed G, Al Alim SA, Sherif GM (2010) Autologous CD34+ and CD133+ stem cells transplantation in patients with end stage liver disease. World J Gastroenterol 16(42):5297–5305. https://doi.org/10.3748/wjg.v16.i42.5297

Salama H, Zekri A-RN, Medhat E, Al Alim SA, Ahmed OS, Bahnassy AA, Lotfy MM, Ahmed R, Musa S (2014) Peripheral vein infusion of autologous mesenchymal stem cells in Egyptian HCV-positive patients with end-stage liver disease. Stem Cell Res Ther 5(3):70. https://doi.org/10.1186/scrt459

Schacher FC, Martins Pezzi da Silva A, Silla LM d R, Álvares-da-Silva MR (2021) Bone marrow mesenchymal stem cells in acute-on-chronic liver failure grades 2 and 3: a phase I-II randomized clinical trial. Can J Gastroenterol Hepatol 2021:3662776. https://doi.org/10.1155/2021/3662776

Semeraro R, Cardinale V, Carpino G, Gentile R, Napoli C, Venere R, Gatto M, Brunelli R, Gaudio E, Alvaro D (2013) The fetal liver as cell source for the regenerative medicine of liver and pancreas. Ann Transl Med 1(2):13. https://doi.org/10.3978/j.issn.2305-5839.2012.10.02

Sharma AD, Cantz T, Vogel A, Schambach A, Haridass D, Iken M, Bleidissel M, Manns MP, Schöler HR, Ott M (2008) Murine embryonic stem cell-derived hepatic progenitor cells engraft in recipient livers with limited capacity of liver tissue formation. Cell Transplant 17(3):313–323. https://doi.org/10.3727/096368908784153896

Shi M, Zhang Z, Xu R, Lin H, Fu J, Zou Z, Zhang A, Shi J, Chen L, Lv S, He W, Geng H, Jin L, Liu Z, Wang F-S (2012) Human mesenchymal stem cell transfusion is safe and improves liver function in acute-on-chronic liver failure patients. Stem Cells Transl Med 1(10):725–731. https://doi.org/10.5966/sctm.2012-0034

Sipp D (2017) The malignant niche: safe spaces for toxic stem cell marketing. Npj Regen Med 2(1):1–4. https://doi.org/10.1038/s41536-017-0036-x

Sipp D, Robey PG, Turner L (2018) Clear up this stem-cell mess. Nature 561(7724):455–457. https://doi.org/10.1038/d41586-018-06756-9

Sison-Young RL, Lauschke VM, Johann E, Alexandre E, Antherieu S, Aerts H, Gerets HHJ, Labbe G, Hoët D, Dorau M, Schofield CA, Lovatt CA, Holder JC, Stahl SH, Richert L, Kitteringham NR, Jones RP, Elmasry M, Weaver RJ, Hewitt PG, Ingelman-Sundberg M, Goldring CE, Park BK (2017) A multi-center assessment of single-cell models aligned to standard measures of cell health for prediction of acute hepatotoxicity. Arch Toxicol 91(3):1385–1400. https://doi.org/10.1007/s00204-016-1745-4

Song Z, Cai J, Liu Y, Zhao D, Yong J, Duo S, Song X, Guo Y, Zhao Y, Qin H, Yin X, Wu C, Che J, Lu S, Ding M, Deng H (2009) Efficient generation of hepatocyte-like cells from human induced pluripotent stem cells. Cell Res 19(11):1233–1242. https://doi.org/10.1038/cr.2009.107

Stachelscheid H, Wulf-Goldenberg A, Eckert K, Jensen J, Edsbagge J, Björquist P, Rivero M, Strehl R, Jozefczuk J, Prigione A, Adjaye J, Urbaniak T, Bussmann P, Zeilinger K, Gerlach JC (2013) Teratoma formation of human embryonic stem cells in three-dimensional perfusion culture bioreactors. J Tissue Eng Regen Med 7(9):729–741. https://doi.org/10.1002/term.1467

Sumide K, Matsuoka Y, Kawamura H, Nakatsuka R, Fujioka T, Asano H, Takihara Y, Sonoda Y (2018) A revised road map for the commitment of human cord blood CD34-negative hematopoietic stem cells. Nat Commun 9(1):2202. https://doi.org/10.1038/s41467-018-04441-z

Takahashi K, Yamanaka S (2006) Induction of pluripotent stem cells from mouse embryonic and adult fibroblast cultures by defined factors. Cell 126(4):663–676. https://doi.org/10.1016/j.cell.2006.07.024

Takahashi K, Tanabe K, Ohnuki M, Narita M, Ichisaka T, Tomoda K, Yamanaka S (2007) Induction of pluripotent stem cells from adult human fibroblasts by defined factors. Cell 131(5):861–872. https://doi.org/10.1016/j.cell.2007.11.019

Tarlow BD, Pelz C, Naugler WE, Wakefield L, Wilson EM, Finegold MJ, Grompe M (2014) Bipotential adult liver progenitors are derived from chronically injured mature hepatocytes. Cell Stem Cell 15(5):605–618. https://doi.org/10.1016/j.stem.2014.09.008

Terai S, Ishikawa T, Omori K, Aoyama K, Marumoto Y, Urata Y, Yokoyama Y, Uchida K, Yamasaki T, Fujii Y, Okita K, Sakaida I (2006) Improved liver function in patients with liver cirrhosis after autologous bone marrow cell infusion therapy. Stem Cells 24(10):2292–2298. https://doi.org/10.1634/stemcells.2005-0542

Theise ND, Saxena R, Portmann BC, Thung SN, Yee H, Chiriboga L, Kumar A, Crawford JM (1999) The canals of Hering and hepatic stem cells in humans. Hepatology 30(6):1425–1433. https://doi.org/10.1002/hep.510300614

Thorgeirsson SS, Grisham JW (2006) Hematopoietic cells as hepatocyte stem cells: a critical review of the evidence. Hepatology 43(1):2–8. https://doi.org/10.1002/hep.21015

Tsuchida T, Murata S, Matsuki K, Mori A, Matsuo M, Mikami S, Okamoto S, Ueno Y, Tadokoro T, Zheng Y-W, Taniguchi H (2019) The regenerative effect of portal vein injection of liver organoids by Retrorsine/partial hepatectomy in rats. Int J Mol Sci 21(1):E178. https://doi.org/10.3390/ijms21010178

Tsujimoto H, Osafune K (2021) Current status and future directions of clinical applications using iPS cells-focus on Japan. FEBS J. https://doi.org/10.1111/febs.16162

Wang J, Kimura T, Asada R, Harada S, Yokota S, Kawamoto Y, Fujimura Y, Tsuji T, Ikehara S, Sonoda Y (2003) SCID-repopulating cell activity of human cord blood-derived CD34- cells assured by intra-bone marrow injection. Blood 101(8):2924–2931. https://doi.org/10.1182/blood-2002-09-2782

Wiesner R, Edwards E, Freeman R, Harper A, Kim R, Kamath P, Kremers W, Lake J, Howard T, Merion RM, Wolfe RA, Krom R, United Network for Organ Sharing Liver Disease Severity Score Committee (2003) Model for end-stage liver disease (MELD) and allocation of donor livers. Gastroenterology 124(1):91–96. https://doi.org/10.1053/gast.2003.50016

Wilmut I, Schnieke AE, McWhir J, Kind AJ, Campbell KHS (1997) Viable offspring derived from fetal and adult mammalian cells. Nature 385(6619):810–813. https://doi.org/10.1038/385810a0

Wilson HV (1907) A new method by which sponges may be artificially reared. Science 25(649):912–915. https://doi.org/10.1126/science.25.649.912

Xiang Z, Hua M, Hao Z, Biao H, Zhu C, Zhai G, Wu J (2022) The roles of mesenchymal stem cells in gastrointestinal cancers. Front Immunol 13:844001. https://doi.org/10.3389/fimmu.2022.844001

Yanagida A, Ito K, Chikada H, Nakauchi H, Kamiya A (2013) An in vitro expansion system for generation of human iPS cell-derived hepatic progenitor-like cells exhibiting a bipotent differentiation potential. PLoS One 8(7):e67541. https://doi.org/10.1371/journal.pone.0067541

Zanjani ED, Almeida-Porada G, Livingston AG, Flake AW, Ogawa M (1998) Human bone marrow CD34- cells engraft in vivo and undergo multilineage expression that includes giving rise to CD34+ cells. Exp Hematol 26(4):353–360

Zekri A-RN, Salama H, Medhat E, Musa S, Abdel-Haleem H, Ahmed OS, Khedr HAH, Lotfy MM, Zachariah KS, Bahnassy AA (2015) The impact of repeated autologous infusion of haematopoietic stem cells in patients with liver insufficiency. Stem Cell Res Ther 6:118. https://doi.org/10.1186/s13287-015-0106-1

Zhang Z, Lin H, Shi M, Xu R, Fu J, Lv J, Chen L, Lv S, Li Y, Yu S, Geng H, Jin L, Lau GKK, Wang F-S (2012) Human umbilical cord mesenchymal stem cells improve liver function and ascites in decompensated liver cirrhosis patients. J Gastroenterol Hepatol 27(Suppl 2):112–120. https://doi.org/10.1111/j.1440-1746.2011.07024.x

Adv Exp Med Biol - Cell Biology and Translational Medicine (2022) 17: 73–95
https://doi.org/10.1007/5584_2022_716
© Springer Nature Switzerland AG 2022
Published online: 4 July 2022

An Affordable Approach of Mesenchymal Stem Cell Therapy in Treating Perianal Fistula Treatment

Hui-Nee Hon, Pei-Yi Ho, Jing-Wen Lee,
Nur Amalin Amni Mahmud, Hafsa Binte Munir,
Thamil Selvee Ramasamy, Vijayendran Govindasamy ⓘ,
Kong-Yong Then, Anjan Kumar Das, and Soon-Keng Cheong

Abstract

The application of stem cells to treat perianal fistula due to Crohn's disease has attracted a lot of interest in recent decades. Though still a popular procedure, the existing surgical methods may be an ideal form of therapy since the recurrence rate is high, which affects the quality of life badly. Stem cell therapy offers to be a better solution in treating PF, but the utilisation is often restricted because of the manufacturing cost. Hence in this review, the selection of suitable cell sources, the use of bioreactors and preconditioning MSCs as well as modified stem cells will be discussed for a more affordable as compared with the current MSC therapy towards PF. We anticipate that exploring these approaches may give a complete picture in understanding stem cells in order to make them effective and affordable for long-term therapeutic applications.

Keywords

Manufacturing cost · Refractory Crohn's disease · Stem cells genetic manipulation · Surgery

H.-N. Hon, P.-Y. Ho, J.-W. Lee, N. A. A. Mahmud,
H. B. Munir, V. Govindasamy (✉), and K.-Y. Then
Cryocord, 1, Bio X Centre, Persiaran Cyber Point Selatan,
Cyberjaya, Selangor, Malaysia

T. S. Ramasamy
Stem Cell Biology Laboratory, Department of Molecular
Medicine, Faculty of Medicine, Universiti Malaya,
Kuala Lumpur, Malaysia

A. K. Das
Maharaja Agrasen Hospital, Siliguri, West Bengal, India

S.-K. Cheong
Faculty of Medicine & Health Sciences, Universiti Tunku
Abdul Rahman (UTAR), Kajang, Selangor, Malaysia

Abbreviations

ADSCs	Adipose-derived stem cells
AGA	American Gastroenterological Association
bFGF	Basic fibroblast growth factor
BM-MSCs	Bone marrow-derived mesenchymal stem cells
CCL	C–C motif chemokine ligand
CD	Crohn's disease
CD163	Cluster of differentiation 163
CK	Casein kinase
COX2	Cyclooxygenase 2
CPF	Complex perianal fistula
CSF	Colony-stimulating factor
CXCL	Chemokine (C-X-C motif) ligand

CXCL12	The stromal cell-derived factor-1
CXCR	Chemokine (C-X-C motif) receptor
DC	Dendritic cell
DDL4	Delta-like 4
eASCs	Expanded allogeneic adipose-derived stem cells
EGF	Epidermal growth factor
eIF2	Eukaryotic initiation factor 2
EMT	Epithelial-to-mesenchymal transition
EPG	Epidermal growth factor
EV	Extracellular vesicle
FAS	Fas cell surface death receptor
FDA	Food and Drug Administration
FGF	Fibroblast growth factor
FGFR	Fibroblast growth factor receptors
FoxP3+ Treg cells	Forkhead box P3+ regulatory T cells
G1 phase	Growth 1 phase
G5k3β	Glycogen synthase kinase 3 beta
GCN2	General control nonderepressible 2
GDF-15	Growth differentiation factor-15
GM	Granulocyte-macrophage
GM-CSF	Granulocyte-macrophage colony-stimulating factor
GMP	Good Practice Manufacturing
HCAM	Homing cell adhesion molecule
HGF	Hepatocyte growth factor
HLA-G	Human leukocyte antigen G
HUMSCs	Human umbilical cord-derived mesenchymal stem cells
HUVECs	Human umbilical cord vein endothelial cells
IBD	Intestinal bowel disease
ICAM-1	Intercellular adhesion molecule 1
IDO	Indoleamine 2,3-dioxygenase
IEC	Intestinal epithelial cells
IFN-y	Interferon gamma
Ig	Immunoglobulin
IGF-1	Insulin-like growth factor 1
IGFBP	Insulin-like growth factor-binding protein
IL	Interleukin
ILT	Immunoglobulin-like transcript

IRF-1	Interferon regulatory factor-1
ITT	Intention to treat
JAK-STAT 1	Janus kinase and signal transducer and activator of transcription 1
K	Keratin
LIFT	Ligation of the intersphincteric fistula tract
M	Matrix protein
MCP-3	Monocyte chemotactic protein 3
miRNAs	MicroRNAs
MMPs	Matrix metalloproteinases
MSCs	Mesenchymal stem cells
NF-κB	Nuclear factor kappa B
NK	Natural killer
Oct4	Octamer-binding transcription factor 4
PAMPs	Pathogen-associated molecular patterns
PDGF	Platelet-derived growth factor
PF	Perianal fistula
PGE2	Prostaglandin E_2
PI3K	Phosphoinositide 3-kinase
PKA	Protein kinase A
S phase	Synthesis phase
SDF-1α	Stromal cell-derived factor-1
SLUG	Snail family transcriptional repressor 2
SNAIL1	Snail family zinc finger 1
Sox2	SRY-box transcription factor 2
TC	Transitional cells
TGF-β	Transforming growth factor beta
Th	T helper
TLR4	Toll-like receptor 4
TNF-α	Tumour necrosis factor alpha
TRAEs	Treatment-related adverse events
Treg cells	Regulatory T cells
TSG6	Tumour necrosis factor-stimulated gene 6
UB-MSCs	Umbilical blood-derived mesenchymal stem cells
UC-MSCs	Umbilical cord-derived mesenchymal stem cells
VCAM-1	Vascular cell adhesion molecule 1
VEGF	Vascular endothelial growth factor

VLA-4	Very late antigen 4
VWBR	Vertical-Wheel™ Bioreactors
WJ	Wharton's jelly

1 Introduction

Perianal fistula (PF) is a probable consequence of Crohn's disease (CD) since as many as 26% of patients with CD eventually develop PF within 20 years after the diagnosis. This suggests that the CD cases are perhaps a good tracking parameter for PF incidences (Dudukgian and Abcarian 2011; Schwartz et al. 2019). Traditionally, the occurrence and incidence rate of PF is more common and higher in the Western world such as North America, Europe and Scandinavia as compared to the rest of the world (Ng 2014). However, there have been noticeable changes since the last decade wherein several studies on the epidemiology of CD in the Asia Pacific and developing region revealed an increasing trend, while the rate was stable or rather regressive in the Western countries (Ahuja and Tandon 2010). This shift is likely due to changes in diet, stressful lifestyles and industrialisation in developing countries (Ng 2014).

Current treatment options for PF closely follow its anatomical features. The common one is surgical intervention, namely, fistulotomy, advancement flap procedure and ligation of the intersphincteric fistula tract (LIFT) which aims for a complete healing of fistula although they are less effective in complex cases like transsphincteric fistula (Ji et al. 2021; Limura and Giordana 2015). Despite this, surgeries are complicated with a prolonged recovery period that delays patients from returning to normal life (Sanad et al. 2019). The recovery rate of surgery is less impressive, for example, the success rate for LIFT is only 65% (Lehmann and Graf 2013). Furthermore, it is not uncommon for patients to experience fistula recurrences. Emile et al. (2017) reported that 10.3% of patients relapsed and subjugated themselves to second or third surgeries. Besides healing, there is also a need to address patient satisfaction as well. Side effects like bowel incontinence affect 7% of patients

after undergoing surgeries, deteriorating their quality of life by adapting to the adversaries (Dudukgian and Abcarian 2011; Panés and Rimola 2017).

Among the recent revolutionary therapeutic procedures, stem cell therapy showed significant improvement in fistula treatment. Table 1 summarises ten complete clinical trials that were being reported as of 2021, with bone marrow and adipose tissue used as the distributions for the source of mesenchymal stem cells (MSCs). MSCs injected into the tissue surrounding the fistula can restore the damaged tissues primarily through their immunomodulatory effect, stimulating a cascade of immune reactions to foster natural healing (Carvello et al. 2019; Prockop and Oh 2012). The treatment procedure is simpler and takes a shorter time of procedure and hospital stay (Park et al. 2021). Statistically, it boasts a higher success rate, lower recurrences and complete healing with higher patient satisfaction (Ciccocioppo et al. 2019; Herreros et al. 2019).

Despite the superior efficacy of stem cell therapy, its application for PF treatment is underwhelming (Gallo et al. 2020). It suffers from slow industrial growth, yet to be fully realised and made available at a wider scale. High cost of the therapy shrinks its market size, confining stem cell therapy as a last resort for very complex cases only when all other methods fail, thus making it a very niche treatment (Choi et al. 2019). Alofisel, a currently existing medicine specifically for PF based on MSCs derived from allogeneic adipose tissue, costs around $67,000 per dose which some patients need multiple doses for complete fistula closure (Scott 2018). Consequently, the unaffordable cost of treatment is deterring patients from pursuing it, relying on cheaper alternatives. The total cost of treatment is largely dependent on its production cost and numbers of in-process and final release quality assays, as well as on storage (Scott 2018).

A profitable product model is interlinked to the source, isolation and expansion techniques of MSCs. New opportunities like discovering new MSC sources can provide cheaper extraction methods. New advancements in bioreactors previously recruited for bacteria and viruses

Table 1 Clinical studies with stem cell therapy for PF from 2016 to 2021 and their current status

Author (year)/clinical trial number (if applicable)	Study/clinical trial title	Clinical trial phases	Patient enrolment and injection dosage	Type of cell and its source	Status	Results (if applicable)
Park et al. (2016)	Allogeneic adipose-derived stem cells for the treatment of perianal fistula in Crohn's disease: a pilot clinical trial	Phase 1	Group 1, n = 3 (10×10^6 cells/mL); group 2, n = 3 (30×10^6 cells/mL)	Allogeneic ADSCs	Completed	At month 8: Complete closure was observed in group 1, 2/3 (67%); group 2, 1/3 (33%)
Panés et al. (2016)/NCT01541579	Expanded allogeneic adipose-derived mesenchymal stem cell (Cx601) for complex perianal fistulas in Crohn's disease: A phase 3 randomized, double-blind controlled trial	Phase 3	Cx601, n = 107 (120×10^6 cells/mL); placebo, n = 105	Allogeneic ADSCs	Completed	At week 24: 57/107 (53%) Cx601 vs. 43/105 (41%) placebo achieved clinical remission
Dietz et al. (2017)/NCT01915927	Autologous mesenchymal stem cells, applied in a bioabsorbable matrix, for treatment of perianal fistula in patients with Crohn's disease	Phase 1	12 (20×10^6 cells/mL)	Autologous ADSCs	Completed	At 6 months: 10/12 (83%) achieved complete clinical healing
Panés et al. (2018a, b)/NCT01541579	Long-term efficacy and safety of stem cell therapy (Cx601) for complex perianal fistulas in patients with Crohn's disease	Phase 3	Cx601, n = 107 (120×10^6 cells/mL); placebo, n = 105	Allogeneic ADSCs	Completed	At week 52: 61/103 (59%) Cx601 vs. 42/101 (42%) placebo achieved clinical remission in modified ITT group
Wainstein et al. (2018)	Stem cell therapy in refractory perianal Crohn's disease: Long-term follow-up	Observational pilot study	9 (100×10^6–120×10^6 cells/mL)	Autologous ADSCs	Completed	At median follow-up of 31 months: 8/9 (89%) patients achieved complete healing
Serrero et al. (2017)/NCT02520843	Long-term safety and efficacy of local microinjection combining autologous microfat and adipose-derived stromal vascular fraction for the treatment of refractory perianal fistula in Crohn's disease	Phase 1	10 (10.9×10^6–47.8×10^6 cells/mL)	Autologous ADSCs	Completed	At week 48: 80% of patients had clinical responses; 60% of patients had combined remission

Laureti et al. (2020)/ NCT03555773	Refractory complex Crohn's perianal fistulas: A role for autologous microfragmented adipose tissue injection	Prospective pilot study	15 (20 mL microfragmented ADSCs)	Autologous ADSCs	Completed	At 24 weeks: 10/15 (67%) patients achieved combined remission that includes clinical and radiographic assessment; 4/15 (27%) patients had improved condition
Dige et al. (2019)/ NCT03803917	Efficacy of injection of freshly collected autologous adipose tissue into perianal fistula in patients with Crohn's disease	Phase 1	21 (18 mL–104 mL)	Autologous ADSCs	Completed	At 6 months: 12/21 (57%) patients had complete fistula healing; 3/21 (14%) patients had ceased fistula secretion; 1/21 (5%) patients had reduced fistula secretion
Zhou et al. (2020)/ ChiCTR1800014599	Autologous adipose-derived stem cells for the treatment of Crohn's fistula-in-ano: An open label, controlled trial	Phase 2	ADSCs, n = 11 (5 × 10^6 cells/ mL); placebo, n = 11	Autologous ADSCs	Completed	At 12 months: Healing rates were observed in 7/11 (64%) patients in the ADSCs group vs. 6/11 (55%) patients in the placebo group
Barnhoorn et al. (2020)/ NCT01144962	Long-term evaluation of allogeneic bone marrow-derived mesenchymal stromal cell therapy for Crohn's disease perianal fistulas	Phase 1	Cohort 1, n = 5 (10 × 10^6 cells/ mL); cohort 2, n = 5 (30 × 10^6 cells/mL); cohort 3, n = 5 (90 × 10^6 cells/mL); placebo, n = 6	Allogeneic BM-MSCs	Completed	At 4 years: Cohort 1, 3/4 (75%); cohort 2, 4/4 (100%); cohort 3, 1/5 (20%); achieved complete clinical fistula closure vs. placebo, 0/3 (0%)
2020/NCT04519671	A phase IB/IIA study of adult allogeneic bone marrow derived mesenchymal stem cells for the treatment of perianal Fistulizing Crohn's disease	Phases 1 and 2	40 (75 × 10^6 cells/mL)	Allogeneic BM-MSCs	Recruiting	–
2021/NCT04791878	A phase I study of adult allogeneic bone marrow derived mesenchymal stem cells for pediatric perianal Fistulizing Crohn's disease	Phase 1	10 (75 × 10^6 cells/mL)	Allogeneic BM-MSCs	Recruiting	–

(continued)

Table 1 (continued)

Author (year)/ clinical trial number (if applicable)	Study/clinical trial title	Clinical trial phases	Patient enrolment and injection dosage	Type of cell and its source	Status	Results (if applicable)
2021/NCT04939337	Study to assess the safety and efficacy of allogeneic umbilical cord-derived mesenchymal stem cells (TH-SC01), for treatment of complex perianal fistulas in perianal Crohn's disease	Phase 1	24 (120×10^6 cells/mL)	Allogeneic UC-MSCs	Enrolling by invitation	–
2021/NCT05039411	A phase I study of the safety of allogeneic human umbilical cord mesenchymal stem cells (UC-MSCs) for perianal fistulas in patients with Crohn's disease	Phase 1	7 ($125–150 \times 10^6$ cells/mL)	Allogeneic UC-MSCs	Not yet recruiting	–

Numbers of clinical trials have been conducted to identify the efficiency and the efficacy of MSCs towards PF based on the injection dosage and type of cell source. There are still several clinical trials up until now to prove the safety of MSCs against PF with Crohn's disease expansion can be modified to sustain a biological environment for the various techniques that exist for cell culture production, with limitations for MSCs (McKee and Chaudhry 2017; Damasceno et al. 2020). Hence, in this review, the market demand and the details of finding a practical and cost-effective approach to transfer MSCs from various sources for treating PF using shorter time are explored.

2 Pathophysiology of Perianal Fistula

A fistula represents a tunnel under the skin which connects two epithelial surfaces. The most prevalent among the fistulas is PF, which typically connects the rectum and drains out to the skin around the anus. Apart from CD, PF also occurs due to infection (Scharl et al. 2016). Multifactorial changes in physical behaviours and biological functions in the rectum area cause a surge in inflammation which induces an inflammatory response, such as activation of macrophages, monocytes and neutrophils (Chen et al. 2018). This leads to the secretion of pro-inflammatory cytokines, chemotactic and cell-activating peptides as well as tissue-degrading enzymes and reactive oxygen radicals, which induce local tissue injury (Scharl et al. 2016).

The release of the pro-inflammatory cytokine, tumour necrosis factor (TNF), stimulates the expression of transforming growth factor-beta (TGF-β) which leads to the production of β-integrin that act as a catalyser to the onset of epithelial-to-mesenchymal transition (EMT) (Scharl and Rogler 2014). EMT redifferentiates epithelial cells located on the inner lining of the rectum into fibroblastic-like cells with migratory capability and penetrates adjacent tissue (Panés and Rimola 2017; Scharl and Rogler 2014). TGF-β also triggers interleukin-13 (IL-13) and increases the secretion of matrix metalloproteinases (MMPs) which are associated with cell-invasive aid (Scharl and Rogler 2014). The overactivation of β-integrin and MMPs marks

the birth of the fistula. Once the wastes from the body pile up, during defecation, the intraluminal pressure drives the wastes into subcutaneous tissues (Bataille et al. 2004). The accumulation of wastes leads to the formation of abscesses which are potentially the source of bacterial infection. The luminal pressure makes the tunnel become longer until an external opening is formed (de Zoeten et al. 2013). As a result of this mechanism, the deep penetrating tract develops into PF. Based on the pathophysiology and the severity of fistulas, precise classification of PF in patients has been highlighted in order to come out with the best clinical strategy (Marzo et al. 2015). There are three classifications of PF, which are Parks classification, St James University Hospital classification and American Gastroenterological Association (AGA) classification shown in Fig. 1 (Panés and Rimola 2017).

3 Mechanism of Action of MSCs

The main mechanism of action of MSC entirely is attributed to its paracrine factors such as cytokines, chemokines and growth factors (Park et al. 2018). MSCs release secretomes that help to suppress inflammation, increase cell proliferation and repair damaged tissue (Park et al. 2018). Here we briefly explain the mechanism that is likely to happen upon the injection of MSCs to the side of the fistula. The mechanism of action is also shown in Fig. 2.

3.1 Homing Ability

After injection into the fistula tissue, MSCs enter the vascular system and migrate into the injured site through a homing mechanism (Li et al. 2019; Ullah et al. 2019). Several cell signalling molecules play a crucial role for efficient homing (Spees et al. 2016). For example, the presence of pro-inflammatory cytokines like TNF-α activates endothelial cells in blood vessels to induce intercellular adhesion molecule 1 (ICAM-1), vascular

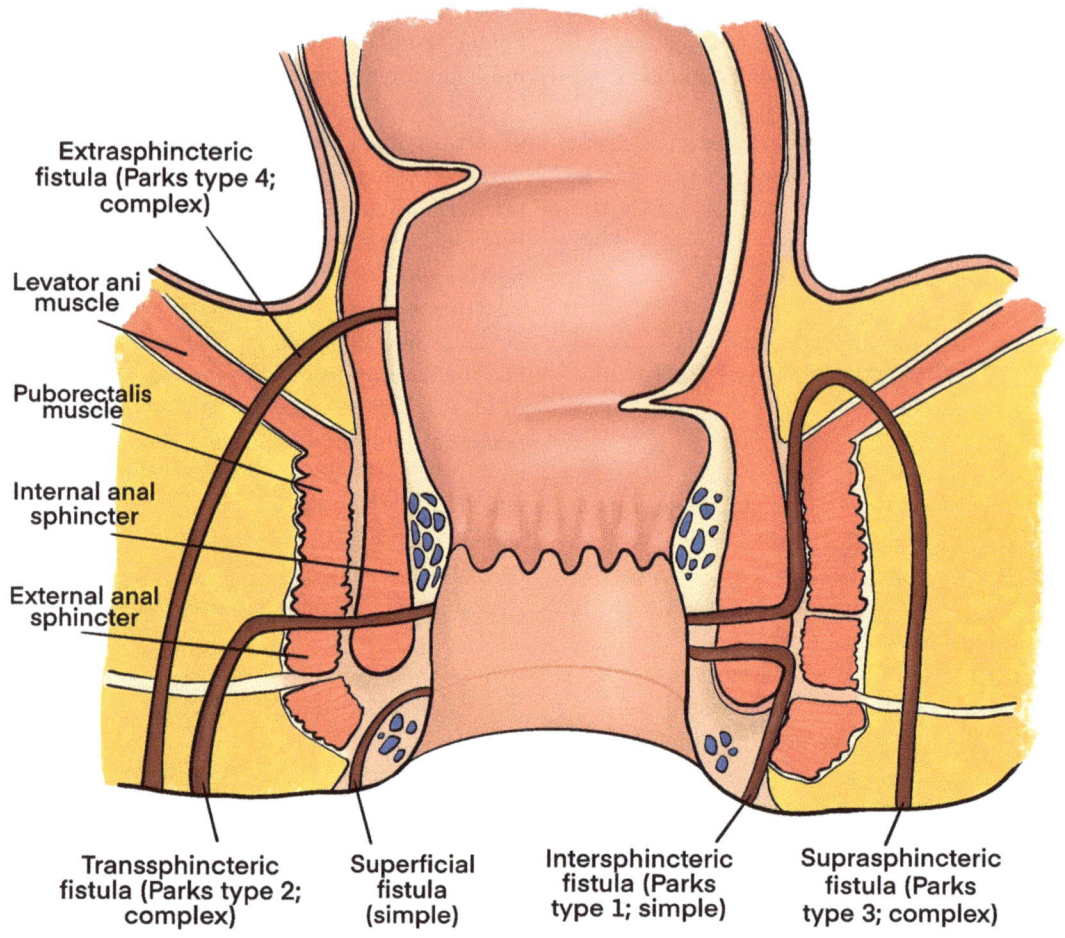

Fig. 1 Classification of PF. Superficial fistula: A fistula without interception with sphincter or muscular structure (Agha et al. 2013). Intersphincteric fistula: Located in intersphincteric plane. Fistula located in between internal anal sphincter until the external anal sphincter (Agha et al. 2013). Transsphincteric fistula: A fistula that will travel through the external sphincter (high or low) to pass some distance distally in the intersphincteric plane before going through the external sphincter (Agha et al. 2013). Suprasphincteric fistula: Fistula that perforates through the intersphincteric plane before piercing the levator ani and descending through the ischioanal fossa (Agha et al. 2013). Extrasphincteric fistula: Fistula starts from the external anal sphincter that tracks across the levator ani to the perineum (Agha et al. 2013)

cell adhesion molecule 1 (VCAM-1) and P-selectin activation for increasing adhesion of MSCs to the endothelial cells (Teo et al. 2012). The movement of MSCs on vascular cell surfaces prompts the upregulation of ligands such as cluster of differentiation 44 (CD44), homing cell adhesion molecule (HCAM) and CD49d (Andreas et al. 2014). Integrins like very late antigen 4 (VLA-4) are formed by galectin-1 to regulate the adhesion of MSCs. Platelet- and neutrophil-derived growth factors such as fibroblast growth factor (FGF) as listed in Table 2 are released by MSCs to interact with basic fibroblast growth factor (bFGF) in endothelial cells to regulate the adhesiveness of galectin-1 to P-selectin (Langer et al. 2009).

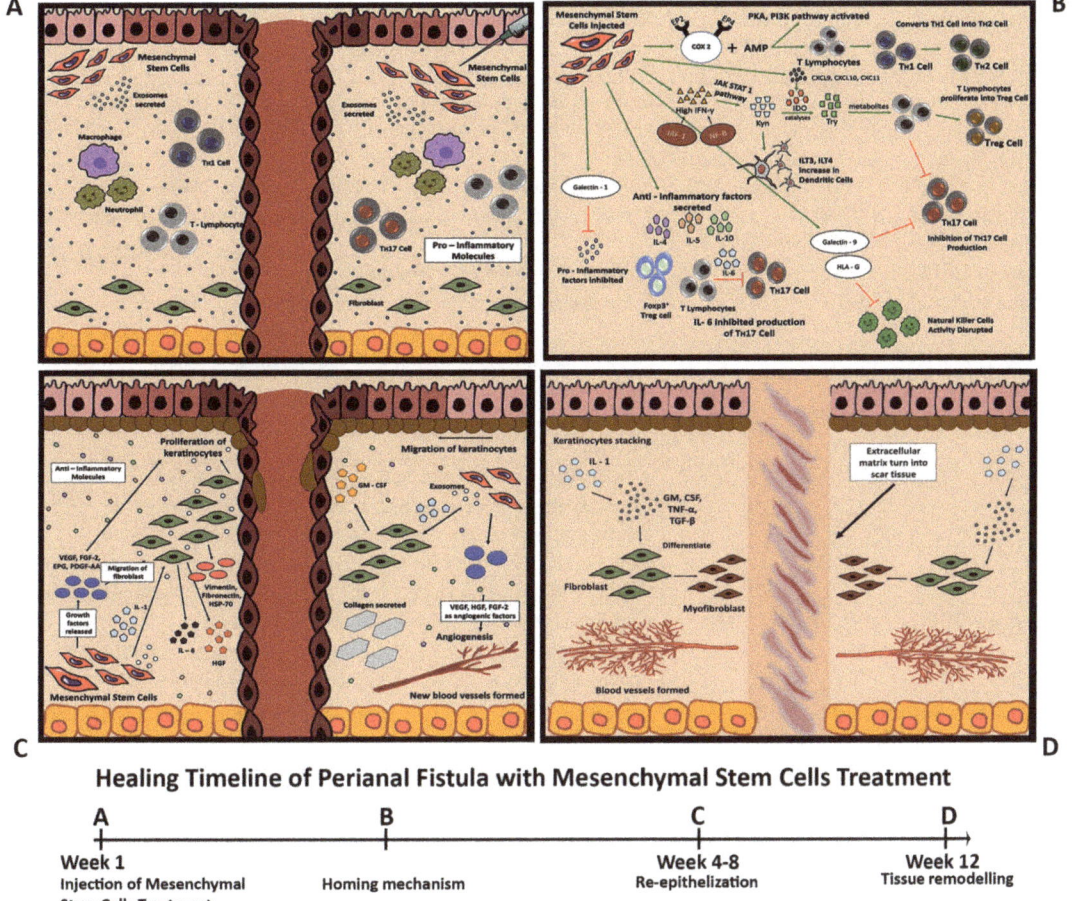

Fig. 2 Healing time of PF with MSCs treatment. (**a**) MSCs were injected into the active inflammation site. The active inflammation site contains an abnormal amount of pro-inflammatory molecules such as TNF-α, IL-1β, TGF-β and others including TH1 and TH17. (**b**) Mechanism of action of MSCs to the active inflammation site including suppressing the production of pro-inflammatory factors and anti-inflammatory factors which are secreted. The production of TH17 is inhibited by the anti-inflammatory factors and Th1 is converted into Th2. Pathway protein kinase (PKA), phosphoinositide 3-kinase (PI3K) and JAK STAT1 activated. (**c**) Re-epithelisation of tissue around the fistula. The abundance of anti-inflammatory molecules balances out the pro-inflammatory molecules. Growth factors such as VEGF, FGF-2, EPG and PDGF-AA are released to assist the migration of fibroblast and the proliferation of keratinocytes to the damaged tissue. The growth factors also contributed to the angiogenesis process, allowing the secretion of collagen. (**d**) Tissue remodelling leaving the extracellular matrix turns into scar tissue and the fistula is healed. New blood vessels completely formed

3.2 Anti-Inflammatory Mechanism

Once it has migrated to the injured site, microenvironment of the injured tissues which consist of pro-inflammatory cytokines secreted during inflammation such as interferon gamma (IFN-y), TNF, IL-1 and IL-17 activates the immunomodulation role of MSCs (Waterman et al. 2010;

Sangiorgi and Panepucci 2016). Several immunoregulatory factors like IL-10, prostaglandin E_2 (PGE2), human leukocyte antigen G (HLA-G), indoleamine 2,3-deoxygenate (IDO) 1 and IDO2 and chemokines like chemokine C-X-C motif receptor (CXCL) 9, CXCL10 and CXCL11 will be secreted by MSCs as listed in Table 3 (Li et al. 2018). Once the MSC receptors bind to the

Table 2 Growth factors secreted by MSCs

Author (year)	Growth factors secreted by MSCs	Function
Joel et al. (2019)	HGF	Exert anti-inflammatory signals by causing MSCs to inhibit the proliferation and/or activities of CD4$^+$ Th1, Th17, CD8$^+$ T cell and NK cells
Wang et al. (2014)	bFGF	Proliferates and promotes differentiation of fibroblasts
Delafontaine et al. (2004)	IGF-1	Tissue growth factor with effects to influence blood glucose level
Zhao et al. (2013)	PDGF	Promotes migration of fibroblasts
Tamama et al. (2010)	EGF	Promotes cell regeneration by stimulating cell proliferation
Panek-Jeziorna and Mulak (2020)	FGF-19	Growth factor with potential anti-inflammatory properties
Langer et al. (2009)	FGFR	Interact with bFGF in endothelial cells to arbitrate the adhesiveness of galectin-1 to P-selectin
Zhao et al. (2013)	FGF-2	Attracts leukocyte recruitment to the inflammation site
		Increases the migration of fibroblasts and the functional roles of keratinocytes. Initiates vascularization in the damaged tissue
Ho et al. (2012)	GDF-15	Antiapoptotic, antihypertrophic and anti-inflammatory properties in response to oxidative stress or pro-inflammatory signalling molecules
Zhao et al. (2013)	VEGF	Angiogenesis factor which involves immunology
		Increases the migration of fibroblasts and the functional roles of keratinocytes

The growth factors secreted by MSCs and its function are described. Most of the growth factors show immunomodulatory and anti-inflammatory properties which may benefit in PF treatment

cytokines, chemokines and growth factors in the microenvironment, production of the anti-inflammatory paracrine factors is initiated (Waterman et al. 2010). The paracrine factors will tune the function of the T lymphocytes, macrophages, neutrophils, natural killer (NK) cells, dendritic cells (DC) and B lymphocytes for immunosuppression activities (Wang et al. 2014).

IDO production is activated by the Janus kinase and signal transducer and activator of transcription 1 (JAK-STAT 1) signalling pathway (Ji et al. 2017). In this pathway, the nuclear factor-beta (NF-β) and interferon regulatory factor-1 (IRF-1) bind to upstream IFN-γ-responsive elements of the IDO gene, thus promoting gene expression (Sohni and Verfaillie 2013). IDO behaves as a switch that initiates a cascade of reactions to promote immunosuppression (Luz-Crawford et al. 2013). It directs the monocyte to differentiate into anti-inflammatory and immunosuppressive type

2 macrophage (Francois et al. 2010). Moreover, CXCL12 and CXCR4 secreted by MSCs draw near the T cells to enable IDO to catabolise the tryptophan in T cells (Sohni and Verfaillie 2013). Tryptophan, which is a necessary molecule for T-cell survival, is then broken down into metabolites like kynurenine, quinolinic acid and picolinic acids (Weber et al. 2006). The deficit in tryptophan numbers retards the T-cell multiplication and growth, thus stunting their numbers (Moffett and Namboodiri 2003). This forces a change in the metabolic pathway of ATP production from glycolysis to oxidative phosphorylation and activates a stress response in the immune cells eukaryotic initiation factor 2 (eIF2) and general control nonderepressible 2 (GCN2) through the accumulation of uncharged tRNA (Zhu et al. 2011). Consequently, the arrested cell growth declines their physiological roles, mediating fas cell surface death receptor (FAS)-regulated lymphocyte apoptosis. On the contrary,

Table 3 Paracrine factors secreted by MSCs in PF treatment

Author (year)	Anti-inflammatory factors secreted by MSCs	Function
Schraufstatter et al. (2015)	Complement component C5	Anti-inflammatory cytokine, increases survival of MSCs under oxidative stress
Weiss and Dahlke (2019)	COX2	Anti-inflammatory cytokine, involves in reducing pain
Solodeev et al. (2018)	Fas ligand	Improves cell survival during inflammation tissue damage
Hartung (1998)	G-CSF	Reduces inflammatory activity by inhibiting the secretion or function of the main inflammatory mediators
Luz-Crawford et al. (2013)	IDO 1 & 2	Catabolic enzyme with immunosuppressive properties through kynurenine pathway
Miguel-Hidalgo et al. (2007)	ICAM-1/CD54	Modifies the immunosuppressive functions of MSCs by facilitating the interaction between pro-inflammatory macrophages and MSCs
Lee et al. (2018)	IL-1	Anti-inflammatory cytokine that can inhibit Th17 polarisation
Amorin et al. (2014)	IL-4	Treg cell differentiation and Th2 cells
Pripp and Stanišić (2014)	IL-5	Anti-inflammatory cytokine, increases the development of new nerve cells in the hippocampus and lowered the quantity of potentially damaging inflammation in the brain
Amorin et al. (2014)	IL-10	The most potent anti-inflammatory cytokine, high levels are predicted to involve to the ageing secretome Treg cell differentiation and Th2 cells
Nilsson et al. (2019)	IL-13	Anti-inflammatory cytokine, involves in reversing ageing
Liu et al. (2018)	MCP-3 (CCL7)	Participates in anti-inflammatory responses through binding to its receptors to facilitate the recruitment of immune cells
Park et al. (2006)	PGE2	Vasodilator, reduces inflammation
Luz-Crawford et al. (2013); Mallis et al. (2018)	HLA-G	Inhibits T-cell differentiation into Th1 and Th17. Allows apoptosis of T and B cells
		Stops cytolysis of $CD8^+$ cells and induces development of $CD4^+$, $CD25^+$, $FoxP3^+$ Treg cells
		Disrupts activities of NK cells
Sohni and Verfaillie (2013)	CXCL9, CXCL10, CXCL11, CXCL12	Immunomodulatory roles in MSCs wound healing
Sohni and Verfaillie (2013)	CXCL12, CXCR4	Draws near T cell towards MSCs
Akiyama et al. (2012)	TGF-β	Development and maturation of Treg cell differentiation
Wang et al. (2014)	TSG6	Anti-inflammatory properties
Gieseke et al. (2010)	Galectin-1	Downregulates the pro-inflammatory cytokines TNF-α, IFN-y, IL-2, IL-10
Pripp and Stanišić (2014)	Galectin-9	Anti-inflammatory cytokine, increases the development of new nerve cells in the hippocampus and lowered the quantity of potentially damaging inflammation in the brain
Whelan et al. (2020)	CCL2	Regulates macrophage recruitment, accelerates wound healing

The paracrine factors secreted by MSCs that may involve in PF treatment are stated and its role in PF treatment is also described based on several studies

kynurenine upregulates inhibitory receptors like immunoglobulin-like transcript (ILT) 3 and ILT4 in the DC, while co-stimulating cytokines are suppressed in parallel. HLA-G expressed by MSCs also induce T and B immune cells to undergo apoptosis (Mallis et al. 2018). They further end the cytolysis of antigen-activated $CD8^+$ cells and encourage the development of $CD4^+$, $CD25^+$ and forkhead box $P3^+$ regulatory T cells ($FoxP3^+$ Treg cells).

Galectin-1 secretion by MSCs downregulates the pro-inflammatory cytokines such as IFN-y, IL-2, TNF-α, IL-10 and so on (Gieseke et al. 2010). Together with galectin-9, biological activities of T cells are controlled to reduce T and B lymphocyte numbers (Mallis et al. 2018). Activated MSCs also regulate immunoglobulin (lg) E and IgG concentration by retarding the growth of B cells (Wang et al. 2014). MSCs downregulate IL-2 and IL-15 to keep the natural killer cells dormant, while CD14$^+$ is inhibited by MSCs to stop DC differentiation (Spaggiari and Moretta 2013).

3.3 Cell Proliferation Stage

Activation of GCN2 plays a role in promoting differentiation of Treg cells and downregulates IL-6. Without adequate IL-6, T helper (Th) 17 cells' activities are suppressed (Liu et al. 2020). The tryptophan metabolites, kynurenine, induce the production of tolerogenic DC (Regmi et al. 2019). All the metabolites overall exhibit toxic effects towards CD4$^+$ Th1 and CD8$^+$ T cells but are substantially harmless towards immunosuppressing cells like Th2. This impact prompts T helper cells to start developing into Th2 cells and decrease the development into Th1 cells (Weiss and Dahlke 2019). On the other hand, MSCs block TNF-α secretion by promoting IL-10, and IL-4 secretion also helps in increasing the Treg cell differentiation and Th2 cells (Amorin et al. 2014).

IDO, HLA-G, galectins and other secretomes carry out an extra role of inhibiting the differentiation and development of T cells into Th1 and Th17 (Luz-Crawford et al. 2013). This inhibition forces the macrophages to express growth factors, TGF-β which plays a vital part in the development and maturation of Treg cell differentiation (Akiyama et al. 2012). At the same time, HLA-G also disrupts the protoplasmic activities of NK cells (Mallis et al. 2018).

Furthermore, MSC exosomes can polarise macrophages from pro-inflammatory matrix protein 1 (M1) into anti-inflammatory M2 phenotypes once triggered by the availability of pro-inflammatory cytokines such as chemokines, IL-1β, IL-12 and TNF-α (Murray 2017). It is found that miR-223 in exosomes abates inflammation and helps to speed up healing through macrophage M2 polarisation, while M2 macrophage is brought about by Th2 cytokines and chemokines like TGF-β, IL-10 and M2 markers such as IL-1ra, CD163 and C-C motif chemokine 22 (Murray 2017; Zhuang et al. 2012).

3.4 Re-epithelisation and Tissue Remodelling

MSCs help to remodel the damaged tissues by accelerating wound healing (Nie et al. 2011; Whelan et al. 2020). MSCs will differentiate to fibroblasts and express vimentin, fibronectin and heat shock protein 47 once embedded into the injured site IL-1 secreted by MSCs influences gene expression for other chemokines involved in the metabolic cascade chain like granulocyte-macrophage (GM), colony-stimulating factor (CSF) and TNF-α (Hamilton 2008; Shingyochi et al. 2015). MSCs also differentiate into keratinocytes, producing keratin (K) 5 and K14, integrins, cytokeratin 5, cytokeratin 14, cytokeratin 19, desmoglein 3 and cytokeratin 6α proteins for keratinocyte assembly (Shingyochi et al. 2015). Other than that, they can also differentiate endothelial cells into blood vessel walls (Ebrahimian et al. 2009). Growth factors such as vascular endothelial growth factor (VEGF), epidermal growth factor (EPG), fibroblast growth factor 2 (FGF-2) and PDGF-AA that are secreted by MSCs increase the migration of fibroblasts and the functional roles of keratinocytes (Zhao et al. 2013). VEGF, HGF and FGF-2 also act as angiogenic factors by initiating vascularisation in the damaged tissue.

New blood vessels help in transporting nutrients and oxygen for the cell proliferation of fibroblasts and keratinocytes. At the same time, MSCs secrete IL-1 which directs the migration of fibroblast to the wound, excreting IL-6, HGF and granulocyte-macrophage colony-stimulating factor (GM-CSF) (Zhao et al. 2013).

Lastly, during remodelling, keratinocytes stack up in alignment through epidermal stratification (Santoro and Gaudino 2005). The mechanical tension from the tissue activates TGF-β, and splice variant fibronectin triggers proto-fibroblast differentiation into smoother myofibroblasts, increasing proliferation. When the wound gap closes, the excess capillaries slowly disappear, leaving behind a completely remodelled tissue (Sorg et al. 2017; Hinz et al. 2001).

4   A more Affordable MSC Therapy in Treating PF

Stem cell therapy was widely investigated in various kinds of diseases, including PF. As compared to the other treatments, stem cell therapy has a relatively high healing rate, which is up to 59% with a low complication rate, 13%–18.5% (Fig. 3). Hence, stem cell therapy is a good alternative for treatment that gives a long-term outcome with fewer relapses (Georgiev-Hristov et al. 2018). Table 4 shows the current conventional treatments and its respective cost. Unfortunately, the high cost appears to be a major bottleneck, causing the limited industrial growth towards a wider market base, especially in the low-income countries. The entire process from the manufacturing process until the delivery of the services requires a high amount of cost and resources, thus causing MSCs as a premium therapy that only PF patients with a higher tier of income could be able to afford. Therefore, several techniques and alternative approaches are being suggested in the following section, making MSCs a more affordable therapy.

4.1   Cell Source

MSCs can be derived from a variety of sources such as the adipose tissue, bone marrow and umbilical cord. Among these various sources, BM-MSCs and adipose tissue-derived MSCs (ADSCs) are commonly utilised in PF treatment due to their easy accessibility (Musiał-Wysocka et al. 2019). Two approaches are involved in

obtaining adult MSCs, which are either autologous or allogeneic. Autologous MSCs have the advantage in preventing transplant rejection and performing better cell tolerability; however it is not readily available where it needs to be further isolated and expanded, which may not be suitable in emergency treatment (Molendijk et al. 2015). This leads to allogeneic MSCs which could be pre-cultured in advance and are ready to use anytime. Nonetheless, a healthy donor can be selected to obtain functional and normal MSCs, and these cells may be cryopreserved and readily available for future use (Molendijk et al. 2015).

Cheng et al. (2020) reported that a majority of the reported clinical trials used BM-MSCs and ADSCs in treating PF, with an authorised product available using the latter source. Although these well-studied findings reported a visible healing rate towards PF, several challenges remain unsolved. Firstly, both BM-MSCs and ADSCs recorded low proliferation rates, which were proven by Amable et al. (2014). Besides, age becomes a limiting factor in obtaining high-quality adult tissue-derived MSCs. According to Bustos et al. (2014), using aged BM-MSCs reported a decrease in immunomodulatory activity which was caused by the lowered expression level of chemokine and cytokine receptors that are involved in cell migration. Further, additional steps are needed to extract MSCs from adipose tissue and bone marrow, for example, by bone aspiration and liposuction, and these may increase the risk of contamination if mishandling of samples occurs throughout the workflow (Mazini et al. 2021). With the limitation of adult tissue-derived MSCs, the umbilical cord seems to be an ideal cell source selection in PF treatment. UC-MSCs proved to have a better proliferation rate and cellular migration ability while possessing similar immunomodulatory characteristics with adult tissue-derived MSCs (Omar et al. 2014). For example, WJ-MSCs are reported to have higher expression levels of IL-10, TGF-β and VEGF compared with adult tissue MSCs, which perform better immunosuppressive ability against diseases, and it has been proven since the last decade (Weiss et al. 2008).

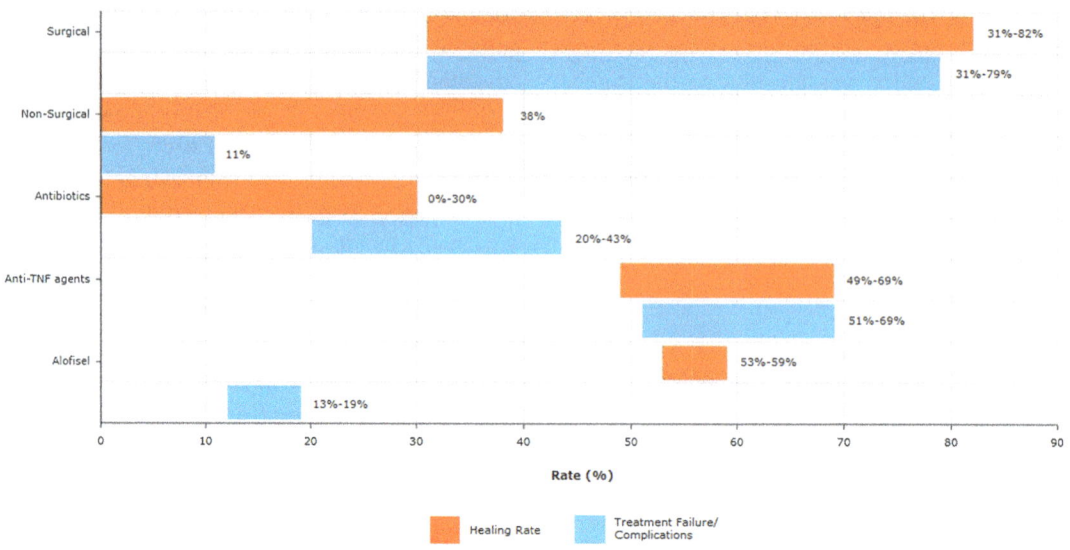

Fig. 3 Graph of treatment success and complication rate of current treatment of PF. Stem cell therapy (Alofisel) has a relatively high healing rate, which is up to 59% with a low complication rate, 13%–19%. Treatment using anti-TNF agents has both high healing and complication rate, which are up to 69%. The healing rate of treatment using antibiotics is low, 0%–30% with high complication rate, which is up to 43%. The healing rate of non-surgical treatment is up to 38% with low complication rate, 0%–11%, while surgical treatment gives highest healing rate, 31%–82% with the highest complication rate, which is up to 79%

4.2 Manufacturing Scales

To produce MSCs in a large quantity as a commercial product for treating PF, a Good Manufacturing Practice (GMP)-compliant production process using a bioreactor is crucial. Traditionally, MSCs are cultured in monolayer flasks, have a simple handling process and are generally low in cost (Rodrigues et al. 2011). However, monolayer culture technology has a higher risk of contamination due to the open system coupled with low cell yield (Mizukami and Swiech 2018). This has prompted for the use of bioreactor for cell expansion to upsize the scale of cell manufacture. Since MSCs are anchorage-dependent, the microcarriers that are present in the bioreactor are used for cell attachment and cell growth by providing higher surface area to volume, allowing a higher yield of MSCs in a shorter time, thus reducing the cost of production (Panchalingam et al. 2015). Moreover, stirred tank bioreactors with impellers achieved a homogenous culture system as it mixes the culture more regularly. As compared with traditional culture using flasks, stirred-tank bioreactors are closed systems that reduce the risk of contamination and with parameter control and monitoring systems which are able to monitor the cultured cells easier (Nienow et al. 2014). According to Mizukami et al. (2017), hollow fibre bioreactor was used in clinical trials testing for intestinal bowel disease (IBD), and successful expansion of up to 11-fold of MSCs in 5 days was reported.

4.3 Preconditioning of MSCs

Preconditioned MSCs under modified culture conditions would help in preserving their therapeutic effects, and this approach has been well studied in stem cell therapy in recent years (Ocansey et al. 2020). For instance, preconditioning of MSCs using IFN-γ seems to be a promising approach in conferring anti-inflammatory effect which was proven by Noone et al. (2013). IFN-γ will specifically target T cells or NK cells and suppress its activity by the release of prostaglandin E2. The suppression of

Table 4 Current conventional treatments (surgical, non-surgical, medical, stem cells) and their cost

Author (year)	Category		Estimated average cost (USD)	No. of patients	Healing rate (%)	Follow-up (weeks)	Complication
van Koperen et al. (2009)	Surgical	Fistulotomy	$2,855	28	82	343	Fistula recurrence (18%)
							Faecal incontinence (61%)
							Total: 79%
Galis-Rozen et al. (2010)		Seton drainage	$2,688	17	59	104	Fistula recurrence (40%)
							Faecal incontinence (6%)
							Total: 46%
Bessi et al. (2019)		Advancement flap	$3,015	34	68	58	Treatment failure (33%)
Gingold et al. (2014)		LIFT	$2,876	15	60	8	Treatment failure (40%)
Senéjoux et al. (2016)		Fistula plug	$2,965	54	31	12	Abscesses formation (7%)
							Plug avulsions (9%)
							CD flare (2%)
							Abdominal pain (2%)
							Miscellaneous (11%)
							Total: 31%
Grimaud et al. (2010)	Non-surgical	Fibrin glue	$2,627	36	38%	8	Abscess formation (11%)
Medical treatment							
Thia et al. (2009)	Antibiotic	Metronidazole	$9.8 for ten tablets (500 mg) (8)	7	0	10	Abscess (42.9%)
Thia et al. (2009)		Ciprofloxacin	$19 for ten tablets (500 mg) (9)	10	30	10	Abscess (20%)
Bouguen et al. (2013)	Anti-TNF agents	Infliximab	$1,229 (100 mg) (11)	156	69	250	Fistula recurrence (40%)
							Abscess formation (29%)
							Total: 69%
Castaño-Milla et al. (2015)		Adalimumab	$3,120 (40 mg) (13)	46	49	52	Treatment failure (51%)
Medical treatment							
Panés et al. (2016)	Stem cells	Alofisel (Cx601)	$67,000	107	53	24	TRAEs:
							Anal abscess (6%)
							Proctalgia (5%)
							Procedural pain (1%)
							Fistula discharge (1%)
							Total: 13%
Panés et al. (2017)				107	59	52	TRAEs:
							Anal abscess (13%)
							Proctalgia (5%)
							Procedural pain (1%)
							Total: 19%
Ramezankhani et al. (2020)		Cupistem	$5,000	43	80.8	96	–

immune cells activity will then prevent it from presenting cytotoxicity characteristics (Noone et al. 2013). Also, according to a recent study by Yu et al. (2019), preconditioning of MSCs with IFN-γ and IL-1β upregulates the production of PGE2 and IDO, thus improving the immunomodulatory property of MSCs and increasing its efficacy towards treatment.

Nicotinamide (NAM) can also be used in cell culture to act as a cell supplement. NAM belongs to the family under vitamin B3 and has been used widely for treatment of various diseases such as diabetes, Alzheimer's disease and cancer (Meng et al. 2018). However, studies have shown that using NAM for cell culturing could enhance cells' performance (Meng et al. 2018). This may be due to the role of NAM as a direct effector on ROCK signalling pathway inhibition, thus improving cell survival and inducing cell differentiation (Watanabe et al. 2007; Zhang et al. 2021). Not only that, studies have also proved the role of NAM to inhibit casein kinase 1 (CK1), which involves in CK1 signalling pathway to induce apoptosis (Janovská et al. 2020). Besides this, several studies have proposed that preconditioning of MSCs under hypoxic conditions presented an improvement in proliferation rate. For example, Haque et al. (2013) stated that the proliferation rate and the cell doubling time that cultured for BM-MSCs under hypoxic conditions reported a significant increase as compared to ambient oxygen concentration.

4.4 Genetically Modified MSCs

Genetically modified MSCs have been used widely in recent years in treating various diseases (Varkouhi et al. 2020). Although limited studies are available on using engineered MSCs against PF treatment, there are still several clinical trials reported showing better wound healing rates and improved immunomodulatory effects against certain diseases. These techniques would be useful as a reference to further examine the efficacy of PF treatment. Table 5 summarises the use of genetically modified methods and their effect on

immunoregulatory function. Here we have briefly explained the therapeutic benefits of genetically modified MSCs as well as the cost that needs to be taken into consideration before introducing this approach to the clinical.

4.4.1 Improving Migration

The cellular migratory ability is crucial in MSCs treatment where the transplanted or injected MSCs would need to migrate to the injured tissues. The improvement of cell migration can be enhanced by modifying miRNAs. According to the in vitro study in a rat model, overexpression of miR-9-5p would help in upregulating the β-catenin signalling pathway, which takes part in most of the cellular regulatory processes in MSCs, such as differentiation and migration (Li et al. 2017; Pai et al. 2017). Overexpressing miR-9-5p would help to suppress the activity of glycogen synthase kinase 3 beta (G5k3β) and casein kinase 1 alpha (CK1α), which act as inhibitors towards the production of β-catenin. The inactive inhibitors will thus help in preventing the degradation of β-catenin, promote the formation of β-catenin and enhance the signalling pathway at a higher level (Pai et al. 2017). The overall process will thus be promoting sufficient MSCs to migrate towards injured tissues. Modification of miR-9-5p of MSCs in PF treatment may provide similar results with the study mentioned, where further studies need to be conducted to test its efficacy of it.

Enhancement of the cellular migratory process could also be achieved by modifying CXC chemokine receptors, in specific CXCR4 and CXCR7. Based on a study by Devetzi et al. (2018), overexpressing both genes will induce the activity of stromal cell-derived factor-1 (SDF-1) which acts as key chemokines in mediating homing of transplanted MSCs to the injured tissues. This results in promoting paracrine signalling to the nearby cells, thus maximising the efficacy of the wound healing process (Devetzi et al. 2018). According to a study by Du et al. (2013), an animal model has been used to determine the effect of CXCR4 and CXCR7 overexpression in BM-MSCS towards early liver regeneration. The results showed

Table 5 Types of genetic modification process

Author (year)	Genetic modification	Source of MSCs	Experimental model	Transfection method	Transfection efficiency	Purpose	Target mechanism
Du et al. (2013)	Overexpression of CXCR4 and CXCR7	Rat BM-MSCs	Rat with liver injury	Viral (adeno-associated virus)	Approximately 98%	Improves cell migration	Increase level of SDF-1α, VEGF and HGF
Han et al. (2014)	Overexpression of Sox2 and Oct4 gene	Human ADSCs	Human ADSCs	Non-viral (liposomal)	Approximately 98%	Reduces premature senescence of MSCs	Extending S phase in cell cycle by increasing the production of cyclin D1
Li et al. (2017)	Overexpression of miR-9-5p	Rat BM-MSCs	Artificial made wound cells	Non-viral (liposomal)	High	Improves cell migration	Induce β-catenin signalling pathway by suppressing activity of CK1α and G5k3β
Fang et al. (2016)	Exosomal miR-21, miR-23a, miR-125b and miR-145	WJ-MSCs	Mouse model with skin injury	Non-viral (microinjection)	High	Improves wound healing process	Induce the translocation of β-catenin to the injured tissue
Li et al. (2016)	Exosomal miR-181c	UC-MSCs	Mouse model with burn injury	Non-viral (microinjection)	High	Anti-inflammatory	Suppresses macrophage activity to produce IL-6 / Downregulates the secretion of inflammatory cytokines IL-1β and TNF-α / Increase of IL-10 production / Inhibits TLR4 expression
Liang et al. (2016)	Overexpression of exosomal miR-125a	Human ADSCs	HUVECs	Non-viral (microinjection)	Approximately 90%	Promotes re-epithelisation	Downregulating the expression of DLL4 thus promotes the formation of epithelial tip cells

Different types of genetic modification approaches of MSCs are described together with its purpose and target mechanism. This may help as a guide to involve genetic modification techniques in PF treatment, where exosomal microRNAs are mostly used improving cellular migration into rat liver graft, promoting regeneration of hepatocytes in the rat's liver (Du et al. 2013). Although the study was only conducted in animal models, altering CXC chemokine receptors may be a promising technology in PF treatment and should be further examined.

4.4.2 Reduced Premature/Replicative Senescence

Modifying octamer-binding transcription factor 4 (Oct4) and SRY-Box transcription factor 2 (Sox2) genes would help in reducing premature senescence of MSCs. Based on a study by Han et al. (2014), the overexpression of these two genes was transfected in human ADSCs with the help of a PB-CA vector. The findings concluded that overexpressing of Sox2 and Oct4 will prolong the duration of the synthesis (S) phase in the cell cycle, where DNA replication occurs. It is important to monitor the processes in each phase of the cell cycle, as the overall events will eventually affect the cell proliferation performance (Resnitzky et al. 1994). High expression levels of Sox2 and Oct4 will trigger the production of cyclin D1 in high amount, thus accelerating the transition from non-proliferative growth 1 (G1) phase into proliferative S phase, prolonging the duration in DNA replication and resulting in extended MSCs growth and expansion (Han et al. 2014). This may be also included in PF treatment, where the yield of MSCs will be increased and may be suitable in a larger scale of production.

4.4.3 Cost

The cost of genetically modified MSCs is still unknown in disease treatment as it has only been used in smaller-scale experiments, consisting only of phase I or phase II clinical trials (Damasceno et al. 2020). Nevertheless, exosome therapies are still yet to be approved by FDA, which means that all the experimental studies reported are still not being commercialised in the market (An Introduction to Exosome Therapy and Its Costs 2020). Further studies are needed to identify its cost-effectiveness on large-scale production, and the cost should be affordable for every patient, especially in PF treatment.

4.5 Cell-Free Therapy

Cell-free therapy involves the utilisation of exosomes derived from MSCs, and this approach is gaining much interest in recent years mainly in human disease treatment. Several research have shown the potential of involving exosomes derived from MSCs in the wound healing process, which may benefit PF treatment. Facilitation of the wound healing process could be achieved by transfecting WJ-MSCs exosomes into the injured tissues. Fang et al. (2016) stated that WJ-MSCs contain exosomal miR-21, miR-23a, miR-125b and miR-145, and these miRNAs were being proved in promoting myofibroblast and scar formation, accelerating the overall wound healing process. Not only that, treating injured tissues by using WJ-MSCs exosomes improves the re-epithelialising of tissues by promoting translocation of β-catenin into the injured wound to enhance the formation of skin cells and promotes migration (Fang et al. 2016).

Exosomes released from WJ-MSCs would also contribute to the regulation of the immune system by suppressing the activity of macrophages to secrete IL-6, which reduce the inflammatory response (Song et al. 2020). Li et al. (2016) also mentioned that miR-181c-expressed exosomes secreted from WJ-MSCs can downregulate the release of inflammatory cytokines IL-1β and TNF-α while promoting IL-10 production, leading to anti-inflammation of cells. Nevertheless, exosomes derived from ADSCs were reported to have a great contribution towards the re-epithelialisation process. This is due to the presence of overexpressed miR-125a in the exosomes acting as a pro-angiogenic factor, where it encourages the development of endothelial tip cells by lowering the expression level of angiogenic inhibitor delta-like 4 (DLL4) (Liang et al. 2016).

5   Remaining Challenges and Conclusion

Several limitations and challenges remained to be addressed in MSC therapy for PF treatment. In normal circumstances, MSCs will differentiate into fibroblast and myofibroblast in the phase of wound healing, forming scar tissue on the site of injury. However, excessive proliferation of myofibroblast may occur during tissue remodelling, thus causing fibrotic disease or excessive scarring on the remodelled tissue (Darby et al. 2014). Although most of the patients may not be affected, some may feel discomfort or pain at the scarring site (Dwarkasing and Schouten 2013). Regardless of MSC source, the production of stem cells requires GMP compliance to ensure the products are safe to be used for the patients. However, maintaining a lower cost while practising GMP is a crucial, yet challenging, aspect for manufacturing companies. Besides that, implementing a bioreactor system in MSCs production remains unclear. Detailed financial planning should be performed by considering several financial aspects, such as the cost of the bioreactor as well as on the maintenance. On the other hand, the efficacy of using genetically modified MSCs in PF treatment remains uncertain due to limited studies available in the industry. Therefore, further detailed examination and clinical trials would need to be conducted if genetically modified MSCs are to be introduced in treating PF for both its safety and efficacy index. In conclusion, developing affordable PF therapy is important to satisfy the current market demand, especially targeting the Asia Pacific market with the increase of fistulising CD incidence rate in the next 20 years.

References

Agha ME, Eid M, Mansy H et al (2013) Preoperative MRI of perianal fistula: Is it really indispensable? Can it be deceptive? Alexandria J Med 49:133–144

Ahuja V, Tandon RK (2010) Inflammatory bowel disease in the Asia–Pacific area: a comparison with developed countries and regional differences. J Dig Dis 11:134–147

Akiyama K, Chen C, Wang D et al (2012) Mesenchymal-stem-cell-induced immunoregulation involves FAS-ligand-/FAS-mediated T cell apoptosis. Cell Stem Cell 10:544–555

Amable PR, Teixeira MV, Carias RB et al (2014) Protein synthesis and secretion in human mesenchymal cells derived from bone marrow, adipose tissue and Wharton's jelly. Stem Cell Res Ther 5:53

Amorin B, Alegretti AP, Valim V et al (2014) Mesenchymal stem cell therapy and acute graft-versus-host disease: a review. Hum Cell 27:137–150

An Introduction to Exosome Therapy and Its Costs (2020) https://bioinformant.com/exosome-therapy/. Accessed 8 Sept 2021

Andreas K, Sittinger M, Ringe J (2014) Toward in situ tissue engineering: chemokine-guided stem cell recruitment. Trends Biotechnol 32:483–492

Barnhoorn MC, Wasser MNJM, Roelofs H et al (2020) Long-term evaluation of allogeneic bone marrow-derived mesenchymal stromal cell therapy for Crohn's disease perianal fistulas. J Crohn's Colitis 14:64–70

Bataille F, Klebl F, Rümmele P et al (2004) Morphological characterisation of Crohn's disease fistulae. Gut 53:1314–1321

Bessi G, Siproudhis L, Merlini l'Héritier A et al (2019) Advancement flap procedure in Crohn and non-Crohn perineal fistulas: a simple surgical approach. Color Dis 21(1):66–72

Bouguen G, Siproudhis L, Gizard E et al (2013) Long-term outcome of perianal fistulizing Crohn's disease treated with infliximab. Clin Gastroenterol Hepatol 11(8):975-981.e4

Bustos ML, Huleihel L, Kapetanaki MG et al (2014) Aging mesenchymal stem cells fail to protect because of impaired migration and antiinflammatory response. Am J Respir Crit Care Med 189:787–798

Carvello M, Lightner A, Yamamoto T et al (2019) Mesenchymal stem cells for perianal Crohn's disease. Cell 8:764

Castaño-Milla C, Chaparro M, Saro C et al (2015) Effectiveness of adalimumab in perianal fistulas in Crohn's disease patients naive to anti-TNF therapy. J Clin Gastroenterol 49(1):34–40

Chen L, Deng H, Cui H et al (2018) Inflammatory responses and inflammation-associated diseases in organs. Oncotarget 9

Cheng F, Huang Z, Li Z (2020) Efficacy and safety of mesenchymal stem cells in treatment of complex perianal fistulas: a meta-analysis. Stem Cells Int 8816737

Choi S, Jeon BG, Chae G et al (2019) The clinical efficacy of stem cell therapy for complex perianal fistulas: a meta-analysis. Tech Coloproctol 23:411–427

Ciccocioppo R, Klersy C, Leffler DA et al (2019) Systematic review with meta-analysis: safety and efficacy of local injections of mesenchymal stem cells in perianal fistulas. JGH Open 3:249–260

Damasceno PKF, de Santana TA, Santos GC et al (2020) Genetic engineering as a strategy to improve the therapeutic efficacy of mesenchymal stem/stromal cells in regenerative medicine. Front Cell Dev Biol 8:737

Darby IA, Laverdet B, Bonté F, Desmoulière A (2014) Fibroblasts and myofibroblasts in wound healing. Clin Cosmet Investig Dermatol 7:301

de Zoeten EF, Pasternak BA, Mattei P et al (2013) Diagnosis and treatment of perianal Crohn disease: NASPGHAN clinical report and consensus statement. J Pediatr Gastroenterol Nutr 57:401–412

Delafontaine P, Song Y, Li Y (2004) Expression, regulation, and function of IGF-1, IGF-1R, and IGF-1 binding proteins in blood vessels. Arterioscler Thromb Vasc Biol 24:435–444

Devetzi M, Goulielmaki M, Khoury N et al (2018) Genetically-modified stem cells in treatment of human diseases: tissue kallikrein (KLK1)-based targeted therapy (review). Int J Mol Med 41:1177–1186

Dietz AB, Dozois EJ, Fletcher JG et al (2017) Autologous mesenchymal stem cells, applied in a bioabsorbable matrix, for treatment of perianal fistulas in patients with Crohn's disease. Gastroenterology 153:59–62

Dige A, Hougaard HT, Agnholt J et al (2019) Efficacy of injection of freshly collected autologous adipose tissue into perianal fistulas in patients with Crohn's disease. Gastroenterology 156:2208–2216

Du Z, Wei C, Yan J, Han B et al (2013) Mesenchymal stem cells overexpressing C-X-C chemokine receptor type 4 improve early liver regeneration of small-for-size liver grafts. Liver Transpl 19:215–225

Dudukgian H, Abcarian H (2011) Why do we have so much trouble treating anal fistula? World J Gastroenterol 17:3292–3296

Dwarkasing RS, Schouten WR (2013) Chronic anal and perianal pain resolved with MRI. Am J Roentgenol 200:1034–1041

Ebrahimian TG, Pouzoulet F, Squiban C et al (2009) Cell therapy based on adipose tissue-derived stromal cells promotes physiological and pathological wound healing. Arterioscler Thromb Vasc Biol 29:503–551

Emile SH, Elfeki H, Thabet W et al (2017) Predictive factors for recurrence of high transsphincteric anal fistula after placement of seton. J Surg Res 213:261–268

Fang S, Xu C, Zhang Y et al (2016) Umbilical cord-derived mesenchymal stem cell-derived exosomal microRNAs suppress myofibroblast differentiation by inhibiting the transforming growth factor-β/SMAD2 pathway during wound healing. Stem Cells Transl Med 5:1425–1439

Francois M, Romieu R, Li M et al (2010) IDO expression in human mesenchymal stromal cells mediates T cell suppression and leads to monocyte differentiation into IL-10 secreting immunosuppressive CD206⁺ M2 macrophages. Blood 116:2784

Galis-Rozen E, Tulchinsky H, Rosen A et al (2010) Long-term outcome of loose seton for complex anal fistula: a two-Centre study of patients with and without Crohn's disease. Color Dis 12(4):358–362

Gallo G, Tiesi V, Fulginiti S et al (2020) Mesenchymal stromal cell therapy in the management of perianal fistulas in Crohn's disease: an up-to-date review. Medicina 56:563

Georgiev-Hristov T, Guadalajara H, Herreros MD et al (2018) A step-by-step surgical protocol for the treatment of perianal fistula with adipose-derived mesenchymal stem cells. J Gastrointest Surg 22:2003–2012

Gieseke F, Böhringer J, Bussolari R et al (2010) Human multipotent mesenchymal stromal cells use galectin-1 to inhibit immune effector cells. Blood 116:3770–3779

Gingold DS, Murrell ZA, Fleshner PR (2014) A prospective evaluation of the ligation of the intersphincteric tract procedure for complex anal fistula in patients with Crohn's disease. Ann Surg 260(6):1057–1061

Grimaud JC, Munoz-Bongrand N, Siproudhis L et al (2010) Fibrin glue is effective healing perianal fistulas in patients with Crohn's disease. Gastroenterology 138(7)

Hamilton JA (2008) Colony-stimulating factors in inflammation and autoimmunity. Nat Rev Immunol 8:533–544

Han S-M, Han S-H, Coh Y-R et al (2014) Enhanced proliferation and differentiation of Oct4- and Sox2-overexpressing human adipose tissue mesenchymal stem cells. Exp Mol Med 46:e101

Haque N, Rahman MT, Abu Kasim NH, Alabsi AM (2013) Hypoxic culture conditions as a solution for mesenchymal stem cell based regenerative therapy. Sci World J 632972

Hartung T (1998) Anti-inflammatory effects of granulocyte colony-stimulating factor. Curr Opin Hematol 5: 221–225

Herreros MD, Garcia-Olmo D, Guadalajara H et al (2019) Stem cell therapy: a compassionate use program in perianal fistula. Stem Cells Int 6132340

Hinz B, Mastrangelo D, Iselin CE et al (2001) Mechanical tension controls granulation tissue contractile activity and myofibroblast differentiation. Am J Pathol 159: 1009–1020

Ho JE, Mahajan A, Chen M-H et al (2012) Clinical and genetic correlates of growth differentiation factor 15 in the community. Clin Chem 58:1582–1591

Janovská P, Normant E, Miskin H, Bryja V (2020) Targeting casein kinase 1 (Ck1) in hematological cancers. Int J Mol Sci 21:1–19

Ji J, Wu Y, Meng Y et al (2017) JAK-STAT signaling mediates the senescence of bone marrow-mesenchymal stem cells from systemic lupus erythematosus patients. Acta Biochim. Biophys. Sin. Shanghai 49:208–215

Ji L, Zhang Y, Xu L et al (2021) Advances in the treatment of anal fistula: a mini-review of recent five-year clinical studies. Front Surg 7:172

Joel MDM, Yuan J, Wang J et al (2019) MSC: immunoregulatory effects, roles on neutrophils and evolving clinical potentials. Am J Transl Res 11:3890–3904

Langer HF, Stellos K, Steingen C et al (2009) Platelet derived bFGF mediates vascular integrative mechanisms of mesenchymal stem cells in vitro. J Mol Cell Cardiol 47:315–325

Laureti S, Gionchetti P, Cappelli A et al (2020) Refractory complex Crohn's perianal fistulas: a role for autologous microfragmented adipose tissue injection. Inflamm Bowel Dis 26:321–330

Lee MN, Hwang H-S, Oh S-H et al (2018) Elevated extracellular calcium ions promote proliferation and migration of mesenchymal stem cells via increasing osteopontin expression. Exp Mol Med 50:1–16

Lehmann J, Graf W (2013) Efficacy of LIFT for recurrent anal fistula. Colorectal Dis 15:592–595

Li X, Liu L, Yang J, Yu Y et al (2016) Exosome derived from human umbilical cord mesenchymal stem cell mediates miR-181c attenuating burn-induced excessive inflammation. EBioMedicine 8:72–82

Li X, He L, Yue Q et al (2017) MiR-9-5p promotes MSC migration by activating β-catenin signaling pathway. Am J Physiol Cell Physiol 313:C80–C93

Li H, Rong P, Ma X et al (2018) Paracrine effect of mesenchymal stem cell as a novel therapeutic strategy for diabetic nephropathy. Life Sci 215:113–118

Li Y, Zhang D, Xu L et al (2019) Cell-cell contact with proinflammatory macrophages enhances the immunotherapeutic effect of mesenchymal stem cells in two abortion models. Cell Mol Immunol 16:908–920

Liang X, Zhang L, Wang S, Han Q, Zhao RC (2016) Exosomes secreted by mesenchymal stem cells promote endothelial cell angiogenesis by transferring miR-125a. J Cell Sci 129:2182–2189

Limura E, Giordana P (2015) Modern management of anal fistula. World J Gastroenterol 21:12–20

Liu Y, Cai Y, Liu L, Wu Y, Xiong X (2018) Crucial biological functions of CCL7 in cancer. Peer J 6

Liu S, Liu F, Zhou Y et al (2020) Immunosuppressive property of MSCs mediated by cell surface receptors. Front Immunol 11:1076

Luz-Crawford P, Kurte M, Bravo-Alegría J et al (2013) Mesenchymal stem cells generate a CD4+ CD25+ Foxp3+ regulatory T cell population during the differentiation process of Th1 and Th17 cells. Stem Cell Res Ther 4:65

Mallis P, Boulari D, Michalopoulos E et al (2018) Evaluation of HLA-G expression in multipotent mesenchymal stromal cells derived from vitrified Wharton's Jelly tissue. Bioengineering 5:–95

Marzo M, Felice C, Pugliese D et al (2015) Management of perianal fistulas in Crohn's disease: an up-to-date review. World J Gastrointest Pathophysiol 21:1394–1403

Mazini L, Ezzoubi M, Malka G (2021) Overview of current adipose-derived stem cell (ADSCs) processing involved in therapeutic advancements: flow chart and regulation updates before and after COVID-19. Curr Stem Cell Res Ther 12:1–17

McKee C, Chaudhry GR (2017) Advances and challenges in stem cell culture. Colloids Surf B Biointerfaces 159:62–77

Meng Y, Ren Z, Xu F et al (2018) Nicotinamide promotes cell survival and differentiation as kinase inhibitor in human pluripotent stem cells. Stem Cell Rep 11:1347–1356

Miguel-Hidalgo JJ, Nithuairisg S, Stockmeier C, Rajkowska G (2007) Distribution of ICAM-1 immunoreactivity during aging in the human orbitofrontal cortex. Brain Behav Immun 21:100–111

Mizukami A, Swiech K (2018) Mesenchymal stromal cells: from discovery to manufacturing and commercialization. Stem Cells Int 4083921

Mizukami A, de Abreu Neto MS, Moreira F et al (2017) A fully-closed and automated hollow fiber bioreactor for clinical-grade manufacturing of human mesenchymal stem/stromal cells. Stem Cell Rev Rep 14:141–143

Moffett JR, Namboodiri MA (2003) Tryptophan and the immune response. Immunol Cell Biol 81:247–265

Molendijk I, Bonsing BA, Roelafs H (2015) Allogeneic bone marrow-derived mesenchymal stromal cells promote healing of refractory perianal fistulas in patients with Crohn's disease. Gastroenterology 149:918–927

Murray PJ (2017) Macrophage polarization. Annu Rev Physiol 79:541–566

Musiał-Wysocka A, Kot M, Majka M (2019) The pros and cons of mesenchymal stem cell-based therapies. Cell Transplant 28:801–812

Ng SC (2014) Epidemiology of inflammatory bowel disease: focus on Asia. Best Pract Res Clin Gastroenterol 28(3):363–372

Nie C, Yang D, Xu J et al (2011) Locally administered adipose-derived stem cells accelerate wound healing through differentiation and vasculogenesis. Cell Transplant 20:205–216

Nienow AW, Rafiq QA, Coopman K, Hewitt CJ (2014) A potentially scalable method for the harvesting of hMSCs from microcarriers. Biochem Eng J 85:79–88

Nilsson MI, Bourgeois JM, Nederveen JP et al (2019) Lifelong aerobic exercise protects against inflammaging and cancer. PLoS One 14

Noone C, Kihm A, English K, O'Dea S, Mahon BP (2013) IFN-γ stimulated human umbilical-tissue-derived cells potently suppress NK activation and resist NK-mediated cytotoxicity in vitro. Stem Cells Dev 22:3003–3014

Ocansey DKW, Pei B, Yan Y et al (2020) Improved therapeutics of modified mesenchymal stem cells: an update. J Transl Med 18:1–14

Omar RE, Beroud J, Stoltz JF, Menu P, Velot E, Decot V (2014) Umbilical cord mesenchymal stem cells: the new gold standard for mesenchymal stem cell-based therapies? Tissue Eng Part B Rev 20:523–544

Pai SG, Carneiro BA, Mota JM et al (2017) Wnt/beta-catenin pathway: modulating anticancer immune response. J Hematol Oncol 10

Panchalingam KM, Jung S, Rosenberg L, Behie LA (2015) Bioprocessing strategies for the large-scale production of human mesenchymal stem cells: a review. Stem Cell Res Ther 6:1–10

Panek-Jeziorna M, Mulak A (2020) An inverse correlation of serum fibroblast growth factor 19 with abdominal pain and inflammatory markers in patients with ulcerative colitis. Gastroenterol Res Pract 2389312

Panés J, Rimola J (2017) Perianal fistulizing Crohn's disease: pathogenesis, diagnosis and therapy. Nat Rev Gastroenterol Hepatol 14:652–664

Panés J, Garcia-Olmo D, Assche GV et al (2016) Expanded allogeneic adipose-derived mesenchymal stem cells (Cx601) for complex perianal fistulas in Crohn's disease: a phase 3 randomised, double-blind controlled trial. Lancet 388:1281–1290

Panés J, García-Olmo D, Assche GV et al (2017) Long-term efficacy and safety of stem cell therapy (Cx601) for complex perianal fistulas in patients with Crohn's disease. Gastroenterology 154(5):1334-1342.e4

Panés J, Reinisch W, Rupniewska E et al (2018a) Burden and outcomes for complex perianal fistulas in Crohn's disease: systematic review. World J Gastroenterol 24: 4821–4834

Panés J, Garcia-Olmo D, Assche GV et al (2018b) Long-term efficacy and safety of stem cell therapy (Cx601) for complex perianal fistulas in patients with Crohn's disease. Gastroenterology 154:1334–1342

Park JY, Pillinger MH, Abramson SB (2006) Prostaglandin E2 synthesis and secretion: the role of PGE2 synthases. Clin Immunol 119:229–240

Park KJ, Ryoo SB, Kim JS et al (2016) Allogeneic adipose-derived stem cells for the treatment of perianal fistula in Crohn's disease: a pilot clinical trial. Color Dis 18:468–476

Park WS, Ahn SY, Sung SI et al (2018) Strategies to enhance paracrine potency of transplanted mesenchymal stem cells in intractable neonatal disorders. Pediatr Res 83:214–222

Park MY, Yoon YS, Lee JL et al (2021) Comparative perianal fistula closure rates following autologous adipose tissue-derived stem cell transplantation or treatment with anti-tumor necrosis factor agents after seton placement in patients with Crohn's disease: a retrospective observational study. Stem Cell Res Ther 12:401

Pripp AH, Stanišić M (2014) The correlation between pro- and anti-inflammatory cytokines in chronic subdural hematoma patients assessed with factor analysis. PLoS One 9:e90149

Prockop DJ, Oh JY (2012) Mesenchymal stem/stromal cells (MSCs): role as guardians of inflammation. Mol Ther 20:14–20

Ramezankhani R, Torabi S, Minaei N et al (2020) Two decades of global progress in authorized advanced therapy medicinal products: an emerging revolution in therapeutic strategies. Front Cell Dev Biol 8:1358

Regmi R, Pathak S, Kim JO et al (2019) Mesenchymal stem cell therapy for the treatment of inflammatory diseases: challenges, opportunities, and future perspectives. Eur J Cell Biol 98:151041

Resnitzky D, Gossen M, Bujard H, Reed SI (1994) Acceleration of the G1/S phase transition by expression of cyclins D1 and E with an inducible system. Mol Cell Biol 14:1669–1679

Rodrigues CAV, Fernandes TG, Diogo MM, Lobato Da Silva C, JMS C (2011) Stem cell cultivation in bioreactors. Biotechnol Adv 29:815–829

Sanad A, Emile S, Thabet W et al (2019) A randomized controlled trial on the effect of topical phenytoin 2% on wound healing after anal fistulotomy. Colorectal Dis 21:697–704

Sangiorgi B, Panepucci RA (2016) Modulation of immunoregulatory properties of mesenchymal stromal cells by toll-like receptors: potential applications on GVHD. Stem Cells Int 9434250

Santoro MM, Gaudino G (2005) Cellular and molecular facets of keratinocyte reepithelization during wound healing. Exp Cell Res 304:274–286

Scharl M, Rogler G (2014) Pathophysiology of fistula formation in Crohn's disease. World J Gastrointest Pathophysiol 5:205–212

Scharl M, Bruckner RS, Rogler G (2016) The two sides of the coin: similarities and differences in the pathomechanisms of fistulas and stricture formations in irritable bowel disease. United European Gastroenterol J 4:506–514

Schraufstatter IU, Khaldoyanidi SK, DiScipio RG (2015) Complement activation in the context of stem cells and tissue repair. World J Stem Cells 7:1090–1108

Schwartz DA, Tagarro I, Carmen Díez M, Sandborn WJ (2019 Oct 18) Prevalence of fistulizing Crohn's disease in the United States: estimate from a systematic literature review attempt and population-based database analysis. Inflamm Bowel Dis 25:1773–1779

Scott LJ (2018) Darvadstrocel: a review in treatment-refractory complex perianal fistulas in Crohn's disease. Bio Drugs 32:627–634

Senéjoux A, Siproudhis L, Abramowitz L et al (2016) Fistula plug in fistulising ano-perineal Crohn's disease: a randomised controlled trial. J Crohns Colitis 10(2): 141–148

Serrero M, Philandrianos C, Visee C et al (2017) OP008 an innovative treatment for refractory perianal fistulas in Crohn's disease: local micro reinjection of autologous fat and adipose derived stromal vascular fraction. J Crohn's Colitis 11:S5

Shingyochi Y, Orbay H, Mizuno H (2015) Adipose-derived stem cells for wound repair and regeneration. Expert Opin Biol Ther 15:1285–1292

Sohni A, Verfaillie CM (2013) Mesenchymal stem cells migration homing and tracking. Stem Cells Int 130763

Solodeev I, Meilik B, Volovitz I et al (2018) Fas-L promotes the stem cell potency of adipose-derived mesenchymal cells. Cell Death Dis 9:1–13

Song A, Wang J, Tong Y et al (2020) BKCa channels regulate the immunomodulatory properties of WJ-MSCs by affecting the exosome protein profiles during the inflammatory response. Stem Cell Res Ther 11:1–16

Sorg H, Tilkorn DJ, Hager S et al (2017) Skin wound healing: an update on the current knowledge and concepts. Eur Surg Res 58:81–94

Spaggiari GM, Moretta L (2013) Interactions between mesenchymal stem cells and dendritic cells. Adv Biochem Eng Biotechnol 130:199–208

Spees JL, Lee RH, Gregory CA (2016) Mechanisms of mesenchymal stem/stromal cell function. Stem Cell Res Ther 7:125

Tamama K, Kawasaki H, Wells A (2010) Epidermal growth factor (EGF) treatment on multipotential stromal cells (MSCs). Possible enhancement of therapeutic potential of MSC. J Biomed Biotechnol 795385

Teo GSL, Ankrum JA, Martinelli R et al (2012) Mesenchymal stem cells transmigrate between and directly through tumor necrosis factor-α-activated endothelial cells via both leukocyte-like and novel mechanisms. Stem Cells 30:2472–2486

Thia KT, Mahadevan U, Feagan BG et al (2009) Ciprofloxacin or metronidazole for the treatment of perianal fistulas in patients with Crohn's disease: a randomized, double-blind, placebo-controlled pilot study. Inflamm Bowel Dis 15(1):17–24

Ullah M, Liu DD, Thakor AS (2019) Mesenchymal stromal cell homing: mechanisms and strategies for improvement. iScience 15:421–438

van Koperen PJ, Safiruddin F, Bemelman WA, Slors JFM (2009) Outcome of surgical treatment for fistula in ano in Crohn's disease. Br J Surg 96(6):675–679

Varkouhi AK, Paula A, Monteiro T et al (2020) Genetically modified mesenchymal stromal/stem cells: application in critical illness. Stem Cell Rev Rep 16:812–827

Wainstein C, Quera R, Fluxá D et al (2018) Stem cell therapy in refractory perineal Crohn's disease: long-term follow-up. Color Dis 20:O68–O75

Wang Y, Chen X, Cao W et al (2014) Plasticity of mesenchymal stem cells in immunomodulation: pathological and therapeutic implications. Nat Immunol 15:1009–1016

Watanabe K, Ueno M, Kamiya D et al (2007) A ROCK inhibitor permits survival of dissociated human embryonic stem cells. Nat Biotechnol 25:681–686

Waterman RS, Tomchuck SL, Henkle SL et al (2010) A new mesenchymal stem cell (MSC) paradigm: polarization into a pro-inflammatory MSC1 or an immunosuppressive MSC2 phenotype. PLoS One 5:e10088

Weber WP, Feder-Mengus C, Chiarugi A et al (2006) Differential effects of the tryptophan metabolite 3-hydroxyanthranilic acid on the proliferation of human CD8+ T cells induced by TCR triggering or homeostatic cytokines. Eur J Immunol 36:296–304

Weiss ARR, Dahlke MH (2019) Immunomodulation by mesenchymal stem cells (MSCs): mechanisms of action of living, apoptotic, and dead MSCs. Front Immunol 10:1191

Weiss ML, Anderson C, Medicetty S et al (2008) Immune properties of human umbilical cord Wharton's jelly-derived cells. Stem Cells 26:2865–2874

Whelan DS, Caplice NM, Clover AJP (2020) Mesenchymal stromal cell derived CCL2 is required for accelerated wound healing. Sci Rep 10:2642

Yu Y, Yoo SM, Park HH et al (2019) Preconditioning with interleukin-1 beta and interferon-gamma enhances the efficacy of human umbilical cord blood-derived mesenchymal stem cells-based therapy via enhancing prostaglandin E2 secretion and indoleamine 2,3-dioxygenase activity in dextran sulfate sodium-induced colitis. J Tissue Eng Regen Med 13:1792–1804

Zhang Y, Xu J, Ren Z et al (2021) Nicotinamide promotes pancreatic differentiation through the dual inhibition of CK1 and ROCK kinases in human embryonic stem cells. Stem Cell Res Ther 12:1–12

Zhao J, Hu L, Liu J et al (2013) The effects of cytokines in adipose stem cell-conditioned medium on the migration and proliferation of skin fibroblasts in vitro. Biomed Res Int 2013:578479

Zhou C, Li M, Zhang Y et al (2020) Autologous adipose-derived stem cells for the treatment of Crohn's fistula-in-ano: an open-label, controlled trial. Stem Cell Res Ther 11:1–13

Zhu Y, Yao S, Chen L (2011) Cell surface signalling molecules in the control of immune responses: a tide model. Immunity 34:466–478

Zhuang G, Meng C, Guo X et al (2012) A novel regulator of macrophage activation: miR-223 in obesity-associated adipose tissue inflammation. Circulation 125:2892–2903

Adv Exp Med Biol - Cell Biology and Translational Medicine (2022) 17: 97–162
https://doi.org/10.1007/5584_2022_715
© Springer Nature Switzerland AG 2022
Published online: 4 July 2022

Virus, Exosome, and MicroRNA: New Insights into Autophagy

Javid Sadri Nahand, Arash Salmaninejad,
Samaneh Mollazadeh, Seyed Saeed Tamehri Zadeh,
Mehdi Rezaee, Amir Hossein Sheida, Fatemeh Sadoughi,
Parisa Maleki Dana, Mahdi Rafiyan, Masoud Zamani,
Seyed Pouya Taghavi, Fatemeh Dashti,
Seyed Mohammad Ali Mirazimi, Hossein Bannazadeh Baghi,
Mohsen Moghoofei, Mohammad Karimzadeh,
Massoud Vosough, and Hamed Mirzaei

Abstract

Autophagy is known as a conserved self-eating mechanism that contributes to cells to degrade different intracellular components (i.e., macromolecular complexes, aggregated proteins, soluble proteins, organelles, and foreign bodies). Autophagy needs formation of a double-membrane structure, which is composed of the sequestered cytoplasmic contents, called autophagosome. There are a variety of internal and external factors involved in initiation and progression of

J. Sadri Nahand
Infectious and Tropical Diseases Research Center, Tabriz University of Medical Sciences, Tabriz, Iran

A. Salmaninejad
Department of Medical Genetics, Faculty of Medicine, Guilan University of Medical Sciences, Guilan, Iran

Department of Medical Genetics, Faculty of Medicine, Mashhad University of Medical Sciences, Mashhad, Iran

S. Mollazadeh
Natural Products and Medicinal Plants Research Center, North Khorasan University of Medical Sciences, Bojnurd, Iran

S. S. Tamehri Zadeh
School of Medicine, Tehran University of Medical Sciences, Tehran, Iran

M. Rezaee
Department of Anesthesiology, School of Medicine, Shahid Madani Hospital, Alborz University of Medical Sciences, Karaj, Iran

A. H. Sheida, F. Sadoughi, P. M. Dana, M. Rafiyan, M. Zamani, S. P. Taghavi, F. Dashti, and S. M. A. Mirazimi
School of Medicine, Kashan University of Medical Sciences, Kashan, Iran

Student Research Committee, Kashan University of Medical Sciences, Kashan, Iran

H. Bannazadeh Baghi
Infectious and Tropical Diseases Research Center, Tabriz University of Medical Sciences, Tabriz, Iran

Department of Virology, Faculty of Medicine, Tabriz University of Medical Sciences, Tabriz, Iran

M. Moghoofei
Department of Microbiology, Faculty of Medicine, Kermanshah University of Medical Sciences, Kermanshah, Iran

M. Karimzadeh
Department of Virology, Faculty of Medicine, Iran University of Medical Sciences, Tehran, Iran

autophagy process. Viruses as external factors are one of the particles that could be associated with different stages of this process. Viruses exert their functions via activation and/or inhibition of a wide range of cellular and molecular targets, which are involved in autophagy process. Besides viruses, a variety of cellular and molecular pathways that are activated and inhibited by several factors (e.g., genetics, epigenetics, and environment factors) are related to beginning and developing of autophagy mechanism. Exosomes and microRNAs have been emerged as novel and effective players anticipated in various stages of autophagy. More knowledge in these pathways and identification of accurate roles of them could help to provide better therapeutic approaches in several diseases such as cancer. We highlighted the roles of viruses, exosomes, and microRNAs in the autophagy processes.

Keywords

Autophagy · Cancer · Chemoresistance · Exosome · MicroRNA · Viral infection

1   Autophagy

Although autophagy was recognized around 50 years ago in mammalian cells, its molecular function was revealed vastly in the past decade. Autophagy usually occurs as an evolutionary conserved mechanism in all eukaryotic cells for sustaining cell homeostasis. Recent studies have revealed that autophagy is one of the vital biological mechanisms, which is related to health, longevity, differentiation, starvation, homeostasis, cell survival, adaptation, elimination of microorganisms, and cell death (Shafabakhsh et al. 2021). This process begins with the formation of double-membrane

vesicles (DMVs), which is generally termed autophagosome, as well as by various processes, such as fusing with lysosomes. This event leads to degradation/recycling of components which exist in cytoplasmic lysosomes (Cuervo 2004). Critical roles of autophagy process in longevity, homeostasis, and cell death have been recently demonstrated (Mizushima 2007). In eukaryotic cells, autophagy includes microautophagy, macroautophagy, and chaperone-mediated autophagy (CMA), three key intracellular pathways. Core molecular machinery of autophagy referred to subset of autophagy-related (ATG) proteins is essential for autophagosome formation (Fig. 1) (Mizushima 2007). P53 and Bcl-2 protein\families with dual regulatory properties play significant roles in autophagy induction (Yoon et al. 2012; Singletary and Milner 2008). Autophagy is involved in various pathologies, such as neurodegenerative and age-related disorders, infections, and inflammatory/immunity diseases, and especially in invasion and cancer progression (Yang and Klionsky 2010). Increasing evidences show the importance of autophagy in cancer and support the concept when it gets disturbed, it can lead to an accelerated tumorigenesis. Also, comparative evidences have shown that degradation of autophagy or proteolysis in tumors is less than normal cells (Yang et al. 2011). Anticancer role of autophagy is due to elimination of damaged cell component and inhibition of tumor growth. However, autophagy can cause tumor cells withstand stress in undesirable conditions leading to survival. Stress-induced autophagy may result in resistance to treatment and result in the progression of tumor cells (Yang et al. 2011; Kondo et al. 2005). Additionally, the efficacy of autophagy inhibitors, along with chemotherapy, in preventing tumor growth and inducing cell death is far better than the chemotherapy alone. Recent investigations have shown that autophagy may play an important role in drug resistance. It means that autophagy may contribute to increase tumor cells resistance to chemotherapeutic

M. Vosough
Department of Regenerative Medicine, Cell Science
Research Center, Royan Institute for Stem Cell Biology
and Technology, ACECR, Tehran, Iran

H. Mirzaei (✉)
Research Center for Biochemistry and Nutrition in
Metabolic Diseases, Institute for Basic Sciences, Kashan
University of Medical Sciences, Kashan, Iran
e-mail: mirzaei-h@kaums.ac.ir

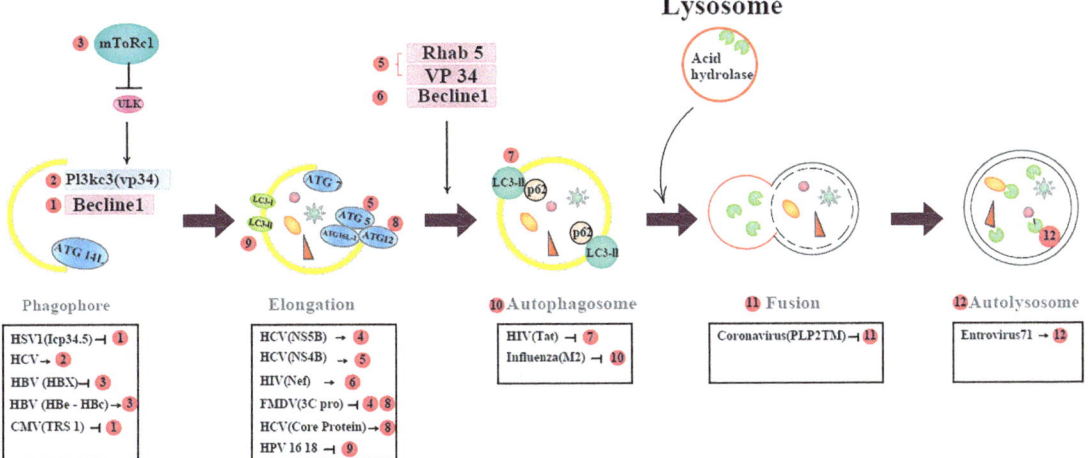

Fig. 1 A schema various stages of autophagy. Viruses, through the production of different proteins, can affect different stages of autophagy, such as the early stages of autophagy (phagophore and elongation), and the ending stages (Autophagosome and autolysosome), in order to survive in the host cell longer. Some of the viruses that can affect the autophagy process are hepatitis C and B viruses and HIV. For example, the hepatitis C virus by producing NS5B protein influences the ATG5 autophagy regulatory protein, which proceeds autophagy in the elongation phase, or, by producing the NS4B protein, affects the Rhab5 factor and drives autophagy from the elongation phase to autophagosome formation. Hepatitis B virus targets and impedes mToR protein kinase through the production of HBX protein, increases the efficiency of the ULK protein, and leads to the initial phase of autophagy, the formation of phagophore. It has also been observed that some of the proteins of the virus (HBe and HBc) inhibit the early stage of autophagy by increasing the efficiency of the mToR protein kinase. HIV produces nef and thus affects one of the major proteins in autophagy called BECN1, which causes autophagy to progress from the elongation stage to the formation of autophagosome, or by producing TAT stops a factor necessary for the formation of autophagosome (LC3-II-PE complex), losing the autophagosome form. There are other viruses that can apply their effects on the autophagy process. For example, coronaviruses prevent lysosomal incorporation with the PLP2TM protein and impede the formation of autolysosomes. Enterovirus 71 affects and disables autolysosomes. Influenza virus via M2 protein destroys autophagosome. HSV prevents the formation of phagophore by producing Icp34.5 and inhibiting BECN1

and anticancer agents. Therefore, autophagy regulation can be considered as an appropriate therapeutic target in the therapy of cancer. Thus, various autophagy-modulating approaches may be assumed to circumvent chemoresistance (Huang et al. 2016; YiRen et al. 2017). Mounting evidences have revealed that autophagy along with chemotherapy and its association with chemoresistance can be a new therapeutic goal to succeed in cancer treatment.

2 MicroRNA and Autophagy

2.1 Regulation of Autophagy by MicroRNAs

MicroRNAs (miRNAs) are a group of noncoding small RNA molecules (~19–22 nucleotides long) which regulate protein-coding genes (Mollazadeh et al. 2019; Neshati et al. 2018; Letafati et al. 2022; Mousavi et al. 2022; Balandeh et al. 2021; Razavi et al. 2021; Mirzaei and Hamblin 2020). The main miRNA mechanisms are translational repression and mRNA degradation. In the nucleus, RNA polymerase II (RNAPII) produces long primary transcripts (pri-miRNAs), which acts as a substrate for RNase III enzymes and Drosha-DGCR8 complex (a microprocessor that is essential for miRNA maturation) to produce precursor miRNAs (pre-miRNAs). Then, pre-miRNA is exported from the nucleus into the cytoplasm by exportin-5 and Ran-GTP. In the cytoplasm, pre-miRNA is cleaved by another RNase III enzyme, Dicer, into miRNA duplexes approximately 19–22 nucleotides long. Mature miRNA is incorporated into RNA-induced silencing complex (RISC) where it remains stable and binds to its complementary target mRNA. miRNAs are involved in many major biological functions such as intracellular signaling, cellular metabolism, differentiation, pathological processes, and regulation of gene expression (Su et al. 2015). Some miRNAs are only expressed in specific cell types. Expression patterns of miRNAs are unique to individual tissues and differ between cancer and normal tissues (Jafari et al. 2018). Aberrant expression of miRNAs is associated with multiple human diseases, such as metabolic disease, neurological disorders (Tavakolizadeh et al. 2018), cardiovascular complications, viral diseases (Keshavarz et al. 2018), immune-related diseases, and especially malignancies (Bartels and Tsongalis 2010).

Autophagy-related protein 7 (ATG7) was recently considered as a potential target of miR-96-5p. The aberrant expression of this miRNA reduces autophagy activity (Yu et al. 2018a). Based on the current data, miR-20a-5p inhibits cell proliferation and autophagy and promotes apoptosis through negative regulation of ATG7 (Yu et al. 2018b). Moreover, the overexpression of miR-140-5p/miR-149 inhibits apoptosis and promotes autophagy by downregulating fucosyltransferase1 (FUT1) (Wang et al. 2018a). According to the investigation conducted by Liu et al. (2017a) miR-20a negatively relates to autophagy/lysosome pathway. They reported that miR-20a inhibited autophagy and lysosomal proteolytic activity through targeting several key regulators of autophagy, including BECN1, ATG16L1, and sequestosome 1 (SQSTM1) (Liu et al. 2017a). Various molecular components involve in autophagy cascade, including Atg1/unc-51-like kinase (ULK) complex, Beclin-1/class III phosphatidylinositol 3-kinase (PI3K) complex, Atg9 and vacuole membrane protein 1(VMP1), two ubiquitin-like protein (Atg12 and Atg8/LC3) conjugation systems, and proteins which mediate fusion between autophagosomes and lysosomes (Kroemer et al. 2010). Some of these core components of autophagy pathway are direct targets of miRNAs (such as miR-30a, miR-23a, and miR-129-5p) and have key roles in the inhibition/induction of autophagy process (Fig. 2) (Xiao et al.

2015; Guo et al. 2017a; Zhu et al. 2009; Sadri Nahand et al. 2021; Pourhanifeh et al. 2020a, b; Rezaei et al. 2020; Jamali et al. 2020). In the following, the role of miRNAs in the regulation of autophagy and their potential molecular mechanisms has been reported in some disorders.

Meng and colleagues revealed the clinical significance of miR-138 in patients with malignant melanoma, which inhibits cell proliferation and induces apoptosis. Overexpression of miR-138 increases cell autophagy by LC3 protein induction as well as the suppression of PI3K/AKT/mTOR and PDK1 (Meng et al. 2017). It was exhibited that the upregulation of miR-18a-5p in melanoma cell lines and tissues had promising role in melanoma pathogenesis mediated by EPHA7 silence leading to tumor development as well as apoptosis and autophagy blockage (Guo et al. 2021).

Long et al. reviewed the association between miRNAs and autophagy in colorectal cancer (CRC) and concluded that miRNA-regulated autophagy could be up- or downregulated in various CRC conditions associated with the tumor microenvironment. In this context, it can referrer to the roles of miR-140-5p and miR-502 in inhibition of autophagy in chemotherapy of CRC stem cells; miR-214, miR-183-5p, and miR-31 in inhibition of autophagy in radiotherapy of CRC; and miR-124, miR-18a, and miR-210 in promotion of autophagy in metabolism and hypoxia of CRC. Also, blockage of autophagy in inflammatory bowel disease could be mediated via miR-142-3p, miR-143, miR-130a, etc. (Long et al. 2020).

In hepatocellular carcinoma (HCC), autophagy could be reduced via miR-490-3p/ATG7 (Ou et al. 2018) or microRNA-181a/Atg5 axis, suggestive of the profounding value of autophagy deficiency in HCC (Yang et al. 2018). Jin et al. showed that miR-513b-5p attenuated tumorigenesis of liver cancer cells in HCC via inactivation of PIK3R3-mediated autophagy (Jin et al. 2021). Zhang et al. demonstrated that downregulation of miR-638 in human liver cancer led to a noticeable reduction in malignancy of liver cancer cell accompanied by increase of autophagosomes and

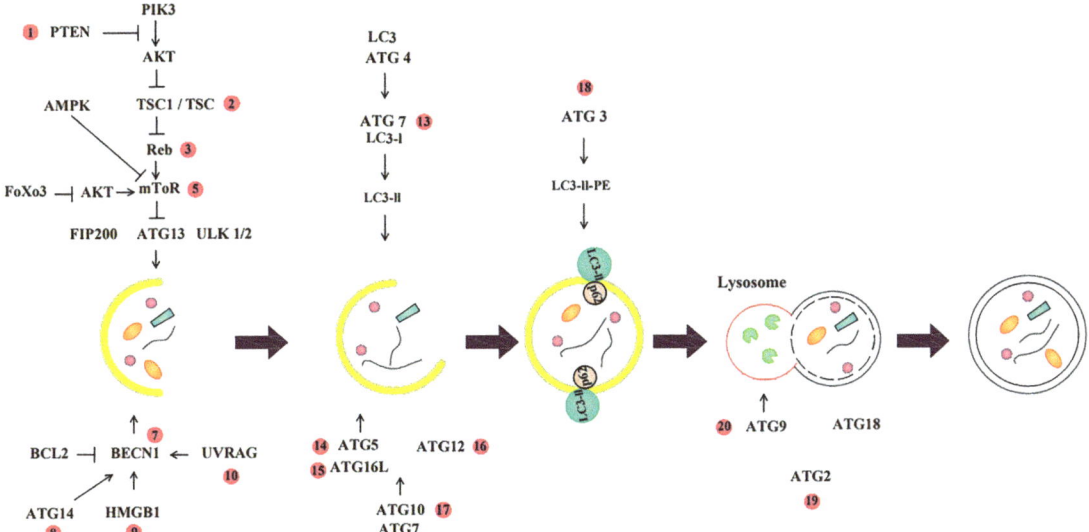

Fig. 2 Various factors involved in the formation of the autophagy mechanism, each of which is affected by different microRNAs that somehow regulate the autophagy steps. The PTEN protein by inhibiting the PIK3-akt pathway paves the way for autophagy to start, increasing the expression of mir-21 that inhibits this protein, thus activating the pik3-akt pathway, and preventing the onset of autophagy (1). Autophagy starts/miR-193b-3p declines, and this protein is most produced and autophagy occurs more (2). Reb has an incremental effect on the mTOR protein that activates this pathway and prevents the formation of the initial autophagy phase. The expression of miR-199a-5p decreases, the inhibitory effect on Reb is inactivated, and autophagy is inhibited (3). Foxo3 disables Akt and causes the MPT pathway to be deactivated/miR-27a decreases its expression, and the foxo3 protein is further produced and autophagy continues to function (4). mTOR, one of the important pathways involved in the autophagy mechanism, has an inhibitory effect on this process and does not allow autophagy to begin and applies its effect on the ULK1/2 factor/miR-7 declines, and its inhibitory effect on this pathway is removed, and the autophagy does not start (5). ULK1/2 is one of the important factors for the onset of autophagy and phagophore formation, declining miR-26b and its inhibitory effect on ULK2, and autophagy begins its pathway, but the expression of miR-290-295 cluster is increased, and the ULK1 protein level is reduced, and the phagophore is not formed, so autophagy does not occur (6). Beclin-1 is somehow one of the important proteins in the development of phagophore and the onset of autophagy. The expression of mir-20a increases, and its inhibitory effect on the gene does not allow the formation of proteins and, accordingly, autophagy does not begin, but miR-30a expression reduces, the Beclin-1 gene is more expressed, and autophagy starts (7). ATG14 has an increased effect on Beclin-1 and makes phagophore more likely to form miR-135a expression increases, thereby inhibiting ATG14 gene and autophagy formation (8). HMGB1

stimulates the Beclin-1 gene and causes the autophagy to start its first phase/miR-34a expression decreases, and its inhibitory effect on the HMGB1 gene is removed, and Beclin-1expression increases (9). UVRAG interferes somehow behind the initial pathway of autophagy and reaches the formation of autophagosome/the miR-183 which disrupts the process by targeting and inhibiting the gene (10). FIP200, present in ULK complex and is effective in the formation of phagophore/miR-224-3p expression, is increased, and an inhibitory effect on this gene is increased, and the initial phase of autophagy does not occur, but miR-20b, which declines, causes an increase in the expression level of FIP200, and phagophore is formed (11). AMK with inhibitory effect on MTOR pathway and TSC1/TSC2 stimulation inhibits autophagy. The expression of miR-185 is reduced, AMK is more expressed, and autophagy is more active (12). ATG7 is a factor accelerating the conversion of lc3-I to Lc3-II, which is an important process for the onset of autophagosome formation. miR-490-3p expression declines and further stimulates its target and ATG7, and autophagy continues (13). ATG5 is a protein that causes autophagy to evolve from the phagophore formation phase to the next formation of the process. miR-181a is increased, and most of the ATG5 gene is inhibited, and this functional trend is disrupted (14). ATG16L is a factor that accelerates the formation of autophagosomes. The expression of miR-130a is increased, the level of the ATG16L protein decreases, and the autophagy is inhibited/expression of the miR-410 decreases, and this process continues (15). The activity of ATG12 is similar to that of ATG16L. The expression of MIR-23a is reduced, its inhibitory effect on this gene is reduced, and autophagy continues its process. miR-378 inhibits the autophagy process by inhibiting the gene (16). ATG10 is a protein that stimulates the activity of ATG5, ATG16L, and ATG12 proteins and accelerates the process of autophagosome formation. miR-20 has an inhibitory effect on this protein, which can disrupt this activity (17). ATG3 is a factor to stimulate the formation autolysosomes, suggestive of tumor-suppressive role of miR-638 via silence of EZH2 (Zhang et al. 2021a).

In osteosarcoma (OS), miR-210-5p induced epithelial-mesenchymal transition (EMT) and oncogenic autophagy via PIK3R5/AKT/mTOR axis (Liu et al. 2020a). Also, upregulation of miR-22 in OS suppressed autophagy and induced apoptosis resulted in increased sensitivity to cisplatin (Meng et al. 2020). In prostate cancer (PC) cells, overexpression of miR-381 increased cellular autophagy and apoptosis, while decreased cell proliferation mediated by reelin (RELN) suppression (Liao and Zhang 2020). Deng et al. recognized that miR-493 respectively activated cytotoxic autophagy and reduced invasion of PC cells via up-modulation of BECN1 and ATG7 (Deng et al. 2020).

In cervical cancer cells, miR-211 overexpression targeted autophagy and apoptosis through Bcl-2 regulation (Liu et al. 2020b). Besides, aberrant expression of miR-106a in cervical squamous cell carcinoma (CSCC) was related to malignancy parameters of CSCC tissues. Based, overexpression of miR-106a elevated CSCC growth and suppressed autophagy via binding to 3UTR of LKB1 in human papilloma virus (HPV) 16-positive CSCC (Cui et al. 2020). Consistently, miR-378 has a potential impact on cervical cancer progression via binding to ATG12-regulated autophagy (Tan et al. 2018). In the ovarian cancer (OC), increased expression of miR-34 activates apoptosis and autophagy followed by significant reduction in the proliferation of cancerous cells (OVACAR-3 cells) via silencing Notch 1 (Jia et al. 2019). Shao et al. identified that miR-1251-5p upregulation had oncogenic effects on human ovarian cancer via preventing TBCs (negative modulator of autophagy) (Shao et al. 2019).

In bladder cancer, reduced expression of miR-221 facilitated autophagy through increasing TP53INP1 levels, indicative of the valuable importance of miR-221 as therapeutic targets in this malignancy (Liu et al. 2020c). Dai et al. represented the *tumorigenic* capacity of miR-130 in bladder cancer cells as it was proved by autophagy induction through blocking CYLD (Dai et al. 2020). Also, Zhang et al. displayed that upregulation of miR-21 in bladder tumor cells (T24 cells) promoted T24 cells progression alongside with apoptosis and autophagy obstruction via downregulation of, Beclin-1, PTEN, caspase-3, LC3-II, and E-cadherin (Zhang et al. 2020a). Similarly, Rezaei et al. focused on the impacts of up-/downregulation of miRNAs in the different lung diseases including lung cancer either in in vitro and in vivo conditions or human. In this regard, up- and downregulation of respectively miR-210 and miR-181 inactivated autophagy, while down- and upregulation of respectively miR-3127-5p and miR-21 activated autophagy (Rezaei et al. 2020).

In esophageal squamous cell carcinoma (ESCC), autophagy is triggered by miR-503 via PKA/mTOR pathway followed by inhibition of ESCC invasiveness (Wu et al. 2018). In another study, Li et al. focused on the effect of miR-126 on apoptosis and autophagy of ESCC cells and found that miR-126 expression was increased in ESCC followed by enhancement of apoptosis and autophagy; however, miR-126 inhibition reversed current trend via suppression of STAT3 (Li et al. 2020a). Phatak et al. (2021) clarified that miR-141-3p could act as an oncogene in esophageal cancer cells via binding to TSC1 mRNA which led to tumor progression as well as autophagy reduction (Phatak et al. 2021).

In gastric cancer (GC) cells, miR-let-7a/Rictor/ Akt-mTOR axis modulates autophagy activity

Fig. 2 (continued) of LC3-PE, which causes the LC3 protein binding to phosphatidylethanolamine and the formation of autophagosome and maintains its stability. The expression of miR-1 is reduced, this factor is further developed, and autophagosome is formed (18). ATG2 protein is effective in the formation of autolysosome. The expression of miR-143 is increased, the ats2 gene is suppressed, and this process is disrupted (19). ATG9 is an agent for stimulating the formation of autolysosome and accelerating the process of lysosome fusion with autophagosome/miR-29a expression which is decreased and the level of atg9 increased, and this trend continues (20). Akt is a stimulant factor for the mTOR pathway and prevents the formation of autophagy/miR-185 which targets this gene and inhibits autophagy (21)

(Fan et al. 2018). Among another regulators of autophagy in GC, it can mention miR-183 which its downregulation blocks apoptosis and autophagy via interacting with MALAT1 and SIRT1 through PI3K/AKT/mTOR pathway (Li et al. 2019a). Li et al. evidenced that miR-133a-3p could strengthen autophagy and proliferation of GC cells via downregulation of FOXP3 (Li et al. 2020b). In breast cancer cells, transfection of MCF-7 with miR-26b mimic reduced autophagy dependent to irradiation through silence of DRAM1 (Meng et al. 2018). Ai et al. (2019) clarified that overexpression of miR-107 in breast cancer cell lines (MDA-MB-231 and MDA-MB-453 cells) causes significant reduction in cellular autophagy, proliferation, and metastasis via silencing HMGB1. ULK1 and lysosomal protein transmembrane 4 beta (LAPTM4B), autophagy-related mediators, have also been identified as direct targets of miR-489 which is downregulated in the most of breast cancer cells and several drug resistant breast cancer cell lines (Soni et al. 2018a).

In the metabolic diseases such as osteoporosis, the condition can be exacerbated via miR-15 overexpression which modulates osteoblast genesis and autophagy alongside with downregulation of USP7 (Lu et al. 2021). Wang et al. provided evidences that in osteoarthritis (OA), joint disease, miR-140-5p/miR-149 could affect autophagy, apoptosis, and proliferation of chondrocytes via their potential target, FUT1 (Wang et al. 2018a). Also, miR-20 has a pivotal impact on OA evidenced by inhibition of autophagy and chondrocytes proliferation through ATG10/PI3K/AKT/mTOR axis (Vojtechova and Tachezy 2018). Besides, He et al. (2018) assigned that the inhibition of miR-20 promoted proliferation and autophagy machinery in articular chondrocytes by targeting ATG10 via PI3K/AKT/mTOR signaling pathway (He and Cheng 2018). Furthermore, pathogenesis of intervertebral disc degeneration (IDD) can be influenced by miRNA-regulated autophagy including decreased autophagy facilitated by upregulation of miR-210 and miR-202-5p via targeting ATG7 (Lan et al. 2020). Yun et al.

(2020) highlighted the promising role of miR-185 in preventing IDD via improving cell survival and suppressing apoptosis and autophagy of nucleus pulposus cell via blockage of galectin-3/Wnt/β-catenin pathway (Yun et al. 2020). Similar results have been achieved by miR-142-3 overexpression in controlling and inhibiting IDD (Xue et al. 2021).

Moreover, evidences are in a favor of miR-145-3p in exerting autophagic flux in multiple myeloma (MM) via *HDAC4* inhibition (Wu et al. 2020). In the neurodegenerative disorders such as Parkinson's disease (PD) defined by dopaminergic neurons apoptosis, Wen et al. (2018a) confirmed that AMPK/mTOR-regulated autophagy and apoptosis could be a potential therapeutic platform as this axis can be inhibited by miR-185 overexpression leading to prevention of dopaminergic cells death in PD model (Wen et al. 2018a). Similarly, Li et al. (2018a) observed that autophagy in PD could be triggered by miR-181b/PTEN/Akt/mTOR axis in a way that overexpression of miR-181b is associated with increased cell viability. Also, Lu et al. (2020) conducted similar research on PD model and reached to the findings that upregulation of miR-133a in a PD cell model increased cell proliferation and inhibited autophagy and apoptosis by binding to 3 UTR of RAC1 (Lu et al. 2020). Wen and colleagues demonstrated that overexpression of miR-185 inhibited autophagy and apoptosis through regulating the AMPK/mTOR signaling pathway in PD (Wen et al. 2018b). In Alzheimer's disease (AD), the amounts of miRNA-101a was significantly decreased in patients as well as in vivo model and resulted in autophagy regulation through the MAPK pathway (see Table 1) (Li et al. 2019b). Another novel therapeutic option in AD could be proposed by upregulation of miR-16-5p or downregulation of BTG2, which inhibit neuronal damage and autophagy (Dong et al. 2021). Yang et al. (2020) pinpointed that melatonin could reduce neuronal death and autophagy in cerebral ischemia-reperfusion injury (CIRI) mechanistically through regulation of miR-26a-5p/NRSF as well as JAK2-STAT3

Table 1 microRNAs and autophagy

miRNA	Expression	Target	Inhibition/induction	Disease	Note	Ref
miR-26b	Down	ULK2	Induction	Prostate cancer	Downregulation of mTOR	Clotaire et al. (2016)
miR-21	Up	Rab11	Inhibition	Renal ischemia-reperfusion	Reduction of Beclin-1andLC3-II expression and upregulation of p62	Liu et al. (2015a)
miR-185	Down	mTOR AMPK	Induction	Parkinson	Increase of neuronal apoptosis through elevating AMPK/mTOR signaling pathway activity, upregulation of Beclin-1, LC3-I/LC-II	Wen et al. (2018a)
miR-96-5p	Up	FOXO1	Inhibition	Breast cancer	Increase of migration, invasiveness, and proliferation by decreasing apoptosis	Doan et al. (2017)
miR-502	Down	Rab1B DHODH	Induction	Colon cancer	Increase of cell proliferation and metastasis	Zhai et al. (2013)
miR-100	Down	mTOR IGF-1R	Inhibition	HCC	Decrease of LC3B-II and Akt proteins enhance tumor growth	Ge et al. (2014)
miR-30a	Down	Beclin-1	Induction	Breast cancer Lung cancer Glioma	–	Zhu et al. (2009)
miR-143	Down	ATG2B HK2	Induction	Non-small-cell lung cancer (NSCLC)	Promotion of cell proliferation, metastasis and Warburg effect	Wei et al. (2015)
miR-23a	Down	ATG12	Induction	Melanoma	Increase of the expression of RUNX2 reduces miR-23a Increase of metastasis and invasion via blocking AMPK-RhoA pathway	Guo et al. (2017a)
miR-130a	Up	ATG16L	Inhibition	COPD	Enhancement of apoptosis and increase of the development of COPD	Li et al. (2016a)
miR-193b-3p	Down	TSC1	Induction	Amyotrophic lateral sclerosis (ALS)	Increase of cell survival by increase of TSC1 expression, and decrease of mTORC1 activity, apoptosis	Dhital et al. (2017)

(continued)

Table 1 (continued)

miRNA	Expression	Target	Inhibition/induction	Disease	Note	Ref
let-7i	Up	IGF-1R	Induction	Ankylosing spondylitis (AS)	Protection of T cell from apoptosis through (PI3 K)/Akt and MAPK signaling pathways, LC3B-II increase in T cell and p62 decline, inhibition of mTOR	Hou et al. (2014a)
miR-20a-5p	Down	ATG16L1	Induction	Ischemic kidney injury	The hypoxia downregulated HIF-1α and miR-20a-5p expression, increase of LC3-II	Wang et al. (2015a)
miR-20a	Up	THBS2	Induction	Cervical cancer tissue	The miR-20a deficiency led to trigger the decrease of autophagic activity in cervical cancer cell lines	Zhao et al. (2015a)
miR-338-5p	Up	PIK3C3	Inhibition	Colorectal cancer	Promotion of metastasis and cell migration, decline of ATG14, LC3-II and Beclin-1 expression	Ju et al. (2013)
miR-301a	Up	NDRG2	Induction	Prostate cancer	Hypoxia-induced miR-301a expression, increase of cell viability and decrease of cell apoptosis, promotion of PTEN expression	Guo et al. (2016)
miR-301b	Up	NDRG2	Induction	Prostate cancer	Hypoxia-induced miR-301b expression, increase of cell viability and decrease of cell apoptosis, promotion of PTEN expression	Guo et al. (2016)
miR-290-295 cluster	UP	Atg7 ULK1	Inhibition	Melanoma	Promotion of cell Proliferation and migration Increase of melanoma cells resistance to glucose deficiency	Cheng et al. (2012)

(continued)

Table 1 (continued)

miRNA	Expression	Target	Inhibition/induction	Disease	Note	Ref
miR-185	Down	AKT1 RICTOR RHEB	Inhibition	HCC	Increase of cell proliferation by overexpression of mTOR Decrease of apoptosis via Bcl-2, upregulation of cyclin D1	Zhou et al. (2017)
miR-96-5p	Up	ATG7	Inhibition	Liver fibrosis	TGF-β1 promotes miR-96-5p expression, inverse cell proliferation, inhibition of mRNA, and protein levels of α-SMA and Col1α1	Yu et al. (2018b)
miR-101	Down	EZH2	Induction	HCC	Increase of chemoresistance and decline of apoptosis	Xu et al. (2014)
miR-101	–	–	Inhibition	Liver ischemia/reperfusion injury (LIRI)	miR-101 can inhibit autophagy and reduce LIRI by activating the mTOR pathway	Song et al. (2019)
miR-101a	Down	–	Inhibition	Alzheimer's disease (plasma)	miRNA-101a could regulate autophagy by targeting the MAPK pathway	Li et al. (2019b)
miR-129-5p	Up	Beclin-1	Inhibition	Prostate cancer	Increase of resistance to the Norcantharidin (NCTD)	Xiao et al. (2016)
miR-140-5p/miR-149	Down	FUT1	Inhibition	Osteoarthritis	Decrease of chondrocyte proliferation, overexpression of IL-1β, and promotion of apoptosis	Wang et al. (2018a)
miR-124	Down	Bim	Inhibition	Parkinson	Increase of apoptosis and inhibition of autophagosome accumulation and lysosomal depletion	Wang et al. (2016)
miR-124	–	p62/p38	Induction	Parkinson	miR-124 can suppress neuroinflammation during the Parkinson's disease development via targeting autophagy, p62, and p38	Yao et al. (2019)

(continued)

Table 1 (continued)

miRNA	Expression	Target	Inhibition/ induction	Disease	Note	Ref
miR-224-3p	Up	FIP200	Inhibition	Cervical cancer	Promotion of cell proliferation	Fang et al. (2016)
miR-143	Down	GABARAPL1	Induction	Gastric cancer	Increase of resistance to the quercetin	Du et al. (2015)
miR-22	Up	PTEN	Inhibition	Diabetic nephropathy	Increase of renal tubulointerstitial fibrosis, increase of glucose inducing miR-22 and promoting AKT/mTOR pathway	Zhang et al. (2018a)
miR-130a	Down	ATG2B DICER1	Induction	Chronic lymphocytic leukemia	Promotion of cell proliferation	Kovaleva et al. (2012)
miR-181a	Up	MTMR3	Inhibition	Gastric cancer	Promotion of cell proliferation, metastasis and inhibition of apoptosis	Lin et al. (2017)
miR-21	Up	PTEN	Inhibition	HCC	Increase of resistance to sorafenib, promotion of AKT pathway	He et al. (2015a)
miR-409-3p	Down	Beclin-1	Induction	Colon cancer	Increase of resistance to oxaliplatin	Tan et al. (2016)
miR-30a	Down	Beclin-1	Induction	Renal carcinoma	Increase of resistance to sorafenib, upregulation of ATG5 and decrease of apoptosis	Zheng et al. (2015)
miR-503	Up	PRKACA	Inhibition	Esophageal carcinoma	Promotion of cell proliferation, metastasis, increase of PKA/mTOR signaling pathway activity	Wu et al. (2018)
miR-143	Up	ATG2B	Inhibition	Crohn's disease	Blockage of autophagy in intestinal epithelial cells, decline of autophagosome and autolysosome formation, downregulation of IκBα Promotion of pro-inflammatory cytokine expression: IFN-γ, TNF-α, and IL-8	Lin et al. (2018)

(continued)

Table 1 (continued)

miRNA	Expression	Target	Inhibition/induction	Disease	Note	Ref
miR-423-3p	Up	Bim	Induction	Gastric cancer	Increase of cell proliferation, invasion and migration, upregulation of LC3, and decrease of P62 and apoptosis	Kong et al. (2017)
miR-135a	Up	Atg14	Inhibition	HCC	Factor VII-increased miR-135a, decrease of LC3A/B protein level, promotion of mTOR activation	Huang et al. (2017)
miR-34a	Down	HMGB1	Induction	Acute myeloid leukemia (AML)	Inhibition of apoptosis Increase of LC3 level, enhancement of chemoresistance	Liu et al. (2017b)
miR-34	Down	Notch 1	Induction	Ovarian cancer cell lines	miR-34 can be inhibiting ovarian cancer cells proliferation by triggering apoptosis and autophagy. It suppresses cell invasion through targeting Notch 1	Jia et al. (2019)
miR-142-3p	Down	HMGB1	Induction	Non-small-cell lung cancer (NSCLC)	Promotion of mTOR, AKT, and P13K activation, increase of chemoresistance	Chen et al. (2017a)
miR-142-3p	Down	HMGB1	Induction	Acute myelogenous leukemia (AML)	Increase of drug resistance in AML cells, inhibition of apoptosis	Zhang et al. (2017a)
miR-142-3p	Down	KLF9	Inhibition	Human ectopic endometrial tissues	Upregulation of mir-142-3p levels can restrict autophagy and induce apoptosis of CRL-7566 cells	Ma et al. (2019)
miR-30b	Down	Atg12, Atg5	Induction	Hepatic ischemia-reperfusion injury (IRI)	Upregulation of LC3-II and increase of autophagosomes	Li et al. (2016b)
miR-30b	–	–	Induction	Vascular calcification	Restoring of miR-30b expression can promote autophagy	Xu et al. (2019a)

(continued)

Table 1 (continued)

miRNA	Expression	Target	Inhibition/induction	Disease	Note	Ref
miR-199a-5p	Down	Rheb	Inhibition	Ankylosing spondylitis (AS)	Enhancement of mTOR signaling pathway, decrease of LC3, Beclin-1, and ATG5 expression, Increase of pro-inflammatory cytokines: TNF-α, IL-17, and IL-23	Wang et al. (2017a)
miR-320	Up	HIF-1α	Induction	Retinoblastoma (RB)	Upregulation of HIF-1α and hypoxia, increase of LC3 and Beclin-1 expression, decrease of p62 and p-mTOR	Liang et al. (2017)
miR-32	Up	DAB2IP	Induction	Gastric cancer	Increase of radioresistance in GC, decline of apoptosis and mTOR	Wu et al. (2016a)
miR-224	Up	Smad4	Inhibition	Hepatitis B Virus-associated HCC	Hepatitis B reducesAtg5 and Beclin-1, increase p62, inhibits the formation of autophagosome, blocks TGF-β signaling pathway, and promotes cell proliferation and metastasis	Lan et al. (2014)
miR-181a	Down	p38 JNK	Induction	Parkinson	Enhance p38 MAPK/JNK signaling pathways and apoptosis	Liu et al. (2017c)
miR-410	Down	ATG16L1	Induction	Osteosarcoma	Increase of chemoresistance and inhibition of apoptosis	Chen et al. (2017b)
miR-449a	Down	CISD2	Inhibition	Glioma	Increase level of BCL-2 and cell proliferation, downregulate Beclin-1	Sun et al. (2017a)
miR-378	Up	ATG12	Inhibition	Cervical cancer	Increase migration, invasiveness, proliferation, and metastasis	Tan et al. (2018)
miR-33	Up	ABCA1	Inhibition	Atherosclerosis	Decrease autophagosome formation and LC3 in macrophage	Ouimet et al. (2017)

(continued)

Table 1 (continued)

miRNA	Expression	Target	Inhibition/induction	Disease	Note	Ref
miR-32	Up	DAB2IP	Induction	Prostate cancer	Increase radioresistance in prostate cancer cells, promote autophagy through the mTOR-S6K pathway	Liao et al. (2015)
miR-7	Down	mTOR	Inhibition	HCC	Increase cell proliferation	Wang et al. (2017b)
miR-26b	Down	DRAM1	Induction	Breast cancer	Increase radioresistance in breast cancer cell	Meng et al. (2018)
miR-20a	Up	BECN1 ATG16L1 SQSTM1	Inhibition	Breast cancer	C-myce promotes miR-20a and elevates ROS level and DNA damage	Liu et al. (2017a)
miR-638	Up	TP53INP2	Inhibition	Melanoma	Inhibit apoptosis via block p53, increase methylation at CpG islands, enhance melanoma metastasis	Bhattacharya et al. (2015)
miR-212	Down	SIRT1	Induction	Prostate cancer	Increase angiogenesis and cellular senescence	Ramalinga et al. (2015)
miR224-3p	Down	ATG5 FIP200	Induction	Glioblastoma	Hypoxia inhibits miR224-3p and mTOR activity, increase levels of ATG 16,12,13 and ULK1	Guo et al. (2015)
miR-183	UP	UVRAG	Inhibition	Colorectal cancer	Decrease of apoptosis and autophagosome formation, increase of cell proliferation	Huangfu et al. (2016)
miR-183	Down	SIRT1	Induction	Gastric cancer tissue and cell lines	miR-183 can enhance gastric cancer cell viability and inhibit cell apoptosis by promoting autophagy MALAT1-miR-183-SIRT1 axis and PI3K/AKT/mTOR pathway are involve in autophagy of gastric cancer cells	Li et al. (2019a)
miR-29	Down	–	Induction	Retinal pigment epithelial cells	Overexpression of miR-29 can induce autophagy of ARPE-19 cells and primary human retinal pigment epithelial cells	Cai et al. (2019)

(continued)

Table 1 (continued)

miRNA	Expression	Target	Inhibition/ induction	Disease	Note	Ref
miR-29a	Down	ATG9A TFEB	Induction	Pancreatic cancer	Increase resistance to gemcitabine, autophagy flux, autophagosome formation and autophagosome-lysosome fusion, overexpression of LC3B and decrease p62, promote cell proliferation, migration and invasion	Kwon et al. (2016)
miR-29a	Up	PTEN	Inhibition	Pathological cardiac hypertrophy (rat model)	miR-29a can inhibit autophagy by regulating the PTEN/AKT/mTOR signaling pathway	Shi et al. (2019)
miR-29b-3p	Down	SPARC	Inhibition	Blood samples of heart failure (HF) and hypoxia-induced H9c2 cells	Hypoxia led to downregulation of miR-29b-3p level and induces autophagy and apoptosis of H9c2 cells miR-29b-3p suppresses apoptosis and autophagy by targeting SPARC in hypoxia-induced H9c2 cells	Zhou et al. (2019)
miR-30a	Down	BECN1	Induction	Diabetic cataract	Hgh glucose-promoting apoptosis	Zhang et al. (2017b)
miR-24-3p	Up	DEDD	Induction	Bladder cancer	Increase of cell proliferation, invasion, migration and LC3, decline of apoptosis and p62	Yu et al. (2017a)
miR-138	Down	Sirt1	Induction	Lung cancer	Increase cell proliferation, invasion, metastasis, EMT and AMPK signaling pathway, decrease apoptosis and mTOR activity	Ye et al. (2017)
miR-490-3p	Down	ATG7	Induction	HCC	Increase cell proliferation, decrease apoptosis	Ou et al. (2018)
miR-20a-5p	Down	ATG7	Induction	Neuroblastoma	Increase of LC3-II/ LC3-I and autophagosome formation, decline of apoptosis	Yu et al. (2018b)

(continued)

Table 1 (continued)

miRNA	Expression	Target	Inhibition/induction	Disease	Note	Ref
miR-23a	Down	ATG12	Induction	Melanoma	Increase cell proliferation, invasion and metastasis	Guo et al. (2017b)
miR-138	Down	PDK1	Inhibition	Malignant melanoma	Promote PI3K/AKT/mTOR signaling pathway, decrease levels of LC3, caspase-3 and Bax	Meng et al. (2017)
miR-181a	Up	Atg5	Inhibition	HCC	Decline of apoptosis	Yang et al. (2018)
miR-489	Down	ULK1 LAPTM4B	Induction	Breast cancer	Increase of chemoresistance	Chen et al. (2018)
miR-30a	Down	Beclin-1	Induction	Medulloblastoma	Increase cell proliferation and LC3B level	Singh et al. (2017)
miR-214-3p	Down	Atg12	Induction	Sporadic Alzheimer's disease	Increase levels of LC3bII and Beclin-1, and enhance number of GFP-LC3-positive autophagosome vesicles and apoptosis	Lv et al. (2016)
miR-214	–	PTEN	Induction	Ischemic heart disease (H9c2 cell line)	Oridonin can induce apoptosis and autophagy by regulating PI3K/AKT/mTOR pathway via overexpression of miR-214	Gong et al. (2019)
miR-20a	Down	RB1CC1/FIP200	Induction	Breast cancer	Decrease of mTOR activity	Li et al. (2016c)
miR-20b	Down	RB1CC1/FIP200	Induction	Breast cancer	Decrease of mTOR activity	Li et al. (2016c)
miR-181b	Down	PTEN	Induction	Parkinson	Decrease of PI3 K/Akt/mTOR signaling pathway	Li et al. (2018a)
miR-181b	Up	CREBRF	Induction	Gallbladder cancer	miR-181b inhibits tumor suppression mediated with ginsenoside Rg3 of gallbladder carcinoma through inducing autophagy flux by targeting CREBRF	Wu et al. (2019)
miR-20	Up	ATG10	Inhibition	Osteoarthritis (OA)	Decrease of PI3 K/Akt/mTOR signaling pathway and enhancement of proliferation in chondrocytes	Vojtechova and Tachezy (2018)

(continued)

Table 1 (continued)

miRNA	Expression	Target	Inhibition/induction	Disease	Note	Ref
miR-222	Up	PPP2R2A	Inhibition	Bladder cancer	Increase of Akt/mTOR signaling pathway, enhanced resistance of bladder cancer cells to cisplatin	Zeng et al. (2016)
miR-125b	Up	APC	Induction	Colorectal cancer	CXCL12/CXCR4 promotes miR-125b expression and elevates Wnt/β-catenin signaling pathway, EMT and cell invasion, enhances resistance of colorectal cancer cells to fluorouracil	Yu et al. (2017b)
miR-125a	Up	–	Inhibition	Thyroiditis (mice)	Overexpression of miR-125a can be inhibits autophagy by targeting PI3K/Akt/mTOR signaling pathway in mouse model of thyroiditis	Chen et al. (2019a)
miR-218	Down	YEATS4	Induction	Colorectal cancer	Increase resistance to the oxaliplatin (L-OHP), inhibit of apoptosis	Fu et al. (2016)
miR-22	Down	HMGB1	Induction	Osteosarcoma	Increase drug resistance in osteosarcoma cells, promote cell proliferation, migration, and invasion	Guo et al. (2014)
miR-1	Down	ATG3	Induction	Non-small-cell lung cancer (NSCLC)	Increase drug resistance in NSCLC cells	Hua et al. (2018)
miR-27a	Down	FoxO3a	Induction	Traumatic brain injury (TBI)	Increase level of Beclin-1 and decrease p62	Sun et al. (2017b)
miR-27a	Up	SYK	Inhibition	Melanoma tissues	Depletion of miR-27a lead to induced autophagy and apoptosis of melanoma cells through the activation of the SYK-dependent mTOR signaling pathway	Tang et al. (2019)
miR-31	Down	–	Induction	Colorectal cancer-associated fibroblasts (CAFs)	Elevation of levels of Beclin-1, ATG, DRAM, and LC3, decrease of apoptosis, increase of radioresistance	Yang et al. (2016a)

(continued)

Table 1 (continued)

miRNA	Expression	Target	Inhibition/induction	Disease	Note	Ref
miR-140-5p	Down	Smad2 ATG12	Induction	Colorectal cancer	Decrease of TGF-β signaling pathway and necrosis, increase of cell proliferation, metastasis, and invasion	Zhai et al. (2015)
miR-let-7a	Down	Rictor	Inhibition	Gastric cancer	Upregulation of Akt/mTOR signaling pathway	Fan et al. (2018)
miR-221	Up	TP53INP1	Inhibition	Colorectal cancer	Increase cell proliferation, decrease level of LC3	Liao et al. (2018)
miR-107	Down	HMGB1	Inhibition	Breast cancer tissue and cell line	miR-107 can suppress autophagy, migration and proliferation of breast cancer cells through targeting HMGB1	Ai et al. (2019)
miR-107	Down	TRAF3	Induction	Osteoarthritis chondrocytes	Upregulation of miR-107 can inhibit the activation of NF-κB and AKT/mTOR pathway by targeting TRAF3 genes. Also, miR-107 overexpression suppresses apoptosis and promotes autophagy	Zhao et al. (2019a)
miR-223	Up	Atg16l1	Inhibition	Brain microglial cells (BV2 cells) and	miR-223 can inhibit autophagy and induce CNS inflammation by targeting ATG16L1 Expression level of miR-223 was upregulated in CNS and spleen during experimental autoimmune encephalomyelitis (EAE) progression	Li et al. (2019c)

(continued)

Table 1 (continued)

miRNA	Expression	Target	Inhibition/induction	Disease	Note	Ref
miR-365	–	ATG3	Inhibition	HCC tissue and cell line	Enforced expression of miR-365 led to significant inhibition of the ATG3 expression in hepatocellular carcinoma cells LncRNA PVT1 can promote autophagy by sponging miR-365 in HCC	Yang et al. (2019a)
miR-206	Down	STC2	Induction	Head and neck squamous cell carcinoma (HNSCC) tissue and cell line	Enforced expression of miR-206 can lead to enhanced autophagy of HNSCC cells	Xue et al. (2019)
miR-206	–	–	Inhibition	Osteoarthritis (rat model)	miR-206 can inhibit autophagy and apoptosis of osteoarthritis cells by activating the IGF-1-mediated PI3K/AKT-mTOR signaling pathway	Yu et al. (2019)
miR-93	–	BECN1, SQSTM1, ATG5, ATG4B	Inhibition	Glioblastoma cancer	miR-93 can inhibit autophagy activity by downregulation of autophagy regulatory genes level	Huang et al. (2019a)
miR-99a and miR-449a	Down	Beclin-1	Inhibition	Thrombosis (serum sample)	Upregulation of miR-99a and miR-449a can inhibit beclin-1 expression levels and autophagy	Zeng et al. (2019)
miR-216a	Down	MAP1S	Inhibition	Colorectal cancer	miR-216a can act as a tumor suppressor miRNA and inhibit autophagy through the TGF-β/MAP1S pathway	Wang et al. (2019a)
miR-18a	Up	BDNF	Inhibition	Cardiomyocytes (from an acute myocardial infarction (AMI) rat model)	miR-18a can lead to inhibiting autophagy and promoting senescence of cardiomyocytes after AMI by targeting BDNF	Lin et al. (2019)

(continued)

Table 1 (continued)

miRNA	Expression	Target	Inhibition/induction	Disease	Note	Ref
miR-155	–	–	Inhibition	Human umbilical vein endothelial cells (HUVECs)	Downregulation of miR-155 expression can lead to decreasing oxidant-induced injury and inducing cell proliferation by upregulating autophagy	Chen et al. (2019b)
miR-506-3p	–	SPHK1	Inhibition	Osteosarcoma cancer cell	miR-506-3p can initiate epithelial-to-mesenchymal transition (EMT) and inhibit autophagy of osteosarcoma cancer cells by targeting SPHK1	Wang et al. (2019b)
miR-326	–	XBP1	Induction	Parkinson's disease (mouse model)	miR-326 can inhibit nitric oxide synthase (iNOS) expression and induce autophagy of dopaminergic neurons via targeting XBP1	Zhao et al. (2019b)
miR-1251-5p	Up	TBCC	Induction	Human ovarian cancer cell lines and tissues	miR-1251-5p can induce autophagy and act as an oncogene to suppress TBCC and α-/β-tubulin expression	Shao et al. (2019)
miR-217	–	NAT2	Induction	CCL4-induced liver injury (rat models)	miR-217 can induce apoptosis and autophagy and inhibit proliferation of hepatocytes by targeting NAT2	Yang et al. (2019b)
miR-34a-5p	Up	–	Induction	CIH-induced HCAECs	Overexpression of miR-34a-5p can contribute to chronic intermittent hypoxia (CIH)-induced human coronary artery endothelial cell (HCAEC) autophagy by Bcl-2/Beclin-1 pathway	Lv et al. (2019)

pathway (Yang et al. 2020). It was suggested that neuronal deficit and autophagy in ischemic stroke could be abolished by miR-378 trough targeting GRB2, while lncRNA MEG3 could sponge the miR-378 and activate the expression of GRB2 (Luo et al. 2020).

Shi et al. clarified that miR-126 loss of function could activate myocardial autophagy induced by Beclin-1 and contributed in acute myocardial infarction (AMI) development (Shi et al. 2020). In contrast, miR-18a downregulation had protective effects against AMI via activation of BDNF expression and inhibition of Akt/mTOR axis (Lin et al. 2019). In the Su et al. study (Su et al. 2020), it was manifested that downregulation of miR-30e-3p lessened autophagy and activated apoptosis and injury in cardiomyocytes under ischemia/hypoxia conditions potentially through Egr-1 regulation (Su et al. 2020). MiRNA-regulated abnormal apoptosis and autophagy of cardiomyocyte have a great of importance in heart failure (HF). Alongside with reduced expression of miR-29b-3p in HF patients, the level of this miRNA was decreased in an in vitro HF model under hypoxia condition followed by elevated apoptosis and autophagy via inactivation of SPARC and regulation of TGF-β1/Smad3 cascade (see Table 1) (Zhou et al. 2019).

In the liver complications such as liver fibrosis characterized by hepatic stellate cell (HSC) activation, the regulation of HSC autophagy has attracted research interests. There is line of evidence shown that introduction of miR-96-5p into HSCs (LX-2 cells) is accompanied by repressing autophagy in the cells via ATG7 regulation (Yu et al. 2018c).

In the renal problems including renal tubulointerstitial fibrosis (TIF) as a main result of diabetic nephropathy (DN), accumulating data implicated the major role of miRNAs in the autophagy regulation. Zhang et al. findings represented that miR-22 partially targets PTEN-blocked autophagy followed by TIF development (Liu et al. 2018a). Furthermore, p53/miR-214/ULK1 axis affects autophagy dysregulation in diabetic kidney disease (DKD) (Ma et al. 2020). Moreover, Liu et al. disclosed that the expression of miR-25-3p was increased in polycystic kidney

disease (PKD) model via interacting with ATG14-activated autophagy as well as promoting proliferation of renal cell (Liu et al. 2020d). Table 1 lists some miRNAs regulating autophagy in some human cancer cells.

2.2 MiRNAs Interactions in Chemo-Induced Autophagy

Increasing data have reported that autophagy, along with chemotherapy and its association with chemoresistance can be a new therapeutic platform to succeed in cancer treatment. To find the correlation between miRNAs and chemotherapy-induced autophagy, experimental investigations were reviewed. More importantly, the cross talk between miRNAs (modulators of multiple pathways) and autophagy holds promise to overcome chemoresistance in malignancies (Soni et al. 2018b).

Chen and colleagues found that *miR-519a not only plays a role in* glioma by regulating STAT3-mediated autophagy pathway but also affects autophagy in glioblastoma multiforme (GBM) cells and also temozolomide (TMZ) chemosensitivity. The results showed that miR-519a enhanced the sensitivity of GBM cells to TMZ. Also, a significant association was found between miR-519a effects and autophagy. Overall, miR-519a promoted autophagy in glioblastoma through targeting STAT3/Bcl-2 signaling pathway (Li et al. 2018b). Besides, overexpression of miR-29b in GBM cells inhibited cell survival, activated apoptosis and autophagy, and sensitized tested cells to TMZ (Xu et al. 2021). Because of TMZ importance in the treatment of glioblastomas and its ability to induce autophagy, Xu and colleagues assessed the regulatory role of miR-30a in glioblastoma cells treated with TMZ. They revealed that miR-30a increases U251 glioblastoma cells' chemosensitivity to TMZ through direct target of Beclin-1 and inhibition of autophagy (see Table 2) (Xu et al. 2018a). In an in vivo study, Chakrabarti and colleagues proved that antitumor activities of luteolin and silibinin, chemotherapeutic agents, were augmented due to the overexpression of miR-7-1-3p leading to

Table 2 Dysregulated expression of miRNAs in chemotherapy and their function in autophagy cascade

MicroRNA	Expression	Target	Drug/chemotherapy/radiotherapy	Inhibition/induction of autophagy	Type of disease	Ref
miR-519a	Up	STAT3 Bcl-2	Chemotherapy (temozolomide)	Induction	Glioblastoma	Li et al. (2018b)
miR-199a-5p	Down	Beclin-1	Chemotherapy (cisplatin [also known as diamminedichloridoplatinum (II) (DDP)]	Induction	Osteosarcoma (OS)	Li et al. (2016d)
miR-18a	Up	mTORC1	Radiotherapy	Induction	Colon cancer	Qased et al. (2013)
miR-22	Up	HMGB1	Chemotherapy	Inhibition	Osteosarcoma (OS)	Li et al. (2014a)
miR-22	–	MTDH	Chemotherapy (cisplatin)	Inhibition	Osteosarcoma cancer cells	Wang et al. (2019c)
miR-155	Up	–	Chemotherapy	Induction	Osteosarcoma (OS)	Chen et al. (2014a)
miR-152	Up	ATG14	Chemotherapy (cisplatin)	Inhibition	Ovarian cancer	He et al. (2015b)
miR-214	Up	UCP2	Chemotherapy (tamoxifen, fulvestrant)	Inhibition	Breast cancer	Yu et al. (2015a)
miR-30a	Down	Beclin-1	Chemotherapy (temozolomide)	Induction	Glioblastoma	Xu et al. (2018a)
miR-7-1-3p	Up	XIAP	Luteolin Silibinin	Inhibition	Glioblastoma	Chakrabarti and Ray (2016)
miR-199a-5p	Down	ATG7	Chemotherapy (cisplatin)	Induction	HCC	Xu et al. (2012)
miR-199a-5p	Down	DRAM1	Chemotherapy (Adriamycin)	Inhibition	Acute myeloid leukemia (AML)	Li et al. (2019d)
miR-423-5p	Up	p-ERK 1/2	Chemotherapy (sorafenib)	Induction	HCC	Stiuso et al. (2015)
miR-125b	Up	EVA1A	Chemotherapy (oxaliplatin)	Inhibition	HCC	Ren et al. (2018)
miR-216a	Up	Beclin-1	Radiotherapy	Inhibition	Pancreatic cancer	Zhang et al. (2015a)
miR-410-3p	Up	HMGB1	Chemotherapy (gemcitabine)	Inhibition	Pancreatic ductal adenocarcinoma (PDAC)	Xiong et al. (2017)
miR-140-5p	Up	IP3k2	Chemotherapy	Induction	Osteosarcoma	Wei et al. (2016)
hsa-miR-302a-3p	Down	ULK1	Chemotherapy (5-fluorouracil)	Induction	Colon cancer	Hou et al. (2014b)
hsa-miR-548ah-5p	Down	ATG16L1	Chemotherapy (5-fluorouracil)	Induction	Colon cancer	Hou et al. (2014b)

miRNA	Up/Down	Target	Treatment	Induction/Inhibition	Disease/Cell	Reference
hsa-miR-30a-5p	Up	PIK3R2	Chemotherapy (5-fluorouracil)	Induction	Colon cancer	Hou et al. (2014b)
hsa-let-7c-5p	Up	BCL2L1	Chemotherapy (5-fluorouracil)	Induction	Colon cancer	Hou et al. (2014b)
hsa-miR-99b-5p	Up	mTOR	Chemotherapy (5-fluorouracil)	Induction	Colon cancer	Hou et al. (2014b)
hsa-miR-23a-3p	Up	BCL2	Chemotherapy (5-fluorouracil)	Induction	Colon cancer	Hou et al. (2014b)
hsa-miR-195a-5p	Up	BCL2	Chemotherapy (5-fluorouracil)	Induction	Colon cancer	Hou et al. (2014b)
miR-34a	Up	FoxO3	Lipopolysaccharide (LPS)	Induction	Acute lung injury (ALI)	Song et al. (2017)
miR-15a-3p	Up	Bcl-2	Polygonatum odoratum lectin (POL)	Induction	Lung adenocarcinoma	Wu et al. (2016b)
miR-1290	Down	GSK3β	Polygonatum odoratum lectin (POL)	Induction	Lung adenocarcinoma	Wu et al. (2016b)
miR-193b	Up	–	Chemotherapy (5-fluorouracil)	Induction	Esophageal cancer	Nyhan et al. (2016)
miR-193b	Down	FEN1	Chemotherapy (epirubicin)	Induction	Osteosarcoma cells	Dong et al. (2019)
miR-384-5p	Down	Beclin-1	Streptozotocin (STZ)	Induction	Diabetic encephalopathy	Wang et al. (2018b)
miR-101	Up	STMN1 RAB5A ATG4D	Chemotherapy (cisplatin)	Inhibition	HCC	Xu et al. (2013)
miR-30a	Up	Beclin-1	Chemotherapy (doxorubicin)	Inhibition	Osteosarcoma	Xu et al. (2016)
miR-30b	–	ATG5	Chemotherapy (cisplatin)	Inhibition	Gastric cancer	Xi et al. (2019)
miR-221/222	Up	ATG12	Chemotherapy (dexamethasone)	Inhibition	Multiple myeloma	Xu et al. (2018b)
miR-142-3p	Up	ATG5 ATG16L1	Chemotherapy (sorafenib)	Inhibition	HCC	Zhang et al. (2018c)
miR-21	Down	–	Chemotherapy (etoposide, doxorubicin)	Induction	Chronic myeloid leukemia	Seca et al. (2013)
miR-137	–	ATG5	Chemotherapy (doxorubicin)	Inhibition	Pancreatic cancer cells	Wang et al. (2019d)
miR-224-3p	–	ATG5	Chemotherapy (temozolomide)	Inhibition	Glioblastoma and astrocytoma	Huang et al. (2019b)
miR-146a	–	TAF9b/P53 pathway	Doxorubicin (DOX)	Induction	Human AC16 cell line	Pan et al. (2019)

inhibition of autophagy and induction of apoptosis in glioblastoma cells (Chakrabarti and Ray 2016). In addition, miR-224-3p weakened resistance to TMZ in glioblastoma cells (LN229 cells) via abolishing autophagy under hypoxia via ATG5 downregulation (Liu et al. 2020e).

Xiao et al. (2016) investigated the role of miR-199a-5p in reducing chemoresistance to cisplatin or diamminedichloridoplatinum (II) (DDP) in OS. They showed that treatment of OS cells with DDP attenuated the expression level of miR-199a-5p; increased the level of various proteins, such as Beclin-1 and LC3; and induced autophagy machinery, which highlights the relationship between treatment cytotoxicity, autophagy inhibition, and their effects on chemoresistance (see Table 2) (Li et al. 2016d). Chen and colleagues observed that overexpression of miR-155 during chemotherapy induced autophagy leading to mediate chemoresistance in OS (Chen et al. 2014a). Wang et al. noted that upregulation of miR-22 in OS cells (MG-63) increased *sensitivity to cisplatin* mediated via negative regulation of autophagy by down-expression of MTDH (Wang et al. 2019c). Alongside, miR-193b/FEN1 axis ameliorated the epirubicin sensitivity of OS cells through autophagy induction (Dong et al. 2019). miR-375 could be another target to sensitize OS to cisplatin as its overexpression in cisplatin-resistant OS models delayed tumor progression and autophagy via targeting ATG2B (Gao et al. 2020a). Qased et al. investigated the role of miR-18a in autophagy process in HCT116 (human CRC cells). To do so, HCT116 cells were irradiated, and the expression levels of miR-18a were subsequently measured in the cells. The results showed that the radiation led to increased expression level of miR-18a and enhanced autophagy induction (Qased et al. 2013). Li et al. showed that the expression levels of miR-22 are enhanced during chemotherapy and target HMGB1, which results in inhibition of HMGB1-induced autophagy (see Table 2) (Li et al. 2014a).

He et al. reported that miR-152 plays an important role in autophagy regulation and drug resistance in ovarian cancer (OC) (He et al. 2015b). They showed that miR-152 was significantly downregulated in cisplatin-resistant cells. It has been reported that overexpression miR-152 leads to induction of apoptosis in cisplatin-resistant cancer cells as well as a decrease of cisplatin-induced autophagy. In this in vitro study, it was documented that ATG14 downregulation by EGR1-miR-152 sensitizes ovarian cancer cells to cisplatin-induced apoptosis through inhibiting cyto-protective autophagy (He et al. 2015b). Vescarelli et al. verified that miR-200c considerably sensitized chemoresistant OC cells to olaparib via regulating NRP1 (Vescarelli et al. 2020). In addition, miR-29c-3p overexpression inhibited autophagy which in turn reversed cisplatin resistance of OC by downregulation of FOXP1/ATG14 pathway (Hu et al. 2020). Esfandyari et al. (2021) demonstrated that miR-143 overexpression in cervical cancer cells (CaSki cells) could increase cisplatin sensitivity of treated cells via induction of apoptosis and autophagy (Esfandyari et al. 2021). Tamoxifen (TAM) and fulvestrant (FUL) are considered as effective drugs for patients with ER-positive breast cancer, but the rate of response to these therapies is limited because of various barriers, such as endocrine resistance. In this regard, Yu and colleagues found that miR-214 enhanced breast cancer cells sensitivity to TAM and FUL through autophagy inhibition (Yu et al. 2015a). In a comparable study on breast cancer, Soni et al. identified that miR-489 enhanced sensitivity to doxorubicin (Dox) as a result of autophagy inhibition dependent to LAPTM4B downregulation (Soni et al. 2018b).

Xu et al. reported that miR-199a-5p downregulation induced by cisplatin enhances drug resistance through activating autophagy in HCC (Xu et al. 2012). Soni et al. evaluated the role of miR-155-5p on Adriamycin (ADR)-resistant liver carcinoma cells (HepG2/ADR), and their findings indicated the effects of miR-155-5p as sensitizer of ADR, activator of apoptosis, and inhibitor of autophagy via attaching to ATG5 3UTR (Soni et al. 2018b). Also, higher expression of miR-541 inhibited the autophagy in HCC cells by targeting ATG2A and RAB1B leading to promising response to sorafenib (Xu et al. 2020a). In another study, it was revealed that upregulated miR-142-3p increased sensitivity of HCC cells to sorafenib by targeting ATG5 and ATG16L1 as negative modulators of autophagy (Zhang et al. 2018b). Similar findings have been reported for miR-101/RAB5A/STMN1/ATG4D axis in the HCC cells (HepG2) which improved the response to cisplatin due to inhibition of autophagy mechanism (Xu et al. 2013). Consistently, Ren et al. demonstrated that miR-125b/EVA1A axis-mediated autophagy reversed resistance of HCC cells to oxaliplatin (Wei-Wei et al. 2018).

Chemoresistance of nasopharyngeal carcinoma (NPC) has been investigated by Zhao et al. (2020) study in which they verified that miR-1278 expression was decreased in NPC tissues associated with worse chemotherapy response. Nonetheless, upregulation of miR-1278 dramatically raised anticancer effects of cisplatin in NPC cells together with reduced autophagy via inhibiting ATG2B (Zhao et al. 2020).

Yang et al. (2021) showed that miR-136-5p upregulation not only had negative effects on malignant progression of laryngeal squamous cell carcinoma (LSCC) and hypopharyngeal squamous cell carcinoma (HPSCC) cells but also reversed cisplatin resistance in the tested cells via inactivation of ROCK1 Akt/mTOR axis (Yang et al. 2021).

Recently, Xi et al. explored the lncRNA MALAT1/miR-30b/ATG5 axis in cisplatin resistance of GC cells (AGS/CDDP and HGC-27/CDDP) and documented that miR-30b attenuated cisplatin resistance by reduced expression of not only MALAT1-activated autophagy but also ATG5 (see Table 2) (Xi et al. 2019). In another study, Chen et al. (2020a) identified that miR-30a could sensitize gastrointestinal stromal tumors (GISTs) cells to imatinib (IM) via silence of Beclin-1-regulated autophagy (Chen et al. 2020a). Also, He et al. (2020a) discovered that miR-153-5p upregulation in oxaliplatin (L-OHP)-resistant CRC cells could overcome L-OHP resistance via silencing Bcl-2-induced autophagy (He et al. 2020a). Furthermore, Liu et al. (2020f) indicated that lncRNA NEAT1 upregulation sponged miR-34a in CRC. Additionally, NEAT1 inhibition significantly slowed down CRC tumorigenesis and elevated sensitivity of

cells to 5-fluorouracil (5-FU). miR-34a overexpression also showed comparable trends with NEAT1 inhibition via binding to autophagy components (HMGB1, ATG4B, and ATG9A) (Liu et al. 2020f). The role of miRNA in chemoresistance of pancreatic cancer (PC) cells was evaluated by miR-137 overexpression in PANC-1 cell lines. The results indicated that miR-137 chemo-sensitized the cells to Dox via ATG5-triggered autophagy (Wang et al. 2019d).

The main hurdle for the proper treatment of multiple myeloma (MM) is still chemoresistance. Of note, the cross talk between miRNA dysregulation and autophagy illustrated that miR-221/222 could suppress dexamethasone (Dex) sensitivity in MM cells via inhibition of autophagy associated with ATG12/p27-mTOR axis (Xu et al. 2019b). In various studies, drug resistance in non-small-cell lung cancer (NSCLC) has been investigated. In a research conducted by Hua et al., overexpression of miR-1 reversed cisplatin resistance in NSCLC by suppression of ATG3-regulated autophagy (Hua et al. 2018). Therefore, miRNAs have regulatory roles in chemoresistance due to their effects on autophagy induction. These mediators should be further investigated in numerous in vivo and in vitro studies to find the molecular mechanisms related to resistance. Table 2 lists the effects of autophagy-related miRNAs on some human cancer chemotherapy.

3   Exosome and Autophagy

Exosomes, membrane-coated vesicles with 30–120 nm size, are released by several cells, such as lymphocytes, platelets, epithelial cells, mast cells, dendritic cells, neurons, and endothelial cells (Théry et al. 2002; Hashemipour et al. 2021). Exosome has main roles in biological events, including inflammation, tumorigenesis, metastasis, and response to therapy (Kharaziha et al. 2012). Various researches have demonstrated that exosomes can also be considered as diagnostic means and targeted drug delivery system. It has been identified that almost all biological body fluids, including blood, serum, saliva, milk, amniotic fluid, semen, breast milk, and urine contain exosomes (Keller et al. 2011; Lässer 2015).

Exosomes carry diverse unique molecular cargos, including lipids, proteins, and nucleic acid fragment. Some of the proteins are involved in assembly, movement, and organization of exosomes (e.g., annexins, actins, tumor susceptibility gene 101, vesicle-associated membrane protein 8, and fibronectin) and observed in the structures of exosome. Furthermore, a cluster of proteins known as exosome surface markers, such as CD9, CD63, CD81, and CD82, are useful for the detection of exosomes (Zhao et al. 2015b; Barclay et al. 2017). Mounting evidence has established that exosomes have a wide range of roles in human pathological and physiological processes. Since exosomes deliver their constituents into recipient cells, they are able to play a prominent role in cell signaling and local/distant cell-to-cell communication (Lakkaraju and Rodriguez-Boulan 2008; Van Niel et al. 2006). These data demonstrated that exosomal molecular constituents can represent disease conditions (Feng et al. 2013). The idea of the RNAs presence in exosomes has attracted great attention in the research of exosomal RNAs, especially miRNAs as potential diagnostic biomarkers (Taylor and Gercel-Taylor 2008). Recent experiments have demonstrated that exosomal miRNAs are resistant to RNase degradation and thus remain stable in circulating plasma and serum. On the other hand, they are easily evaluated, are minimally invasive, and have high sensitivity and specificity. This evidence indicates that exosomal miRNAs are ideal biomarkers for early clinical diagnostic applications (Lin et al. 2015; Li et al. 2014b).

As cited above, the autophagic process contains five key stages including initiation, nucleation, elongation and maturation, fusion, and degradation (Li et al. 2020c). mTOR acts as the regulator of the initiation stage, and its activation is associated with prohibition of autophagy, whereas its inactivation is able to induce autophagy. It has been revealed that mTOR and the ULK complex (consist of ULK1, FIP200, and autophagy-related protein 13 [Atg13]) is

inactivated and activated, respectively, in stress situations. Beclin-1, an essential component for autophagosome formation, in combination with Vps34 and Atg14L produces a complex, which is necessary for induction autophagy nucleation (Liang et al. 1999; Levine et al. 2015; He and Klionsky 2009; Kihara et al. 2001). In the elongation along with maturation stage, two ubiquitin-like conjugation systems are warranted to facilitate autophagosome membrane expansion. The first system involves the microtubule-associated protein light chain 3 (LC3)-phosphatidylethanolamine (PE) complex. LC3 is cleaved by Atg4 at its C terminal to produce intracellular LC3-I, which is conjugated with PE in the ubiquitin-like reactions of Atg7 and Atg3. The lipid form of LC3 (LC3-II) is attached to the autophagosome membrane (Yu et al. 2015b). The second system involves the Atg12-Atg5-Atg16 complex, in which Atg12 is conjugated with Atg5 via ubiquitin-like reactions of Atg7 and Atg10. The Atg12-Atg5 conjugate interacts noncovalently with Atg16 to form a large complex. While lysosomes bind to autophagosomes to form autolysosomes in the fusion stage, cargo within autolysosomes will be degraded in the degradation stage. Autophagy is tightly modulated to keep homeostasis. Following autophagy initiation, lots of Atg proteins collaborate to manage the next stages of autophagy. It is yet not clear that autophagy conveys protective or detrimental effects in diseases (Saha et al. 2018; Xiong 2015). For example, lack of autophagy is associated with excess amount of tau and synuclein proteins, which induces neurodegenerative disorders. Evidences are in support of the fact that autophagy has a dual effects on cancer cells and initially acts as a tumor inhibitor; however, later it defends tumor cells against the immune system's attacks (Sharma et al. 2021; Hassanpour et al. 2020). Likely, it has been demonstrated that autophagy regulates cardiac and hepatic disorders positively and negatively, respectively. Thus, the control of autophagy via exosomes can have various positive and negative effects on a variety of diseases (Xing et al. 2021).

The role of exosomes in cellular stresses has been evidenced. However, some researches indicate that the interaction between exosomes and autophagy machinery may preserve intracellular protein and homeostasis (Baixauli et al. 2014). In addition, autophagy induction due to nutrient deprivation leads to inhibited exosome secretion (Fader and Colombo 2009). There are some exosomal proteins markers related to autophagy mechanism. Dias et al. showed that PRNP (prion protein gene) is essential to promote the release of exosomes regulating CAV1/caveolin-1-suppressed autophagy (Dias et al. 2016). Moreover, significant levels of autophagy proteins, including WIPI2, LC3, NBR1, and p62, are present in exosomal fractions secreted by apilimod-treated cells (Hessvik et al. 2016). Importantly, different exosomal and autophagic proteins can be applied as potential biomarkers regarding the type of cancer (Salimi et al. 2020).

Also, the role of exosomal miRNAs in autophagy regulation has been demonstrated by various investigations. Yang et al. reported that high serum levels of exosomal miR-30a were observed in AMI patients. Also, they observed that inhibition of miR-30a increased the expression level of Beclin-1, Atg12, and LC3-II/LC3-I known as the regulators of core autophagy machinery and contributed to preserve the hypoxia-induced autophagy (Yang et al. 2016b). Liu and colleagues conducted a study on AMI rat model and in vitro model of hypoxic H9c2 cells to investigate the cardioprotective role of miR-93-5p-encapsulating exosomes released from adipose-derived stromal cells (ADSCs) in ischemia-induced cardiac damage. They found overexpression of inflammatory cytokines as well as miR-93-5p in both patients and rat models with AMI. In addition, the comparison of the protective effects of exosomes on infarction-induced cardiac damage revealed that exosomal treatment containing miR-93-5p derived from ADSCs caused more protection than simple exosomes (Liu et al. 2018b). Also, Li et al. highlighted the impact of bone marrow-derived mesenchymal stem cells (BMMSCs)-derived exosomes enriched in miR-29c on negative regulation of autophagy in cardiac ischemia/reperfusion (I/R) injury through PTEN/Akt/mTOR pathway (Li et al. 2020d). Similarly, human umbilical cord mesenchymal stem cells-exosome (hucMSC-ex) abolished coxsackievirus B3 (CVB3)-activated myocarditis due to upregulation of autophagy function mediated by AMPK/mTOR axis and reduction of cardiomyocyte death (Gu et al. 2020). Santoso et al. (2020) demonstrated that induced pluripotent stem cells and their differentiated cardiomyocyte-delivered exosome (iCM-Ex) treatment had cardioprotective effects against post-MI via improvement of autophagy machinery in vivo and in vitro (Santoso et al. 2020). Besides, Li and colleagues isolated exosomes released by human aortic smooth muscle cells and identified that isolated exosomes contained miR-221/222. They found that miR-221/222 could target 3′UTR of PTEN. Also, overexpression of miR-221/222 downregulated the expression of ATG5, LC3-II and Beclin-1, suggestive of the inhibitor role of exosomal miR-221/222 in autophagy process (Li et al. 2016e).

Yuwen et al. (2017) reported that the expression level of exosomal miR-146a-5p in NSCLC is correlated with chemosensitivity and chemotherapy response to cisplatin. Low levels of miR-146a-5p in serum exosomes were detected in advanced NSCLC patients. In both NSCLC cells and exosomes, the expression level of miR-146a-5p was gradually decreased due to chemoresistance to cisplatin. In addition, miR-146a-5p also inhibited the autophagy through targeting Atg12 (Yuwen et al. 2017). Wang et al. investigated the role of tumor environment such as acute shear stress (ASS) in NSCLC invasion. Their data indicated that ASS activated cell death by exerting the secretion of autophagy and exosome components via SIRT2/TFEB axis (Wang et al. 2020a). In the severe lung injury and respiratory deficit, Wei et al. illustrated that huMSC-ex-delivered miR-377-3p could improve acute lung injury (ALI) induced by lipopolysaccharide through targeting RPTOR followed by autophagy activation (Wei et al. 2020).

Exosomal miR-1910-3p derived from breast cancer cell attenuated metastasis, growth, and autophagy via MTMR3 suppression and NF-κB and wnt/β-catenin signaling induction (Wang et al. 2020b). Since exosomes loaded with miR-1910-3p increased autophagy and breast cancer development via silencing MTMR3 and inducing NF-κB and wnt/β-catenin pathway, it could be considered as a diagnostic biomarker for breast cancer (Wang et al. 2020b). Moreover, hucMSCs-ex transferring miR-224-5p could hamper cellular apoptosis and mount proliferation and autophagy in breast cancer via silence of HOXA5 (Wang et al. 2021a). In another interesting study, Han et al. (2020), showed that exosome-shuttled miR-567 repressed autophagy and chemo-sensitized breast cancer cells to trastuzumab via interacting with ATG5 (Han et al. 2020). Additionally, the anticancer effects of gemcitabine in breast cancer (luminal-b type) could be improved using exosome-overexpressed small interfering RNA (siRNA) MTA1, which suppressed autophagy and EMT/HIF-α pathway (Li et al. 2020e).

In the field of thyroid research, papillary thyroid cancer (PTC) cell exosome-delivered SNHG9 lncRNA could prevent autophagy flux and upregulate apoptosis of human normal thyroid epithelial cell line (Nthy-ori-3 cell) mediated by YBOX3/P21 pathway (Wen et al. 2021).

In cisplatin-resistant GC, Yao et al. manifested that the levels of exosomal circ-PVT1 and miR-30a-5p were respectively upregulated and downregulated, while the silence of Circ-PVT1 reversed cisplatin resistance via reducing autophagy alongside with increasing apoptosis through miR-30a-5p/YAP1 axis (Yao et al. 2021). Comincini et al. evaluated the expression levels of exosomal miR-17 and miR-30a to diagnose celiac disease and discovered that miR-17- and miR-30a-regulated ATG7 and BECN1 known as two key executor of autophagy (Comincini et al. 2017).

Beclin-1 contains three main domains including coiled coil (CCD), evolutionarily conserved (ECD), and Bcl-2-homology-3 (BH3). Several proteins through binding to the various domains of Beclin-1 and forming different complexes regulate autophagy activity (Wirawan et al. 2012). Beclin-1 is encoded by BECN1, which is located on chromosome 17q21 and was shown to be targeted via miR-30a (Zhu et al. 2009). Exosomal miR-30a is capable of prohibiting autophagy via targeting the Beclin-1 pathway and maintains a mandatory role in liver fibrosis and MI. It was revealed by Yang et al. (2016b) that hypoxic cardiomyocytes prohibit autophagy through secreting miR-30a and, thereby, cause cardiomyocyte damage. So, it can be expected that autophagy level can be increased by targeting miR-30a, and, thereby, cardiomyocyte damage will be decreased. In contrast to findings of Yang et al., Zhang et al. found out that epigallocatechin gallate acts as a protective agent for MI through overexpression of exosomal miR-30a and, consequently, prohibiting autophagy and apoptosis (Zhang et al. 2020b). An animal study that was conducted by Xu et al. (2019c) also demonstrated that exosomal miR-30a through prohibiting autophagy decreased the level of cardiomyocyte apoptosis in rats with MI/reperfusion injury. Autophagy becomes active throughout hypoxia and displays protective effects by modifying cell survival. Nevertheless, as myocardial hypoxia continues, excessive autophagy occurs, which causes accumulation of a quite amount of toxic components and, as a consequence, cell death. In Yang et al.'s study, autophagy was inhibited by exosomal miR-30a; hence, there was a lack of protective autophagy in cardiomyocytes, which contributed to cardiomyocyte apoptosis. However, in other studies performed by Zhang and Xu, excessive autophagy was the reason behind cardiomyocytes damage. Exosomal miR-30a is able to decrease the level of cardiomyocyte apoptosis via prohibiting excessive autophagy. Moreover, it has been unveiled that excessive autophagy can induce liver fibrosis. It was shown that in a hepatic fibrosis model that was establish by Chen et al. (2017c), the expression level of exosomal miR-30a, secreting via hepatic stellate cells, was decreased. The upregulation of miR-30a may have the capacity to improve liver fibrosis through prohibiting autophagy mediated by the Beclin-1 pathway.

Li et al. (2021) revealed that osteosarcoma (OS)-secreted exosomal lncRNA OIP5-AS1 regulated autophagy and angiogenesis via reduction of miR-153 and enhancement of ATG5 expressions (Li et al. 2021). In spite of pro-tumor effects of hBMSC-derived exosomes on OS progression via autophagy elevation, knockdown of

ATG5 in OS cells attenuated oncogenic effects of hBMSC exosomes (Huang et al. 2020).

Reportedly, in osteoarthritis (OA) mice model, intra-articular administration of OA exosomes loaded with ATF4 had protective effects against chondrocyte apoptosis via activating autophagy (Wang et al. 2021b). Furthermore, in IVDD model, it was confirmed that normal cartilage end plate stem cell-derived exosomes (N-Exos) had a better therapeutic impact on stopping nucleus pulposus cell apoptosis and delay in IVDD progression in comparison with degenerated cartilage end plate stem cell-derived exosomes (D-Exos) via induction of PI3K/AKT/ autophagy pathway (Luo et al. 2021). Also, the effects of human umbilical cord mesenchymal stem cell-derived exosomes (hucMSC-ex) on tissue damages make them as a promising tool in the regenerative medicine. Based, Jia et al. (2018) discovered that hucMSC-ex enriched with 14-3-3ζ reversed cisplatin-activated nephrotoxicity via interaction with ATG16L-induced autophagy (Jia et al. 2018).

It has been made clear that the levels of anti-inflammatory cytokines and miR-30d-5p are reduced following acute ischemic stroke (AIS). Jiang et al. recognized that exosomes derived from miR-30d-5p-overexpressing ADSCs could overcome autophagy-induced cerebral damage via increasing polarization of M2 microglial/macrophage (Jiang et al. 2018). Chen et al. (2020b) noticed that exosome-delivered circSHOC2 released from ischemic-preconditioned astrocyte (IPAS) potentiated neuronal protective effects against ischemic cerebral injury by affecting autophagy through the miR-7670-3p/SIRT1 axis (Chen et al. 2020b). Recently, Pei et al. verified that astrocyte-released exosomes (AS-Exo) suppressed neuronal autophagy and alleviated neuronal injury and apoptosis in an in vitro model of ischemic injury via overexpression of miR-190b and downregulation of Atg7 (Pei et al. 2020). It has been observed that hucMSC-ex could breakdown blood-brain barrier (BBB) and target substantia nigra leading to protection of dopaminergic neurons via activation of autophagy in a PD model (Chen et al. 2020c). Ma et al. (2021) analyzed the amounts of lncRNA

LINC00470 in glioma-derived exosomes from patients and concluded that overexpressed LINC00470 could abrogate autophagy and raise glioma cells proliferation via binding to miR-580-3p which in turn inactivated WEE1 and induced the PI3K/AKT/mTOR pathway (Ma et al. 2021). Programmed death-ligand 1-containing exosomes (PD-L1-ex) derived from glioblastoma stem cell (GSC) enhanced autophagy and reduced apoptosis via AMPK/ULK1 pathway cascade resulted in enhanced resistance to TMZ, while knockdown of PD-L1 reversed these effects (Zheng et al. 2021). There is line of evidence shown that Schwann cells (SCs) have regenerative role following peripheral nerve injury. In this context, Yin et al. discovered that ADSC-Exos loaded by miR-26b blocked SC autophagy and improved the myelin sheath regeneration in the sciatic nerve injury model via targeting Kpna2 (Yin et al. 2021). Due to the improvement of inflammation secondary to spinal cord injury (SCI) via anti-inflammatory effects of peripheral macrophages (PMs), Zhang et al. represented that PM-derived exosomes (PM-Exos) could promote spinal cord recovery via enhancement of microglial autophagy and anti-inflammatory microglia polarization mediated through PI3K/AKT/mTOR pathway (Zhang et al. 2021b).

In type 2 diabetes mellitus (T2DM) rats, He et al. uncovered that hucMSC-ex promoted hepatic lipid and glucose metabolism potentially by enhancing the autophagosomes via AMPK pathway (He et al. 2020b). Likewise, Zhang et al. reported that liver I-/R-induced injury could be alleviated by huMSC-ex-transmitted miR-20a via regulating apoptotic and autophagic genes including caspase-3, P62, mTOR, and LC3-II (Zhang et al. 2020c). Further, Zhu et al. (2020) verified that ADSC exosome carrying mmu_circ_0000623 inhibited liver fibrosis through autophagy induction (Zhu et al. 2020). Since liver fibrosis can be driven by HSC activation, Wang et al. (Wang et al. 2020c) displayed that natural killer (NK) cell-derived exosome (NK-Exo) attenuated HSC activation via inhibiting TGF-β1 mechanistically through overexpression of miR-223 and inhibition of ATG7-induced autophagy (Wang et al. 2020c).

All in all, studies have recently demonstrated that autophagy has regulatory properties in exosomal production and its release. The link between Atg5 and V1V0-ATPase and their role in induction of exosome production has been documented by Chen et al. (2018). They found that cells with Atg5 and Atg16L1 deficiency exhibit reduced exosome production, but it's not dependent on Atg7 and canonical autophagy. It has been shown that Atg5 affects the production of exosomes by reducing the acidifying of endosomes and disrupting the acidification of V1V0-ATPase. Because of the role of autophagy and exosomes in metastasis, Atg5 is able to induce invasion and metastasis (Guo et al. 2017c). Abdulrahman et al. evaluated the role of autophagy in exosome production and processing. They found that the induction of autophagy by rapamycin, mTOR inhibitor, suppressed the release of exosomal prions; however, the inhibition of autophagy resulted in increased release of both exosomes and prions (Abdulrahman et al. 2018). Totally, further studies were collected in Table 3.

Table 3 Exosome and autophagy

Type of cargo	Exosome source	Effect on autophagy	Type of disease	Note	Ref
miR-146a-5p	Serum	Inhibition	Non-small-cell lung cancer (NSCLC)	miR-146a-5p upregulated and decrease level of Atg12	Yuwen et al. (2017)
miR-93-5p	Adipose-derived stromal cells (ADSCs)	Inhibition	Acute myocardial infarction (AMI)		Liu et al. (2018b)
miR-30d-5p	Adipose-derived stromal cells (ADSCs)	Induction	Acute ischemic stroke (AIS)	Enhancement of M2 microglial/macrophage polarization and reduce of M1 microglial/macrophage polarization. Inhibition ischemia-induced neuronal damage via decreasing of TNF-α, IL-6, and iNOS secretion from M1 microglial cells. Downregulation of Beclin-1 and Atg5. Induction of expression anti-inflammatory cytokines IL-4 and IL-10 from M2 microglial cells	Jiang et al. (2018)
miR-181-5p	Adipose-derived mesenchymal stem cells (ADSCs)	Induction	Liver fibrosis	miR181-5p-ADSC block of STAT3/Bcl-2/Beclin-1-dependent signaling pathway and decrease liver fibrosis	Qu et al. (2017)
miR-30a	Serum H9c2 cell	Inhibition	Acute myocardial infarction (AMI)	Hypoxia promotes expression of miR-30a in cardiomyocytes and increases apoptosis and elevates Atg12 and Beclin-1 protein levels	Yang et al. (2016b)
miR-17	T98G cells	Induction	Celiac disease (CD)	miR-17 downregulated and increase of expression level of ATG7	Comincini et al. (2017)
miR-30a	T98G cells	Induction	Celiac disease (CD)	miR-30a downregulated and increase of expression level of BECN1	Comincini et al. (2017)
miR-221/222	Human aortic smooth muscle cells (HAoSMCs)	Inhibition	–	miR-221/222 upregulation in HUVECs, reduction of PTEN, LC3-II, ATG5, and Beclin-1protein levels. Increase of SQSTM1/p62 level and Akt signaling pathway	Li et al. (2016e)

(continued)

Table 3 (continued)

Type of cargo	Exosome source	Effect on autophagy	Type of disease	Note	Ref
MSC exosome (miR-125b)	Neonatal mice cardiomyocytes (NMCMs) cell	Inhibition	Myocardial infarction (MI)	Decrease of p53/Bnip3 signaling pathway and save myocardial from death	Monaco et al. (2017)
HucMSC exosome (14-3-3ζ)	NRK-52E cells	Induction	Acute kidney injury (AKI)	HucMSC exosome-delivered 14-3-3ζ attached the ATG16L protein and induced autophagosome formation and as a result elevated cisplatin resistance and cell proliferation and reduced apoptosis	Jia et al. (2018)
Exosomes derived from gefitinib-treated (Exo-GF)	PC9 cells	Induction	Non-small-cell lung cancer (NSCLC)	Enhancement cisplatin resistance, overexpression of Bcl-2 and LC3-II protein levels, decrease of Bax and p62 protein levels	Li et al. (2016f)
NA	H9C2 cells	Induction	Myocardial ischemia-reperfusion injury (MIRI)	Exosomes derived from mesenchymal Stem cells enhance cardiomyocyte autophagy, inhibit cell apoptosis and ROS production through H2O2, promote AMPK pathway and decrease Akt and mTOR pathways	Liu et al. (2017d)
HucMSC exosomes	NRK-52E cells	Induction	–	HucMSC exosomes block cisplatin-induced mitochondrial apoptosis and secretion of inflammatory cytokines, decrease of mTOR and NF-KB, increase levels of ATG5 and ATG7	Wang et al. (2017c)

4 Inhibition or Stimulation of Autophagy by the Virus

Viruses are known as intracellular parasites that are highly dependent on the host for their cell cycle. Hence, after entrance, they reprogram the target host cell to meet their basic needs (Fehr and Yu 2013; Bagga and Bouchard 2014). As we cited before, autophagy has a crucial role in preserving cellular hemostasis by participating in different physiological processes, such as, but not limited to, cell differentiation and development, starvation, and degradation of abnormal products. Additionally, it has been shown that autophagy is produced in response to stress conditions such as infection with viral viruses (Senft and Ze'ev 2015; Mizushima and Levine 2010). Also, in response to viral infections, autophagy becomes active by innate immune system to degrade viruses (Deretic et al. 2013).

Additionally, autophagy also takes part in activation of adaptive immune system by accelerating antigen processing (Paludan et al. 2005; Romao et al. 2013). Xenophagy is a type of selective lysosomal degradation pathway that is vital for eliminating pathogens especially bacteria and viruses (Levine 2005). Although autophagosomes potentially are detrimental for invading viruses, several viruses have shown to be able to convert the autophagosome to their home during replication. The autophagosome provides a membrane-bound, protected site to produce their progeny, where their metabolites can be utilized as source of energy for viral replication. Another unique class of autophagy, called lipophagy, targets intracellular lipid droplets, and this process can also be captured by viruses. Lipid droplets are considered as the optimal source for viral assembly since the viruses have the potential to stimulate lipophagy provide the high values of

ATP needed for viral replication (Choi et al. 2018; Heaton and Randall 2011). Taken together, according to recent findings, viruses are developing new strategies to fight or use autophagy to facilitate their replication. Herein, we sought to provide a brief review on how autophagy fights against viral viruses and, thereafter, how the viruses disrupt the autophagic pathway to escape form immune system reactions and prompt their replication.

Recently, several studies have reported that the aim of virus interference with host cell autophagy is to promote the life cycle of virus and avoid detection by the host immune system. The diverse set of viruses are able to dysregulated autophagy machinery (Glick et al. 2010; Jackson 2015). The viral proteins directly or indirectly interact with autophagy components leading to enhance or block autophagy (Mack and Munger 2012). For instance, coronavirus papain-like protease, termed PLP2, induces autophagy via interacting with Beclin-1 (Chen et al. 2014b). Although some viral proteins inhibit the autophagy via interaction with Beclin-1, HIV-Nef and HSV-1 ICP34.5 proteins are capable of inhibiting autophagy-dependent Beclin-1 (Orvedahl et al. 2007; Kyei et al. 2009a; Campbell et al. 2015a). Beclin-1 has Bcl-2 homology 3 (BH3) domain and, through this domain, interacts with anti-apoptotic Bcl-2 family members (Oberstein et al. 2007). This interaction inhibits Beclin-1 assembly to the pre-autophagosomal structure, thereby preventing autophagy (Liang et al. 1998).

The importance of apoptosis and Bcl-2 proteins in immune system regulation and responses to stresses has provided evolutionary pressures on viruses to acquire the genes encoding pro-survival Bcl-2 proteins (Neumann et al. 2015). Large DNA viruses, such as γ-herpesviruses 68 (γ-HV68), adenovirus, Epstein-Barr virus (EBV), and Kaposi's sarcoma-associated herpesvirus (KSAH), mimic the pro-survival Bcl-2 proteins leading to hijack the intrinsic pathway of apoptosis for their purposes (Kvansakul et al. 2017). Liang et al. (2008) reported that murine gamma-herpesvirus 68 (MγHV68) Bcl-2 protected virus-infected cells against apoptosis, also repressed autophagy through its direct binding to Beclin-1 (Liang et al.

2008). In addition to suppressing autophagy by the vBcl2/Beclin-1 complex, KSHV also inhibits this process by viral homolog of cellular FLICE-like inhibitor protein (v-FLIP). Both KSHV v-FLIP and cellular FLIP directly interact with the autophagy-protein ATG3 in competition with LC3 protein. It has been demonstrated that, to suppress autophagic programmed cell death, this interacting ability of KSHV v-FLIP is required (Mack and Munger 2012; Irmler et al. 1997; Thome et al. 1997; Lee et al. 2009). The biochemical evidences show interaction of different HCV and HBV proteins with autophagy machinery components. Nonstructural protein 3 (NS3) of HCV was found to co-localize and associate with the immunity-associated GTPase (IRG) family M that it known autophagy pathway regulator in response to the bacterial infection (Grégoire et al. 2011a; Singh et al. 2006). Core protein of HCV activates autophagy through EIF2AK3and ATF6 UP pathway and/or upregulating Beclin-1 expression (Wang et al. 2014a; Liu et al. 2015b). Moreover, this core protein represses apoptosis and enhances autophagy in hepatocytes through upregulating Beclin-1 (Liu et al. 2015b). Small surface proteins of HBV interact with LC3 and HBV-HBx protein interacts with VPS34 (Sir and Tian 2010; Li et al. 2011a). Sir and colleagues reported that HBx through binding to phosphatidylinositol 3-kinase class III, a critical enzyme in the initiation of autophagy, leads to enhanced activity of this enzyme and thus activates the early autophagic pathway (Sir and Tian 2010).

Espert et al. have shown that autophagy-dependent cell death is activated after binding of HIV envelope glycoprotein to CXCR4 on T cells (Espert et al. 2006; Espert et al. 2007). Bcl-2-associated athanogene 3 (BAG3) is known as a pro-autophagic and anti-apoptotic factor in many normal and neoplastic cells (Rubinstein and Kimchi 2012; Behl 2011; Rosati et al. 2011). Bruno and colleagues reported that transfection of HIV-1 trans-activator (Tat) protein into glioblastoma cells results in increasing BAG3 levels leading to stimulate the autophagic pathway, while silencing of BAG3 results in disrupted balance between autophagy and apoptosis (Bruno et al. 2014). As mentioned earlier, autophagy process involves the formation and maturation of autophagosomes.

Recent studies have showed that interferon-γ (IFN-γ) activates autophagosomes to participate in immunity defense (Deretic 2006). HIV-Tat protein suppresses the formation of autophagosome. In other words, this protein disrupts the IFN-γ signaling pathway through repression of STAT1 phosphorylation and, consequently, inhibits the IFN-γ-induced autophagy (Li et al. 2011b). Additionally, influenza matrix protein 2 and human parainfluenza virus Type 3 phosphoprotein interrupt the maturation of autophagy through blocking autophagosome degradation (Ding et al. 2014; Gannagé et al. 2009).

One of the most important regulators of autophagy is the mammalian target of rapamycin (mTOR), which moderates the balance between autophagy and growth in response to environmental stress and physiological conditions (Cuyàs et al. 2014). Kinase mTOR is the downstream target of PI3K-Akt signaling pathway, which is activated by growth factor receptors and neurotropism as well as promotes cell differentiation, growth, and survival and also reduces apoptosis (Manning and Cantley 2007; Brunet et al. 2001; Hanada et al. 2004). It has been observed that suppression and activation of PI3K/AKT/mTOR pathway lead to promote and inhibit autophagy, respectively (Heras-Sandoval et al. 2014). Surviladze et al. reported that contamination of HaCaT cells with HPV-16 pseudovirions activates thePI3K/Akt/mTOR signaling pathway leading to autophagy inhibition (Surviladze et al. 2013). KSHV-K1, a viral protein, activates thePI3K/Akt/mTOR signaling pathway in endothelial cells and B lymphocytes (Mack and Munger 2012; Tomlinson and Damania 2004; Wang and Damania 2008). Also, HBV induces autophagy in HepG2 cells transfected with HBx through regulating the PI3K/Akt/mTOR pathway (Wang et al. 2013a). It is believed that autophagy plays an important role in the regulation of cancer progression and development and in determining of tumor responses to anticancer treatments. It has been observed that oncolytic viruses (OVs) interact with autophagy in infected tumors to ensure their own survival and replication advantage (Jiang et al. 2011). While an increasing number

of OVs are reported to induce autophagy in infected tumors, some OVs choose to subvert or evade it (Zhang et al. 2006; Moloughney et al. 2011). For instance, Rodriguez-Rocha et al. showed that adenoviruses induce autophagy to promote virus replication and oncolysis in lung cancer A549 and H1299 cells (Rodriguez-Rocha et al. 2011). This concept suggests an insightful indication to OV therapy to improve the quality of life and survival of patients with cancer. Therefore, viruses and viral products can effect on the stimulation or inhibition of autophagy. Searching for using these agents to control stress conditions should be more focused.

HCMV belongs to β-herpesvirus family, which has shown to be transmissible via different body fluids. HCMV is known as one of the biggest viruses since its genome contains of 236 kilobases (Plotkin and Boppana 2019). Albeit it has been demonstrated that primary infection mainly is asymptomatic, the congenital form of the virus can be accompanied by several complications including, but not limited to, disabilities and death. HCMV was shown to have the potential to favor cancer through transformation of infected cells when infecting normal tissues by regulating several signaling pathways (Herbein 2018). The virus modulates autophagy in a dual fashion (Joseph et al. 2017; Nahand et al. 2021). At early phases of infection, it contributes to autophagic vesicle formation. On the contrary, later, it inhibits autophagy via producing some proteins (Chaumorcel et al. 2012). By far, two viral proteins, namely, TRS1 and TRS2, that participate in autophagy prohibition in cooperation with Becline-1 have been explored. It has been demonstrated that simultaneous expression of TRS1 and IRS1 is necessary for prohibition of autophagy in virus infection (Mouna et al. 2016). Recently, viral components with the ability of regulating latency and lytic reactivation, especially those in the uLb' gene region, have been at the center of focus. These viral components are capable of limiting virus replication via moderating immune system response and viral latency through expressing quite a few virus proteins. For instance, a viral protein, namely, UL138, through autophagy machinery, can modulate adaptive immunity of fibroblast when it presents to MHC-1 (Tey and Khanna 2012; Mlera et al. 2020). However, recent evidence clarified that prohibition of autophagy is associated with extreme CD8 + T-cell response because of the internalization of molecules in MHC-I (Loi et al. 2016). Expressing viral proteins derived from HMCV genes 1 and 2 (IE1 and IE2) is essential for immunomodulation and reactivation of host cell virus (Suares et al. 2021; Reddehase and Lemmermann 2019). IE2 is able to modulate gene expression by interacting with UL84 and itself along with a number of cell transcription factors. IE2 protein has a mandatory role in synthesis of viral DNA and was shown to have the potential to counteract host responses (Li et al. 2020f; Møller et al. 2018). Lately, it has been shown that upregulation of IE2 can contribute to autophagy in cells infected with the virus (Zhang et al. 2021c). Briefly, it has been found out that when a cell is infected with HMC, viral proteins result in autophagosomal vesicle formation. Later, the proteins prohibit vesicle-to-lysosome binding, which leads to loss of their degradative capability.

HTLV-1 is a complex type C virus belongs to *Retroviridae* family and contains an envelope which derived from the cell membrane of host (Martin et al. 2016). The virus first was extracted from patients who were suffering from rapidly growing T-cell lymphoma (ATLL) with cutaneous involvement (Martin et al. 2016). Additionally, it has been shown that HTLV-1 has a major role in other diseases including development of poliomyelitis, arthropathy, HTLV-1-associated myelopathy, facial nerve palsy, and infectious dermatitis (Futsch et al. 2018). It has been reported that approximately 5–20 million individuals carry the virus globally; however, a small proportion (3–5%) of them progress secondary ATLL (Gessain and Cassar 2012; Schierhout et al. 2020). Tax is known as a regulatory protein maintaining a crucial role in HTLV-1 replication and, hence, is needed for the virus propagation. It also plays a crucial role in ATLL development since it cooperates with more than 100 cellular proteins to increase cell signaling, inhibit apoptosis, contribute to cell cycle dysregulation, disrupt DNA repair, and

stimulate proto-oncogenes (Mui et al. 2017). It was shown that the virus is able to prohibit the binding between autophagosomes and lysosomes through a mechanism involving tax. As a result, quite a few autophagic vesicles, which are not degraded, appear, and these vesicles are great for virus replication (Tang et al. 2013). Hence, Tax protein combines with the IKK complex to induce NF-kB and Beclin-1 activity. Cell adhesion molecule 1 (CADM1) is a glycoprotein belonging to the type 1 transmembrane cell adhesion family, which is part of immunoglobulin superfamily and is taken into account as a marker of T cells infected with HTLV-1 in (Nakahata et al. 2021; Chen et al. 2015). Tax and NF-kB stimulation and degradation of NF-kB negative regulator, namely, p47, are necessary for CADM1 expression. The main mechanism behind p47 degradation is autophagy, and autophagy can be detected in the majority of HTLV-1 infected ATLL cells (Sarkar et al. 2019). HBZ is another crucial essential viral protein for progression of ATLL (Akram et al. 2017). Recent evidence found out that HBZ can prohibit autophagy as well as apoptosis and, in contrast, stimulate brain-derived neurotrophic factor (BDNF) and its receptor expression (Baratella et al. 2017; Mukai and Ohshima 2014). HBZ can exert different effects based on its location; its expression in cell nucleus and cytoplasm is associated with tumor development and stimulation of inflammation, respectively. Its entry to cytoplasm from nucleus is associated with activation of mTOR via PPP1R15A expression, which is a regulator subunit of protein phosphatase1 (Mukai and Ohshima 2014). Same to other viruses, infection with HTLV-1 is associated with formation of autophagosomes and prohibition of binding to lysosomes so as to inhibit degradation. As a consequence, a great amount of autophagosome vesicles will appear, which provides a suitable environment for the virus formation and, moreover, a physical barrier, which limits the progression of cellular processes (Ren et al. 2015).

Since 2019, the world is witnessing a pandemic caused by a new virus called SARS-CoV-2, causing COVID-19 infection (Khatami et al. 2020). It has been reported that at least

270 million individuals infected with SARS-CoV2 and near 5.3 million people have died because of that (Worldometer 2020). Although its mortality rate is not considerably high, it is highly infectious (Sanche et al. 2020). COVID-19 infection symptoms are broad ranging from fatigue, fever, tiredness, and cough to acute respiratory distress syndrome, MI, stroke, renal injury, and death (Xu et al. 2020b). Albeit some mechanisms have been proposed for sever form of the disease, the exact mechanism behind the diseases pathology is yet not clarified and required more studies (Gorshkov et al. 2020). It has been demonstrated that for the virus replication and transcription, there is a need to DMVs to be formed, indicating the fact that the virus may hijack the autophagosomal machinery to assist DMV formation (Carmona-Gutierrez et al. 2020). Hence, autophagosomes play a crucial role in infection replication by using viral replicase proteins (Cottam et al. 2011). In support of that, also, it was found out that NSP6, a viral replicase protein, colocalized with DMVs positive for LC3, showing a probable correlation between the virus replication and autophagy (Cottam et al. 2011; Bello-Perez et al. 2020). Furthermore, Fulvio et al. designed a study to explore the mechanism that coronaviruses such as mouse hepatitis virus and SARS hijack the formation of EDEMosome, and vesicles participate in the regulation of endoplasmic reticulum degradation, in order to produce the DMVs needed for the virus replication. They declared that mouse hepatitis disrupts two endoplasmic reticulum-associated degradation (ERAD) regulatory proteins, namely, EDEM1 and OS-9, degradation via trapping them into DMVs (Reggiori et al. 2010). This represents that SARS-CoV2 is able to facilitate the virus replication within the infected individual by escaping from autophagy.

Enhanced amount of processed form of LC3B and LC3B-II and an accumulation of SQSTM1, supporting the fact that SARS-CoV2 infection contributed to decreased autophagic flux (Hayn et al. 2021). An experimental study illustrated that although stimulation of autophagy using rapamycin cannot affect the virus considerably, activation of innate immune using interferons keeps the virus sensitive. Therefore, the virus escapes from antiviral mechanism of autophagy. In order to understand the mechanism behind anti-autophagy effects of SARS-CoV2, Hayn et al. (2021) evaluated the effect of 29 of the 30 SARS-CoV-2 proteins on autophagy. They found out that while NSP15 expression is associated with reduced number of autophagosomes positive for LC3B, ORF3a, E, M, and ORF7a expression was associated with accumulation of LC3B. Moreover, the authors showed that E, M, ORF3a, and ORF7a inhibit autophagic flux. It is of importance to note that the reduction of autophagosomes for Nsp15 expression was improved following administration of rapamycin, proposing that possibly Nsp15 impacts mTOR axis. While upon E, ORF3a, and ORF7a expression, the values of processed LC3B-II enhanced, Nsp15 expression led to decrease but not substantial in LC3B-II values. In consistent with this finding, ORF3a, ORF7a, E, and Nsp15 expression is associated with higher values of SQSTM1. Noteworthy, while M expression is associated with higher values of processed LC3B, it was not able to prohibit the degradation of SQSTM1, showing that M cannot inhibit autophagy. Immunofluorescence assay demonstrate that although overexpression of ORF3a, E, and ORF7a is associated with higher numbers of LC3B-positive puncta, M expression is associated with elevated LC3B localization. Also, following Nsp15 expression, decrease in number of autophagosomes was observed. Moreover, the authors proposed that the role of SARS-CoV2 proteins including M, ORF3a, ORF7a, and Nsp15 in autophagy is virtually similar to their function in SARS-CoV-1 and bat coronavirus RaTG13 (Hayn et al. 2021; Koepke et al. 2021). A very recent study showed that ORF3a can intensely prohibit autophagic flux by preventing the fusion of autophagosomes with lysosomes (Zhang et al. 2021d). It was shown that ORF3a colocalized with lysosomes and interacted with VPS39, which is a subunit of the homotypic fusion and protein sorting (HOPS) complex. The interaction between VPS39 and ORF3a contributes to inhibition of -HOPS binding to RAB7, which inhibited the assembly of a fusion machinery, contributing to increase levels of autophagosomes. These findings shed light on the mechanism behind the virus escape degradation, which is disrupting the fusion of autophagosomes with lysosomes (Zhang et al. 2021d). Taken together, the spread of SARS-CoV-2 virus can be limited with using approaches targeting autophagy.

Several drugs, for instance, azithromycin, chloroquine, and hydroxychloroquine, have been considered since these drugs are capable of modulating autophagy signaling pathways (Gao et al. 2020b). The fact that the mentioned medications are able to inhibit endocytic pathway and, thereby, inhibit SARS-CoV2 replication constitute a rational for considering using these drugs in patients who are infected with the virus (Gao et al. 2020b). In clinical settings, inconsistent findings regarding the benefits of these drugs in COVID-19 patients have been reached. Some studies revealed that hydroxychloroquine administration is associated with lower mortality rate in severe COVID-19 patients (Yu et al. 2020; Meo et al. 2020); however, several studies demonstrated that these medications were not able to decrease mortality from infection with SARS-CoV2 (Molina et al. 2020; Singh et al. 2020). Noteworthy, it has been found that these medications are associated with prolonged QT interval, which can lead to cardiac arrhythmia and sudden cardiac death (Chorin et al. 2020; Jankelson et al. 2020). Thus, more investigations are warranted to evaluate the advantageous and disadvantageous of autophagy modulator drugs to limit the virus infection progression. Table 4 lists the effects of viral infection on the regulation of autophagy during some viral diseases.

5   Autophagy Supporting Viral Replication

RNA viruses hijack autophagy for replication. During the autophagy process, DMVs are formed, which maintain a crucial role in poliovirus replication by creating a promising environment for poliovirus replication and keeping polioviruses RNAs away from innate immune receptors recognition and degradation.

Polioviruses, a member of picornavirus family, lack a membrane envelope. Autophagy was shown to be inducer of poliovirus replication, and its inhibition was shown to associated with reduced virus replication (Jackson et al. 2005; Dales et al. 1965). Besides, infection with poliovirus increases the level of LC3 in puncta and expresses two nonstructural poliovirus proteins 2BC and 3A, contributing to lipidation and formation of LC3 and DMVs, respectively, which makes link between the virus replication and autophagy. Similar to polioviruses, foot-and-mouth disease virus and CVB3 exploit autophagy for replication (Berryman et al. 2012; Robinson et al. 2014).

Hepatitis C virus also can trigger autophagy via increasing levels of autophagosomes and using autophagosomal membranes, which is the site for the virus replication (Shrivastava et al. 2011; Dreux and Chisari 2009; Ait-Goughoulte et al. 2008). Nonetheless, the capacity of HCV in stimulating the fusion of lysosome with autophagosomes is still the matter of debate. Several studies have claimed that the virus stimulates autophagosomes and inhibits the autophagosome and lysosome fusion to enhance viral replication and limit virus degradation (Taguwa et al. 2011; Sir et al. 2008a, b). A study stated that HCV enhances the levels of autophagosomes without any change in the levels of autophagy protein degradation, which is (Sir et al. 2008b). Dreux et al. demonstrated that although the autophagy proteins are key components in the translation process of incoming HCV genome, it is not essential for maintenance of the infection (Dreux et al. 2009). However, Ke et al. revealed that the viral replication is totally dependent on the whole autophagic process through complete autolysosome maturation (Ke and Chen 2011b). At early phase of infection with HCV, the interaction between the HCV RNA-dependent RNA polymerase NS5B and ATG5 was observed, which highlights the importance of ATG5 for infection initiation. Blocking ATG5 expression was shown to be associated with the virus replication and maintenance (Guévin et al. 2010).

Table 4 Autophagy and viruses

Virus	Inhibition/induction	Viral product	Target	Sample	Note	Ref
B19	Induction	–	–	In vitro	B19-infected cells survive by cellular autophagy	Nakashima et al. (2006)
Adenoviruses	Induction	–	–	In vitro	Ad E1a and E1b activate LC3 conversion and Atg12-Atg5 complex formation	Rodriguez-Rocha et al. (2011)
PRRSV	Induction	–	–	In vitro	Autophagy is triggered in pulmonary alveolar macrophages by PRRSV infection	Liu et al. (2012)
HP-PRRSV	Induction	–	–	Bystander cells	Induced apoptosis and autophagy in thymi of infected piglets	Wang et al. (2015b)
Mouse norovirus (MNV)	Induction	–	–	In vitro	MNV infection triggers autophagy in host cells and appears to block the downstream degradation of autophagosomes.	O'Donnell et al. (2016)
Rotavirus	Dysregulation	RV-vsRNA1755	IGF-1R	In vitro	RV-vsRNA1755 targets IGF-1R, blockading the PI3K/Akt pathway and triggering autophagy, but it ultimately inhibits autophagy maturation	Zhou et al. (2018)
Influenza A	Inhibition	Matrix protein 2 (M2)		In vitro	M2 protein blocks autophagosome degradation	Gannagé et al. (2009)
Influenza A	Inhibition	M2	LC3		IAV utilizes a mimicry of host protein short linear motifs (SLiMs) to hijack autophagy	Beale et al. (2014)
HSV-1	Inhibition	ICP34.5	Beclin-1	In vitro	Inhibition of Beclin-1-dependent autophagy	Orvedahl et al. (2007)
Human parainfluenza virus type 3 (HPIV3)	Induction	Matrix protein (M)		In vitro	Viral M protein is sufficient to induce mitophagy by bridging autophagosomes and mitochondria	Ding et al. (2017a)
Zika virus	Induction	NS4A and NS4B		In vitro	Akt-mTOR signaling to inhibit neurogenesis and induce autophagy	Liang et al. (2016)
Coronavirus	Induction	PLP2-TM	Beclin-1	In vitro	PLP2-TM activates autophagosome formation but prevents its fusion with lysosomes	Chen et al. (2014b)

(continued)

Table 4 (continued)

Virus	Inhibition/induction	Viral product	Target	Sample	Note	Ref
HIV	Inhibition			CD4+ T cells, U937 cells	The autophagy protein Beclin-1, LC3 II and autophagosomes were found to be markedly decreased	Zhou and Spector (2008)
HIV	Inhibition	Nef	Beclin-1	In vitro	Nef acts as an anti-autophagic maturation factor through interactions with the autophagy regulatory factor Beclin-1	Kyei et al. (2009a)
HIV	Inhibition	Nef	Beclin-1	MOLT-4 cells	During permissive infection, Nef binds BECN1 resulting in mammalian target of rapamycin (MTOR) activation, TFEB phosphorylation and cytosolic sequestration, and the inhibition of autophagy	Campbell et al. (2015a)
Dengue virus	Induction	–	–	Mice	Dengue virus induces amphisome and autophagosome formation as well as the autophagic flux in the brain of infected mice	Lee et al. (2013)
Subgroup J avian leukosis virus (ALV-J)	Inhibition	–	–	In vitro		Liu et al. (2013)
Encephalomyocarditis virus	Induction	2C 3D	–	In vitro	2C and 3D were shown to be involved in inducing autophagy by activating the ER stress pathway	Hou et al. (2014c)
HBV	Induction	HBx	Phosphatidylinositol 3-kinase class III	In vitro Mice	Interestingly, in contrast to starvation-induced autophagy, this enhancement of autophagy by HBV does not lead to an increased autophagic protein degradation rate	Sir and Tian (2010)
HBV	Induction	HBx		HepG2.2.15 cells	Novel function of HBx in increasing autophagy through the upregulation of Beclin-1 expression	Tang et al. (2009a)
HBV	Induction	HBx		HepG2 cells	HBx activates the autophagic lysosome pathway in HepG2 cells through the PI3K-Akt-mTOR pathway	Ju et al. (2013)
HBV	Induction	HBx		Chang cell	HBX induces autophagy via activating DAPK in a pathway related to Beclin-1, but not JNK	Zhang et al. (2014)

Virus	Autophagy	Protein	Target	Cell/System	Mechanism	Reference
HBV		HBx		Huh7 cells	HBx-induced autophagosome formation is MTOR inhibition-independent Repressive effect of HBx on lysosomal function is responsible for the inhibition of autophagic degradation, and this may be critical to the development of HBV-associated HCC	Liu et al. (2014)
HBV	Induction	HBV small surface protein		Huh7 cells	SHBs partially co-localized and interacted with autophagy protein LC3	Li et al. (2011a)
HBV	Induction	HBx	–	Huh7 cells	HBV can be promoting autophagy by the interaction of HBx and c-myc to affect miR-192-3p-XIAP, which in turn regulates Beclin-1	Wang et al. (2019e)
Bluetongue virus (BTV)	Induction	–	–	In vitro	The BTV1-induced inhibition of the Akt-TSC2-mTOR pathway and the upregulation of the AMPK-TSC2-mTOR pathway both contributed to autophagy initiation	Utama et al. (2011)
BTV1	Induction	–	–	BSR cells	BTV1-induced inhibition of the Akt-TSC2-mTOR pathway and the upregulation of the AMPK-TSC2-mTOR pathway both contributed to autophagy initiation	To et al. (2020)
HIV	Induction	Env	CXCR4	In vitro	Autophagy is specifically triggered after Env binding to CXCR4, leading to apoptosis	Espert et al. (2006), Espert et al. (2007)
HIV	Inhibition	Vif	Human autophagy-related protein 8 family proteins	In vitro	The C-terminal part of viral infectivity factor interacts with microtubule-associated protein light chain 3	Borel et al. (2015)
HIV	Induction	Tat	BAG3	Human glial cells	Tat protein is able to stimulate autophagy through increasing BAG3 levels in human glial cells	Bruno et al. (2014)
HIV	Inhibition	Tat		Human primary blood macrophages	HIV-1 Tat protein suppressed IFN-g-induced autophagy processes, including LC3B expression HIV-1 Tat suppressed the induction of autophagy-associated genes and inhibited the formation of autophagosomes	Li et al. (2011b)

(continued)

Table 4 (continued)

Virus	Inhibition/induction	Viral product	Target	Sample	Note	Ref
HCV	Induction		UPR	In vitro	HCV induces the unfolded protein response (UPR), which in turn activates the autophagic pathway	Ke and Chen (2011a)
HCV	Induction	–	–	Huh7	HCV inhibited the AKT-TSC -MTORC1 pathway via ER stress, and the inhibition of the AKT-TSC -MTORC1 pathway contributed to upregulating autophagy	Huang et al. (2013)
HCV	Inhibition	–	–	Monocyte	HCV-positive sera block autophagy during monocyte differentiation LC3 II level increased in monocytes cultured in the presence of HCV-positive sera	Granato et al. (2014)
HCV	–	NS5B	ATG5	Huh7 and C5B cells	HCV utilizes ATG5 as a proviral factor during the onset of viral infection	Guévin et al. (2010)
HBV	Inhibition	HBeHBc		Human ($n = 40$) In vitro	HBe and HBc proteins of HBV activate the mTOR signaling pathway to inhibit autophagy in neutrophils	Hu et al. (2018)
HBV	Induction	HBx				
West Nile virus (WNV)	Induction			In vitro	West Nile Virus-induced LC3 lipidation	Beatman et al. (2012)
Foot-and-mouth disease virus	Inhibition	3C^{pro}	ATG5-ATG12	PK-15 cells	FMDV suppresses NF-κB, IRF3, and autophagy by degradation of ATG5-ATG12 via 3Cpro	Fan et al. (2017)
High-risk HPV	Inhibition	–	–	Tissue sample (uterine cervical cancer, $n = 270$)	Persistent HPV infection may stabilize ATAD3A expression to inhibit cell autophagy and apoptosis as well as to increase drug resistance	Chen et al. (2011)
Influenza A virus	Induction	–	–	Cell lines (A549 cells, MDCK, 293T, WT MEFs, and autophagy-deficient MEFs)	IAV infection increased levels of the autophagosomal marker "microtubule-associated protein light chain 3-II" (LC3-II), at early stage of infection	Khalil (2012)
Human bocavirus (HBoV)	Induction	–	–	Human bronchial epithelial cells	Microtubule-associated protein 1A/1B light chain 3 (LC3) II and autophagy protein 5 were increased in HBoV-transfected HBECs, whereas, the mRNA	Deng et al. (2017)

and protein levels of LC3-I and sequestosome 1 were decreased

Virus	Type	Protein	Cell/model	Findings	References
Porcine hemagglutinating encephalomyelitis virus (PHEV)	Induction	–	Neuro-2a cells	PHEV infection induces atypical autophagy and causes the appearance of autophagosomes but blocks the fusion with lysosomes	Ding et al. (2017b)
Bovine viral diarrhea virus (BVDV)	Induction	–	MDBK cell	BVDV NADL infection triggers autophagosome formation and increases autophagic activities Beclin-1 and ATG14 expression levels were increased as a result of BVDV NADL infection	Fu et al. (2014a)
BVDV	Induction	E^{rns} and E2	MDBK cell	BVDV-NADL infection induced autophagy and significantly elevated the expression levels of autophagy-related genes, Beclin-1 and ATG14, at 12 h postinfection in MDBK cells	Fu et al. (2014b)
Enterovirus 71 (EV71)	Induction	–	Hep2, vero	EV71 infection resulted in the reduction of cellular miR-30a, which led to the inhibition of Beclin-1, a key autophagy-promoting gene that plays important roles at the early phase of autophagosome formation	Fu et al. (2015)
EV-71	Induction	Beclin-1	Human rhabdomyosarcoma, neuroblastoma cells and in vivo	The specific viral proteins encoded contributed to the inhibition of the mTOR/p70S6K pathway and the induction of autophagy	Huang et al. (2009)
EBV	Induction	–	EBV-associated nasal NKTCL ($n = 35$)	Enhanced autophagy and reduced expression of lysosomal enzymes induced regional ACD under EBV infection in natural killer/T-cell lymphomas	Hasui et al. (2011)
Avibirnavirus	Induction	VP2	DF-1 cells, 293T cells	HSP90AA1 binding to the viral protein VP2 resulted in induction of autophagy and AKT-mTOR pathway inactivation	Hu et al. (2015)

(continued)

Table 4 (continued)

Virus	Inhibition/induction	Viral product	Target	Sample	Note	Ref
HCV	Induction	Core protein		QSG-7701	HCV core protein can enhance hepatocytes autophagy through upregulating Beclin-1	Liu et al. (2015b)
HCV	Induction	Core protein		Huh7 hepatoma cell line	Core protein activates autophagy through EIF2AK3 and ATF6 UPR pathway-mediated MAP1LC3B and ATG12 expression	Wang et al. (2014a)
HCV	Induction	-	Class III PI3K-independent pathway	Huh7 hepatoma cell line	-	Sir et al. (2012)
HCV	Induction	NS4B		Huh7.5 cells	Rab5 and Vps34 are involved in NS4B-induced autophagy	Su et al. (2011)
HCV	Induction	-	-	Huh7.5 cells	HCV induces autophagy by upregulating Beclin-1 and activates mTOR signaling pathway	Shrivastava et al. (2012)
HPV16 and 18	Inhibition	-	-	104 cases of cervical cancer tissues	The expression levels of Beclin-1 and LC3B were significantly lower in cervical cancer cells	Wang et al. (2014b)
Flavivirus	Induction	NS4A		Epithelial cells	Expression of Flavivirus NS4A is sufficient to induce PI3K-dependent autophagy and to protect cells against death	McLean et al. (2011)
Simian virus 40	Induction	Small T antigen	-	-	The novel role for the SV40 ST antigen in cancers, where it functions to maintain energy homeostasis during glucose deprivation by activating AMPK, inhibiting mTOR, and inducing autophagy as an alternate energy source	Kumar and Rangarajan (2009)
Varicella-zoster virus	Induction	-	-	Human skin vesicle MRC-5 cells	-	Takahashi et al. (2009)
HHV-8	Induction	RTA	-	RTA-inducible BCBL-1cells (TRExBCBL1-RTA)	Autophagy is involved in the lytic reactivation of HHV-8	Wen et al. (2010)
EBV	Induction	LMP1		B cells		Lee and Sugden (2008)
HCMV	Inhibition	TRS1	Beclin-1	MRC5 cells	The Beclin-1-binding domain of TRS1 is essential to inhibit autophagy	Chaumorcel et al. (2012)

Virus				Cells	Description	Reference
HCMV	Induction	–	–	THP-1 cells	HCMV could induce autophagy, and the capacity of promoting autophagy may be weakened in the latent infection	Liu et al. (2017e)
Porcine circovirus type 2 (PCV2)	Induction	–	–	PK-15 cells	PCV2 might induce autophagy via the AMPK/ERK/TSC2/mTOR signaling pathway in the host cells	Zhu et al. (2012)
HSV-1	Inhibition	Us11	PKR	HeLa cells and fibroblasts		Lussignol et al. (2013)
HPV-16	Inhibition	–	–	HaCaT and 293T cells	The HPV-host cell interaction stimulates the PI3K/Akt/mTOR pathway and inhibits autophagy	Surviladze et al. (2013)
HIV-1	Induction	ASP	LC3	U937 andCOS-7 cells		Torresilla et al. (2013)
Influenza A virus	Induction	NS1	–	CV-1 cells	NS1 stimulates autophagy indirectly by upregulating the synthesis of HA and M2	Zhirnov and Klenk (2013a)
EBV	Induction	Rta	–	293T cells	Autophagic activation is caused by the activation of extracellular signal-regulated kinase (ERK) signaling by Rta	Hung et al. (2014)
HTLV-1	Induction	Tax	–	U251 cells	Tax-triggered autophagy depends on the activation of IKKTax can be degraded via manipulation of autophagy and TRAIL-induced apoptosis	Wang et al. (2014)
HTLV-1	Induction	Tax	–	–	Tax induces Bcl-3 expression Bcl-3 acts as a negative regulator of NF-κB activation and promotes autophagy in HTLV-1-infected cells	Wang et al. (2013c)
HTLV-1	Induction	Tax	–	Human astroglioma cells	HTLV-1 Tax protein induces autophagy via IKK in human astroglioma cells: a protective mechanism against death receptor-mediated apoptosis	Zheng et al. (2014)
EBV	Induction	LMP2A	–	MCF10A cells	MP2A may inhibit anoikis and luminal clearance in acini through induction of autophagy	Fotheringham and Raab-Traub (2015)
DENV	Induction	NS1	–	HMEC-1 cells	–	Chen et al. (2016)
HIV-1	Inhibition	–	–	Monocytic cells		Van Grol et al. (2010)

(continued)

Table 4 (continued)

Virus	Inhibition/induction	Viral product	Target	Sample	Note	Ref
Japanese encephalitis virus	Induction	–	–	N2a cells	HIV-1 impairs autophagy in bystander macrophages/monocytic cells through Src-Akt signaling	Jin et al. (2013)
Japanese encephalitis virus	Induction	C, M and NS3		BHK-21, PK-15 and N2A cells	–	Wang et al. (2015a)
Coxsackievirus A16	Induction	2C and 3C		HeLa cells	CA16 infection inhibited Akt/mTOR signaling and activated extracellular signal-regulated kinase (ERK) signaling, both of which are necessary for autophagy induction	Shi et al. (2015)
Coxsackievirus B3	Induction	–	–	HeLa cells	CVB3 might directly or indirectly induce autophagy via AMPK/MEK/ERK and Ras/Raf/MEK/ERK signaling pathways in the host cells	Xin et al. (2015)
Coxsackievirus B3	Induction	2B		HeLa cells	56 V in the loop region of 2B is critical for the induction of autophagy	Wu et al. 2016c)
γHV68	Inhibition	vBcl2	Beclin-1	NIH3T3 cells	Viral Bcl-2s displays enhanced anti-autophagic activity than cellular Bcl-2	Liang et al. (2008), Liang et al. (2006)
γHV68	Inhibition	M11	Beclin-1	COS7 cells	M11-Beclin-1 BH3 domain binding is required for autophagy inhibition by M11	Sinha et al. (2008)
Bluetongue virus	Induction			BSR cells	BTV- induced disruption of cellular energy metabolism contributes to autophagy	(Lv et al. (2016)
Influenza A virus H5N1	Induction		mTOR signaling	MEF cells	H5N1 causes autophagic cell death through suppression of mTOR signaling	Ma et al. (2011a)
Influenza A virus H5N1	Induction			Mouse lungs and human A549 cells	Autophagy induced by live H5N1 virus in human A549 cells depends on signaling through the Akt-TSC2-mTOR pathway The hemagglutinin protein of H5N1 virus may induce autophagy in A549 cells	Sun et al. (2012)

Virus	Effect	Protein		Cell line	Description	Reference
Newcastle disease virus	Induction			U251 cells		Meng et al. (2012a), Kang et al. (2017)
Newcastle disease virus	Induction	NP and P proteins	–	A549 cells	NDV NP and P proteins induced autophagy through activation of the ER stress-related UPR pathway	Cheng et al. (2016)
Avian reovirus	Induction	–	–	Chicken fibroblast cells and vero cells	The class I PI3K/Akt/mTOR pathway contributes to ARV-triggered autophagy	Meng et al. (2012b)
RABV GD-SH-01	Induction	–	–	Human and mouse neuroblastoma cell lines	Autophagy is induced by GD-SH-01 and can decrease apoptosis in vitro. Furthermore, the M gene of GD-SH-01 may cooperatively induce autophagy	Peng et al. (2016)
HSV-1 HSV-2	Induction	–	–	SIRC cell line		Petrovski et al. (2014)
Enterovirus 71 (EV71)	Induction	–	–	Suckling mouse model	EV71 infection can induce autophagosome, amphisome and autolysosome formation, and the structural protein VP1 and nonstructural protein 2C of EV71 were distributed around the autophagosome and amphisome	Lee et al. (2014)
Pseudorabies virus	Inhibition	–	–	PK-15 cells	Alphaherpesvirus US3 tegument protein may reduce the level of autophagy via activation of the AKT/mTOR pathways in PRV infected cells	Sun et al. (2017c)
Zika virus (ZIKV)	Induction	–	–	Human umbilical vein endothelial cells (HUVEC)	–	Peng et al. (2018)
HIV-1 HIV-2	Induction	–	–	Jurkat and CD4+ T cells	HIV is able to induce the autophagic signaling pathway in HIV-infected host cells, which may be required for HIV infection-mediated apoptotic cell death	Wang et al. (2012)
Sendai virus (HVJ-E)	Induction	–	–	NSCLC cells	HVJ-E could induce autophagy in NSCLC cells via the PI3K/Akt/mTOR/p70S6K signaling pathway	Zhang et al. (2015b)
Rotavirus (RV)	Induction	–	–	HT29 cells, MA104 cells	RV infection can be activating autophagy machinery during RV infection through upregulation and downregulation of miR-99b and let-7g expression levels, respectively	Mukhopadhyay et al. (2019)

HCV dynamically modulates autophagy by expressing ultraviolet radiation resistance-associated gene protein (UVRAG) and Rubicon to increase its replication (Wang et al. 2015c). At the early stages of viral infection, upregulation and downregulation of Rubicon and UVRAG, respectively, by the virus inhibit the autophagosomes maturation and thereby increase the levels of autophagosomes, leading to virus replication (Wang et al. 2015c). Additionally, immunity-related GTPase family M protein (IRGM), an IFN-inducer GTPase, was shown to be able to modulate autophagy process by interacting with several autophagic proteins (Grégoire et al. 2011b). Hansen et al. showed that IRGM by promoting autophagy and Golgi fragmentation induces the virus replication. IRGM stimulates Golgi fragmentation via modulation of Golgi apparatus-specific brefeldin A-resistant guanine nucleotide exchange factor 1 (GBF1) and AMPKα (Hansen et al. 2017). In summary, HCV is able to regulate autophagy process to induce the virus replication.

According to findings of related studies, it can be concluded that flaviviruses take benefits from the close connection between ER and autophagy processes. At first, it was believed that stimulation of autophagy in those infected with flaviviruses is only related to the ER stress-related UPR signaling pathway. On the other hand, it was shown that several nonstructural proteins of West Nile virus (WNV) and DENV are able to stimulate autophagy irrespective of the UPR (Blázquez et al. 2014; Miller et al. 2007). Analyses of neural progenitor cells infected with Zika virus (ZIKV) disclosed that the infection causes a huge remodeling of ER and, moreover, vesicular packet formation, which are assumed to be the spots of ZIKV replication (Offerdahl et al. 2017; Cortese et al. 2017). Infection of skin fibroblast is associated with autophagosomes formation, leading to higher levels of ZIKV replication (Hamel et al. 2015). Furthermore, enhancement in lapidated form of LC3 along with decrement in ATG16L1 expression, a vital autophagy gene, in placentae infected with ZIVK, indicates the fact that autophagy plays a crucial role in vertical transmission of ZIKV (Cao et al. 2017). Liang et al. clarified the mechanism responsible for fetal neurological defects causing by ZIKV. They found out that two proteins exist in ZIKV, namely, NS4A and NS4B, in cooperation with each other inhibit the Akt-mTOR signaling pathway, which contributes to autophagy activation and defective neurogenesis (Liang et al. 2016). Upon early phase of ZIKV and DENV infection, inhibition of FAM134B, which acts as an autophagy receptor, enhances the virus replication. The viruses use their NS3 protease to cleave FAM134B, leading to limit ER and autophagosomes formation (Khaminets et al. 2015; Lennemann and Coyne 2017).

It was shown that two HIV proteins Gag and Nef modulate the autophagy process through interacting with LC3 and Beclin-1, which, finally, causes higher viral replication. During early phase of autophagy, Gag protein interacts with C3, which leads to higher levels of Gag processing and HIV levels in macrophages (Kyei et al. 2009b). Also, during the maturation stage of autophagy, Nef protein of HIV inhibits autophagy maturation via binding to Beclin-1 and, thereby, keeps the virus safe from degradation. Thus, the interaction between the virus and autophagy increases HIV load and replication through inducing early-stage autophagy but prohibits late stages (Kyei et al. 2009b). Nevertheless, it has been detected that during permissive infection, the virus inhibits autophagy so as to prevent the degradation of proteolytic. In the normal situation, mTOR by phosphorylating transcription factor EB (TFEB) limits TFEB translocation. TFEB is able to induce autophagy and lysosomal activation when it transfers to the nucleus. In doing so, TFEB should become dephosphorylated, which is dependent upon mTOR inhibition. For stimulating autophagy within macrophages infected with HIV, the interaction between TLR8 and HIV should be occurred, which is dependent on the dephosphorylation and nuclear translocation of TFEB. The authors also observed that during permissive infection, the interplay between Nef and Beclin-1 contributed to phosphorylation of TFEB, mTOR activation, cytosolic sequestration, and, thereby, autophagy inhibition (Campbell et al. 2015b).

A number of experimental studies have declared that autophagy inhibition causes prohibition of HBV replication, which represents the fundamental role of autophagy in HBV life cycle (Table 3). The studies have utilized cells that were infected with HBV, or transfected with HBV, or exhibiting HBV DNA replication. It was found out by Sir and his colleagues that triggering autophagy by HBV is dependent on the presence of HBx, which increases its activity through binding to PI3KC3. Therefore, autophagy along with PI3KC3 modulates the majority of HBx impacts on HBV replication (Sir et al. 2010a, b). Either inhibition of PI3KC3 or Atg7 contributes to decrease in HBV replication (Sir et al. 2010b). A study found out that autophagy inhibition decreases pgRNA packaging and HBV RNA values to some extent while inhibited HBV DNA replication remarkably (Tang et al. 2009b). Therefore, it can be concluded that this phenomenon indicates that autophagy exerts its effects on HBV replication mainly at the viral DNA replication stage of the viral life (Sir et al. 2010b). Another study similarly found positive effects of autophagy on HBV replication; however, the effects were mostly seen at the stage of envelopment (Rautou et al. 2010). Li et al. designed a study to evaluate the association between autophagy and HBV by suppressing autophagy using 3-methyladenine and siRNA duplexes that suppress fundamental genes need for autophagosome formation. The investigators explored that autophagy inhibition is able to suppress the virus replication notably and stimulating autophagy using starvation and/or rapamycin increases the virus replication (Li et al. 2011c). These inconsistent findings can be explained by using different HBV strains or sublines of Huh7 cells in the relevant studies. Also, a study unveiled ROS HBV capsid assembly in the existence of Hsp90; however, it was observed that ROS without Hsp90 decreases the virus assembly (Kim et al. 2015). Another pathway responsible for HBV-induced autophagy is ROS/JNK signaling pathway. In doing so, ROS/JNK signaling pathway modulates the interaction between Beclin-1 and Bcl-2, which is crucial for activation of autophagy (Zhong et al. 2017). Additionally, it has been shown that HBV has the potential to favor its replication by

subverting autophagy Atg5-12/16L1 complex, without any need for Atg8/LC3 lipidation, which is a vital process for autophagosomes maturation (Döring et al. 2018). At the same time, several studies have claimed that autophagy triggered by HBV inhibits the virus replication. Wu et al. demonstrated that autophagy following infection with HBV is able to degrade envelope proteins (Wu et al. 2016d). For the first time, Lazar et al. demonstrated that HBV decreases the level of envelope protein through that the ERAD signaling pathway. Simultaneous expression of the virus envelope proteins and EDEM1 caused huge envelope protein degradation, which was blocked through EDEM1 inhibition (Lazar et al. 2012). Furthermore, a study revealed that AMPK activation is able to limit the virus production by inducing autophagy, suggesting the therapeutic value of targeting AMPK for HBV management (Xie et al. 2016). Collectively, it can be said that the precise relationship between the virus replication and autophagy merits extra studies.

Also, infection with influenza A virus (IAV) is able to induce enhanced levels of autophagosomes that needed for viral replication (Zhou et al. 2009). A study showed that the virus increases the levels of autophagosomes by inhibition of their fusion with lysosomes, and the presence of matrix 2 (M2) ion-channel protein for prohibition of autophagosomes degradation is pivotal (Gannagé et al. 2009). Another research displayed that M2 escapes from autophagy using its LC3-interacting region (Beale et al. 2014). M2 interacts with LC3 and induces LC3 re-localization to the plasma membrane, and disruption of this interaction downregulates virion budding and stability. The NS1 is another IVA protein that induces autophagy via overexpression of M2 and hemagglutinin (HA) (Zhirnov and Klenk 2013b). Recently, the interplay between M2 protein and MAVS signaling pathway was demonstrated, which leads to MAVS aggregation and, thereby, stimulates MAVS-mediated antiviral innate immunity. Furthermore, it was shown that M2 triggers ROS generation, which is a crucial factor for autophagy activation (Wang et al. 2019f). Additionally, H5N1, a major avian pathogen, has the potential to induce autophagy via prohibiting mTOR (Ma et al. 2011b).

As we mentioned before, Beclin-1 is a fundamental modifier of autophagy process that forms two distinct complexes, one with Atg14 that is needed for autophagosome formation and the other with UVRAG, which is essential for autophagosome maturation (Levine et al. 2015). In a study that was conducted by Qu and his associates, it was demonstrated that infection with SARS-CoV-2 is associated with incomplete autophagy response, which was shown to be needed for effective virus replication. Moreover, the investigators disclosed that although infection with SARS-CoV-2 stimulates autophagosomes formation, the infection contributes to prohibition of autophagosome maturation and block autophagy by inhibiting fundamental genes involved in the virus replication (Qu et al. 2021). They analyzed expression of 26 proteins expressing by the virus and found out that ORF3a expression is associated with incomplete autophagy. The ORF3a interplays with UVRAG to promote and prohibit expression of PI3KC3-C1 (Beclin-1-Vps34-Atg14) and PI3KC3-C2 (Beclin-1-Vps34-UVRAG), respectively. In summary, the authors shed light on how ORF3a inhibits autophagy and, thereby, prompts SARS-CoV-2 replication, which provides a therapeutic potential of targeting autophagy for COVID-19 treatment (Qu et al. 2021).

6 Conclusion

Autophagy is known as conserved intracellular process which transfers cytoplasmic materials to lysosomes for degradation though autophagosomes. This process emerges to be relevant to the pathogenesis of various diseases, and its regulation could have therapeutic value. It has been indicated that a sequence of cellular and molecular signaling pathways by several internal and external factors is involved in initiation and progression of autophagy. Viruses are one of main factors which exert their pathogenesis effects via affecting on autophagy processes.

Besides viruses, a wide range of internal factors including genetic and epigenetic factors could influence on underlying pathways involved in autophagy processes. Very recently, microRNAs and exosomes have been emerged as critical players in the autophagy processes, given that exosomes and microRNAs are able to change behavior of host cells via targeting of a large number of cellular and molecular signaling pathways. Hence, more insights into the various signaling pathways that are targeted by exosomes and microRNAs could pave the way to the finding and designing new therapeutic approaches.

Conflicts of Interest The authors have declared that no competing interest exists.

References

Abdulrahman BA, Abdelaziz DH, Schatzl HM (2018) Autophagy regulates exosomal release of prions in neuronal cells. J Biol Chem 293(23):8956–8968

Ai H, Zhou W, Wang Z, Qiong G, Chen Z, Deng S (2019) microRNAs-107 inhibited autophagy, proliferation, and migration of breast cancer cells by targeting HMGB1. J Cell Biochem 120(5):8696–8705

Ait-Goughoulte M, Kanda T, Meyer K, Ryerse JS, Ray RB, Ray R (2008) Hepatitis C virus genotype 1a growth and induction of autophagy. J Virol 82(5):2241–2249

Akram N, Imran M, Noreen M, Ahmed F, Atif M, Fatima Z et al (2017) Oncogenic role of tumor viruses in humans. Viral Immunol 30(1):20–27

Bagga S, Bouchard MJ (2014) Cell cycle regulation during viral infection. In: Cell cycle control, vol 1170. Springer, pp 165–227

Baixauli F, López-Otín C, Mittelbrunn M (2014) Exosomes and autophagy: coordinated mechanisms for the maintenance of cellular fitness. Front Immunol 5:403

Balandeh E, Mohammadshafie K, Mahmoudi Y, Hossein Pourhanifeh M, Rajabi A, Bahabadi ZR et al (2021) Roles of non-coding RNAs and angiogenesis in glioblastoma. Front Cell Dev Biol 9:716462

Baratella M, Forlani G, Accolla RS (2017) HTLV-1 HBZ viral protein: a key player in HTLV-1 mediated diseases. Front Microbiol 8:2615

Barclay RA, Pleet ML, Akpamagbo Y, Noor K, Mathiesen A, Kashanchi F (2017) Isolation of exosomes from HTLV-infected cells. In: Human T-lymphotropic viruses. Springer, New York, pp 57–75

Bartels CL, Tsongalis GJ (2010) [MicroRNAs: novel biomarkers for human cancer]. Annales de Biologie Clinique 68(3):263–272

Beale R, Wise H, Stuart A, Ravenhill BJ, Digard P, Randow F (2014) A LC3-interacting motif in the influenza A virus M2 protein is required to subvert autophagy and maintain virion stability. Cell Host Microbe 15(2):239–247

Beatman E, Oyer R, Shives KD, Hedman K, Brault AC, Tyler KL et al (2012) West Nile virus growth is independent of autophagy activation. Virology 433(1): 262–272

Behl C (2011) BAG3 and friends: co-chaperones in selective autophagy during aging and disease. Autophagy 7(7):795–798

Bello-Perez M, Sola I, Novoa B, Klionsky DJ, Falco A (2020) Canonical and noncanonical autophagy as potential targets for COVID-19. Cells 9(7):1619

Berryman S, Brooks E, Burman A, Hawes P, Roberts R, Netherton C et al (2012) Foot-and-mouth disease virus induces autophagosomes during cell entry via a class III phosphatidylinositol 3-kinase-independent pathway. J Virol 86(23):12940–12953

Bhattacharya A, Schmitz U, Raatz Y, Schönherr M, Kottek T, Schauer M et al (2015) miR-638 promotes melanoma metastasis and protects melanoma cells from apoptosis and autophagy. Oncotarget 6(5):2966

Blázquez AB, Martín-Acebes MA, Saiz JC (2014) Amino acid substitutions in the non-structural proteins 4A or 4B modulate the induction of autophagy in West Nile virus infected cells independently of the activation of the unfolded protein response. Front Microbiol 5:797

Borel S, Robert-Hebmann V, Alfaisal J, Jain A, Faure M, Espert L et al (2015) HIV-1 viral infectivity factor interacts with microtubule-associated protein light chain 3 and inhibits autophagy. AIDS 29(3):275–286

Brunet A, Datta SR, Greenberg ME (2001) Transcription-dependent and-independent control of neuronal survival by the PI3K–Akt signaling pathway. Curr Opin Neurobiol 11(3):297–305

Bruno AP, De Simone FI, Iorio V, De Marco M, Khalili K, Sariyer IK et al (2014) HIV-1 Tat protein induces glial cell autophagy through enhancement of BAG3 protein levels. Cell Cycle 13(23):3640–3644

Cai J, Zhang H, Zhang Y-f, Zhou Z, Wu S (2019) MicroRNA-29 enhances autophagy and cleanses exogenous mutant αB-crystallin in retinal pigment epithelial cells. Exp Cell Res 374(1):231–248

Campbell GR, Rawat P, Bruckman RS, Spector SA (2015a) Human immunodeficiency virus type 1 Nef inhibits autophagy through transcription factor EB sequestration. PLoS Pathog 11(6):e1005018

Campbell GR, Rawat P, Bruckman RS, Spector SA (2015b) Human immunodeficiency virus type 1 Nef inhibits autophagy through transcription factor EB sequestration. PLoS Pathog 11(6):e1005018

Cao B, Parnell LA, Diamond MS, Mysorekar IU (2017) Inhibition of autophagy limits vertical transmission of Zika virus in pregnant mice. J Exp Med 214(8): 2303–2313

Carmona-Gutierrez D, Bauer MA, Zimmermann A, Kainz K, Hofer SJ, Kroemer G et al (2020) Digesting the crisis: autophagy and coronaviruses. Microb Cell 7(5):119

Chakrabarti M, Ray SK (2016) Anti-tumor activities of luteolin and silibinin in glioblastoma cells: overexpression of miR-7-1-3p augmented luteolin and silibinin to inhibit autophagy and induce apoptosis in glioblastoma in vivo. Apoptosis 21(3):312–328

Chaumorcel M, Lussignol M, Mouna L, Cavignac Y, Fahie K, Cotte-Laffitte J et al (2012) The human cytomegalovirus protein TRS1 inhibits autophagy via its interaction with Beclin 1. J Virol 86(5):2571–2584. https://doi.org/10.1128/JVI.05746-11

Chen T-C, Hung Y-C, Lin T-Y, Chang H-W, Chiang I, Chen Y-Y et al (2011) Human papillomavirus infection and expression of ATPase family AAA domain containing 3A, a novel anti-autophagy factor, in uterine cervical cancer. Int J Mol Med 28(5):689–696

Chen L, Jiang K, Jiang H, Wei P (2014a) miR-155 mediates drug resistance in osteosarcoma cells via inducing autophagy. Exp Ther Med 8(2):527–532

Chen X, Wang K, Xing Y, Tu J, Yang X, Zhao Q et al (2014b) Coronavirus membrane-associated papain-like proteases induce autophagy through interacting with Beclin1 to negatively regulate antiviral innate immunity. Protein Cell 5(12):912–927

Chen L, Liu D, Zhang Y, Zhang H, Cheng H (2015) The autophagy molecule Beclin 1 maintains persistent activity of NF-κB and Stat3 in HTLV-1-transformed T lymphocytes. Biochem Biophys Res Commun 465(4):739–745

Chen H-R, Chuang Y-C, Lin Y-S, Liu H-S, Liu C-C, Perng G-C et al (2016) Dengue virus nonstructural protein 1 induces vascular leakage through macrophage migration inhibitory factor and autophagy. PLoS Negl Trop Dis 10(7):e0004828

Chen X, Broeyer F, de Kam M, Baas J, Cohen A, van Gerven J (2017a) Pharmacodynamic response profiles of anxiolytic and sedative drugs. Br J Clin Pharmacol 83(5):1028–1038

Chen R, Li X, He B, Hu W (2017b) MicroRNA-410 regulates autophagy-related gene ATG16L1 expression and enhances chemosensitivity via autophagy inhibition in osteosarcoma. Mol Med Rep 15(3):1326–1334

Chen J, Yu Y, Li S, Liu Y, Zhou S, Cao S et al (2017c) Micro RNA-30a ameliorates hepatic fibrosis by inhibiting Beclin1-mediated autophagy. J Cell Mol Med 21(12):3679–3692

Chen H, Soni M, Patel Y, Markoutsa E, Jie C, Liu S et al (2018) Autophagy, cell viability and chemo-resistance are regulated by miR-489 in breast cancer. Mol Cancer Res 16(9):1348–1360: molcanres.0634.2017

Chen D, Huang X, Lu S, Deng H, Gan H, Huang R et al (2019a) miRNA-125a modulates autophagy of thyroiditis through PI3K/Akt/mTOR signaling pathway. Exp Ther Med 17(4):2465–2472

Chen H, Liu Gao MY, Zhang L, He FL, Shi YK, Pan XH et al (2019b) MicroRNA-155 affects oxidative damage through regulating autophagy in endothelial cells. Oncol Lett 17(2):2237–2243

Chen W, Li Z, Liu H, Jiang S, Wang G, Sun L et al (2020a) MicroRNA-30a targets BECLIN-1 to inactivate autophagy and sensitizes gastrointestinal stromal tumor cells to imatinib. Cell Death Dis 11(3):1–13

Chen W, Wang H, Zhu Z, Feng J, Chen L (2020b) Exosome-shuttled circSHOC2 from IPASs regulates neuronal autophagy and ameliorates ischemic brain injury via the miR-7670-3p/SIRT1 axis. Mol Ther Nucleic Acids 22:657–672

Chen H-X, Liang F-C, Gu P, Xu B-L, Xu H-J, Wang W-T et al (2020c) Exosomes derived from mesenchymal stem cells repair a Parkinson's disease model by inducing autophagy. Cell Death Dis 11(4):1–17

Cheng M-R, Li Q, Wan T, He B, Han J, Chen H-X et al (2012) Galactosylated chitosan/5-fluorouracil nanoparticles inhibit mouse hepatic cancer growth and its side effects. World J Gastroenterol 18(42):6076

Cheng J-H, Sun Y-J, Zhang F-Q, Zhang X-R, Qiu X-S, Yu L-P et al (2016) Newcastle disease virus NP and P proteins induce autophagy via the endoplasmic reticulum stress-related unfolded protein response. Sci Rep 6:24721

Choi Y, Bowman JW, Jung JU (2018) Autophagy during viral infection—a double-edged sword. Nat Rev Microbiol 16(6):341–354

Chorin E, Wadhwani L, Magnani S, Dai M, Shulman E, Nadeau-Routhier C et al (2020) QT interval prolongation and torsade de pointes in patients with COVID-19 treated with hydroxychloroquine/azithromycin. Heart Rhythm 17(9):1425–1433

Clotaire DZJ, Zhang B, Wei N, Gao R, Zhao F, Wang Y et al (2016) miR-26b inhibits autophagy by targeting ULK2 in prostate cancer cells. Biochem Biophys Res Commun 472(1):194–200

Comincini S, Manai F, Meazza C, Pagani S, Martinelli C, Pasqua N et al (2017) Identification of autophagy-related genes and their regulatory miRNAs associated with celiac disease in children. Int J Mol Sci 18(2):391

Cortese M, Goellner S, Acosta EG, Neufeldt CJ, Oleksiuk O, Lampe M et al (2017) Ultrastructural characterization of Zika virus replication factories. Cell Rep 18(9):2113–2123

Cottam EM, Maier HJ, Manifava M, Vaux LC, Chandra-Schoenfelder P, Gerner W et al (2011) Coronavirus nsp6 proteins generate autophagosomes from the endoplasmic reticulum via an omegasome intermediate. Autophagy 7(11):1335–1347

Cuervo AM (2004) Autophagy: in sickness and in health. Trends Cell Biol 14(2):70–77

Cui X, Wang X, Zhou X, Jia J, Chen H, Zhao W (2020) miR-106a regulates cell proliferation and autophagy by targeting LKB1 in HPV-16–associated cervical cancer. Mol Cancer Res 18(8):1129–1141

Cuyàs E, Corominas-Faja B, Joven J, Menendez JA (2014) Cell cycle regulation by the nutrient-sensing mammalian target of rapamycin (mTOR) pathway. In: Cell cycle control. Springer, New York, pp 113–144

Dai H, Liu C, Liu Y, Zhang Z, Peng C, Wang Z et al (2020) Research on mechanism of miR-130a in regulating autophagy of bladder cancer cells through CYLD. J BUON 25(3):1636–1642

Dales S, Eggers HJ, Tamm I, Palade GE (1965) Electron microscopic study of the formation of poliovirus. Virology 26(3):379–389

Deng YP, Liu YJ, Yang ZQ, Wang YJ, He BY, Liu P (2017) Human bocavirus induces apoptosis and autophagy in human bronchial epithelial cells. Exp Ther Med 14(1):753–758

Deng J, Ma M, Jiang W, Zheng L, Cui S (2020) MiR-493 induces cytotoxic autophagy in prostate cancer cells through regulation on PHLPP2. Curr Pharm Biotechnol 21(14):1451–1456

Deretic V (2006) Autophagy as an immune defense mechanism. Curr Opin Immunol 18(4):375–382

Deretic V, Saitoh T, Akira S (2013) Autophagy in infection, inflammation and immunity. Nat Rev Immunol 13(10):722–737

Dhital S, Sherchand JB, Pokhrel BM, Parajuli K, Shah N, Mishra SK et al (2017) Molecular epidemiology of Rotavirus causing diarrhea among children less than five years of age visiting national level children hospitals, Nepal. BMC Pediatr 17(1):101

Dias MV, Teixeira BL, Rodrigues BR, Sinigaglia-Coimbra R, Porto-Carreiro I, Roffé M et al (2016) PRNP/prion protein regulates the secretion of exosomes modulating CAV1/caveolin-1-suppressed autophagy. Autophagy 12(11):2113–2128

Ding B, Zhang G, Yang X, Zhang S, Chen L, Yan Q et al (2014) Phosphoprotein of human parainfluenza virus type 3 blocks autophagosome-lysosome fusion to increase virus production. Cell Host Microbe 15(5):564–577

Ding B, Zhang L, Li Z, Zhong Y, Tang Q, Qin Y et al (2017a) The matrix protein of human parainfluenza virus type 3 induces mitophagy that suppresses interferon responses. Cell Host Microbe 21(4):538–547.e4

Ding N, Zhao K, Lan Y, Li Z, Lv X, Su J et al (2017b) Induction of atypical autophagy by porcine hemagglutinating encephalomyelitis virus contributes to viral replication. Front Cell Infect Microbiol 7:56

Doan YH, Suzuki Y, Fujii Y, Haga K, Fujimoto A, Takai-Todaka R et al (2017) Complex reassortment events of unusual G9P[4] rotavirus strains in India between 2011 and 2013. Infect Genet Evol 54:417–428

Dong S, Xiao Y, Ma X, He W, Kang J, Peng Z et al (2019) miR-193b increases the chemosensitivity of osteosarcoma cells by promoting FEN1-mediated autophagy. Onco Targets Ther 12:10089

Dong L-X, Bao H-L, Zhang Y-Y, Liu Y, Zhang G-W, An F-M (2021) MicroRNA-16-5p/BTG2 axis affects neurological function, autophagy and apoptosis of hippocampal neurons in Alzheimer's disease. Brain Res Bull 175:254–262

Döring T, Zeyen L, Bartusch C, Prange R (2018) Hepatitis B virus subverts the autophagy elongation complex Atg5-12/16L1 and does not require Atg8/LC3 Lipidation for viral maturation. J Virol 92(7):e01513–e01517

Dreux M, Chisari FV (2009) Autophagy proteins promote hepatitis C virus replication. Autophagy 5(8):1224–1225

Dreux M, Gastaminza P, Wieland SF, Chisari FV (2009) The autophagy machinery is required to initiate hepatitis C virus replication. Proc Natl Acad Sci U S A 106(33):14046–14051

Du F, Feng Y, Fang J, Yang M (2015) MicroRNA-143 enhances chemosensitivity of Quercetin through autophagy inhibition via target GABARAPL1 in gastric cancer cells. Biomed Pharmacother 74:169–177

Esfandyari YB, Doustvandi MA, Amini M, Baradaran B, Zaer SJ, Mozammel N et al (2021) MicroRNA-143 sensitizes cervical cancer cells to cisplatin: a promising anticancer combination therapy. Reprod Sci 28(7): 2036–2049

Espert L, Denizot M, Grimaldi M, Robert-Hebmann V, Gay B, Varbanov M et al (2006) Autophagy is involved in T cell death after binding of HIV-1 envelope proteins to CXCR4. J Clin Invest 116(8): 2161–2172

Espert L, Denizot M, Grimaldi M, Robert-Hebmann V, Gay B, Varbanov M et al (2007) Autophagy and CD4+ T lymphocyte destruction by HIV-1. Autophagy 3(1): 32–34

Fader C, Colombo M (2009) Autophagy and multivesicular bodies: two closely related partners. Cell Death Differ 16(1):70

Fan X, Han S, Yan D, Gao Y, Wei Y, Liu X et al (2017) Foot-and-mouth disease virus infection suppresses autophagy and NF-κB antiviral responses via degradation of ATG5-ATG12 by 3C pro. Cell Death Dis 8(1): e2561

Fan H, Jiang M, Li B, He Y, Huang C, Luo D et al (2018) MicroRNA-let-7a regulates cell autophagy by targeting Rictor in gastric cancer cell lines MGC-803 and SGC-7901. Oncol Rep 39(3):1207–1214

Fang W, Shu S, Yongmei L, Endong Z, Lirong Y, Bei S (2016) miR-224-3p inhibits autophagy in cervical cancer cells by targeting FIP200. Sci Rep 6:33229

Fehr AR, Yu D (2013) Control the host cell cycle: viral regulation of the anaphase-promoting complex. J Virol 87(16):8818–8825. https://doi.org/10.1128/JVI. 00088-13

Feng D-Q, Huang B, Li J, Liu J, Chen X-M, Xu Y-M et al (2013) Selective miRNA expression profile in chronic myeloid leukemia K562 cell-derived exosomes. Asian Pac J Cancer Prev 14(12):7501–7508

Fotheringham JA, Raab-Traub N (2015) Epstein-Barr virus latent membrane protein 2 induces autophagy to promote abnormal acinus formation. J Virol 89(13): 6940–6944

Fu Q, Shi H, Ren Y, Guo F, Ni W, Qiao J et al (2014a) Bovine viral diarrhea virus infection induces autophagy in MDBK cells. J Microbiol 52(7):619–625

Fu Q, Shi H, Shi M, Meng L, Bao H, Zhang G et al (2014b) Roles of bovine viral diarrhea virus envelope glycoproteins in inducing autophagy in MDBK cells. Microb Pathog 76:61–66

Fu Y, Xu W, Chen D, Feng C, Zhang L, Wang X et al (2015) Enterovirus 71 induces autophagy by regulating has-miR-30a expression to promote viral replication. Antivir Res 124:43–53

Fu Q, Cheng J, Zhang J, Zhang Y, Chen X, Xie J et al (2016) Downregulation of YEATS4 by miR-218 sensitizes colorectal cancer cells to L-OHP-induced cell apoptosis by inhibiting cytoprotective autophagy. Oncol Rep 36(6):3682–3690

Futsch N, Prates G, Mahieux R, Casseb J, Dutartre H (2018) Cytokine networks dysregulation during HTLV-1 infection and associated diseases. Viruses 10(12):691

Gannagé M, Dormann D, Albrecht R, Dengjel J, Torossi T, Rämer PC et al (2009) Matrix protein 2 of influenza A virus blocks autophagosome fusion with lysosomes. Cell Host Microbe 6(4):367–380

Gao S, Wang K, Wang X (2020a) miR-375 targeting autophagy-related 2B (ATG2B) suppresses autophagy and tumorigenesis in cisplatin-resistant osteosarcoma cells. Neoplasma 67(4):724–734

Gao J, Tian Z, Yang X (2020b) Breakthrough: chloroquine phosphate has shown apparent efficacy in treatment of COVID-19 associated pneumonia in clinical studies. Biosci Trends 14(1):72–73

Ge Y-Y, Shi Q, Zheng Z-Y, Gong J, Zeng C, Yang J et al (2014) MicroRNA-100 promotes the autophagy of hepatocellular carcinoma cells by inhibiting the expression of mTOR and IGF-1R. Oncotarget 5(15):6218

Gessain A, Cassar O (2012) Epidemiological aspects and world distribution of HTLV-1 infection. Front Microbiol 3:388

Glick D, Barth S, Macleod KF (2010) Autophagy: cellular and molecular mechanisms. J Pathol 221(1):3–12

Gong L, Xu H, Zhang X, Zhang T, Shi J, Chang H (2019) Oridonin relieves hypoxia-evoked apoptosis and autophagy via modulating microRNA-214 in H9c2 cells. Artif Cells Nanomed Biotechnol 47(1): 2585–2592

Gorshkov K, Chen CZ, Bostwick R, Rasmussen L, Xu M, Pradhan M et al (2020) The SARS-CoV-2 cytopathic effect is blocked with autophagy modulators. BioRxiv. https://doi.org/10.1101/2020.05.16.091520

Granato M, Lacconi V, Peddis M, Di Renzo L, Valia S, Rivanera D et al (2014) Hepatitis C virus present in the sera of infected patients interferes with the autophagic process of monocytes impairing their in-vitro differentiation into dendritic cells. Biochim Biophys Acta 1843(7):1348–1355

Grégoire IP, Richetta C, Meyniel-Schicklin L, Borel S, Pradezynski F, Diaz O et al (2011a) IRGM is a common target of RNA viruses that subvert the autophagy network. PLoS Pathog 7(12):e1002422

Grégoire IP, Richetta C, Meyniel-Schicklin L, Borel S, Pradezynski F, Diaz O et al (2011b) IRGM is a common target of RNA viruses that subvert the autophagy network. PLoS Pathog 7(12):e1002422

Gu X, Li Y, Chen K, Wang X, Wang Z, Lian H et al (2020) Exosomes derived from umbilical cord mesenchymal stem cells alleviate viral myocarditis through activating AMPK/mTOR-mediated autophagy flux pathway. J Cell Mol Med 24(13):7515–7530

Guévin C, Manna D, Bélanger C, Konan KV, Mak P, Labonté P (2010) Autophagy protein ATG5 interacts transiently with the hepatitis C virus RNA polymerase (NS5B) early during infection. Virology 405(1):1–7

Guo S, Bai R, Liu W, Zhao A, Zhao Z, Wang Y et al (2014) miR-22 inhibits osteosarcoma cell proliferation and migration by targeting HMGB1 and inhibiting HMGB1-mediated autophagy. Tumor Biol 35(7): 7025–7034

Guo X, Xue H, Guo X, Gao X, Xu S, Yan S et al (2015) MiR224-3p inhibits hypoxia-induced autophagy by targeting autophagy-related genes in human glioblastoma cells. Oncotarget 6(39):41620

Guo Y, Liu J, Guan Y (2016) Hypoxia induced upregulation of miR-301a/b contributes to increased cell autophagy and viability of prostate cancer cells by targeting NDRG2. Eur Rev Med Pharmacol Sci 20(1):101–108

Guo W, Wang H, Yang Y, Guo S, Zhang W, Liu Y et al (2017a) Down-regulated miR-23a contributes to the metastasis of cutaneous melanoma by promoting autophagy. Theranostics 7(8):2231–2249

Guo W, Wang H, Yang Y, Guo S, Zhang W, Zhao T et al (2017b) Down-regulated miR-23a contributes to invasion and metastasis of cutaneous melanoma by promoting autophagy. AACR 7(8):2231–2249

Guo H, Chitiprolu M, Roncevic L, Javalet C, Hemming FJ, Trung MT et al (2017c) Atg5 disassociates the V 1 V 0-ATPase to promote exosome production and tumor metastasis independent of canonical macroautophagy. Dev Cell 43(6):716–730.e7

Guo Y, Shi W, Fang R (2021) miR-18a-5p promotes melanoma cell proliferation and inhibits apoptosis and autophagy by targeting EPHA7 signaling. Mol Med Rep 23(1):79

Hamel R, Dejarnac O, Wichit S, Ekchariyawat P, Neyret A, Luplertlop N et al (2015) Biology of Zika virus infection in human skin cells. J Virol 89(17): 8880–8896

Han M, Hu J, Lu P, Cao H, Yu C, Li X et al (2020) Exosome-transmitted miR-567 reverses trastuzumab resistance by inhibiting ATG5 in breast cancer. Cell Death Dis 11(1):1–15

Hanada M, Feng J, Hemmings BA (2004) Structure, regulation and function of PKB/AKT—a major therapeutic target. Biochim Biophys Acta 1697(1–2):3–16

Hansen MD, Johnsen IB, Stiberg KA, Sherstova T, Wakita T, Richard GM et al (2017) Hepatitis C virus triggers Golgi fragmentation and autophagy through the immunity-related GTPase M. Proc Natl Acad Sci U S A 114(17):E3462–E3e71

Hashemipour M, Boroumand H, Mollazadeh S, Tajiknia V, Nourollahzadeh Z, Borj MR et al (2021) Exosomal microRNAs and exosomal long non-coding RNAs in gynecologic cancers. Gynecol Oncol 161(1):314–327

Hassanpour M, Hajihassani F, Hiradfar A, Aghamohammadzadeh N, Rahbarghazi R, Safaie N et al (2020) Real-state of autophagy signaling pathway in neurodegenerative disease; focus on multiple sclerosis. J Inflamm 17(1):1–8

Hasui K, Wang J, Jia X, Tanaka M, Nagai T, Matsuyama T et al (2011) Enhanced autophagy and reduced expression of cathepsin D are related to autophagic cell death in epstein-barr virus-associated nasal natural killer/T-cell lymphomas: an immunohistochemical analysis of Beclin-1, LC3, mitochondria (AE-1), and cathepsin D in nasopharyngeal lymphomas. Acta Histochem Cytochem 44(3):119–131

Hayn M, Hirschenberger M, Koepke L, Nchioua R, Straub JH, Klute S et al (2021) Systematic functional analysis of SARS-CoV-2 proteins uncovers viral innate immune antagonists and remaining vulnerabilities. Cell Rep 35(7):109126

He W, Cheng Y (2018) Inhibition of miR-20 promotes proliferation and autophagy in articular chondrocytes by PI3K/AKT/mTOR signaling pathway. Biomed Pharmacother 97:607–615

He C, Klionsky DJ (2009) Regulation mechanisms and signaling pathways of autophagy. Annu Rev Genet 43: 67–93

He C, Dong X, Zhai B, Jiang X, Dong D, Li B et al (2015a) MiR-21 mediates sorafenib resistance of hepatocellular carcinoma cells by inhibiting autophagy via the PTEN/Akt pathway. Oncotarget 6(30):28867

He J, Yu JJ, Xu Q, Wang L, Zheng JZ, Liu LZ et al (2015b) Downregulation of ATG14 by EGR1-MIR152 sensitizes ovarian cancer cells to cisplatin-induced apoptosis by inhibiting cyto-protective autophagy. Autophagy 11(2):373–384

He Y, Zhang L, Tan F, Wang L-F, Liu D-H, Wang R-J et al (2020a) MiR-153-5p promotes sensibility of colorectal cancer cells to oxaliplatin via targeting Bcl-2-mediated autophagy pathway. Biosci Biotechnol Biochem 84(8):1645–1651

He Q, Wang L, Zhao R, Yan F, Sha S, Cui C et al (2020b) Mesenchymal stem cell-derived exosomes exert ameliorative effects in type 2 diabetes by improving hepatic glucose and lipid metabolism via enhancing autophagy. Stem Cell Res Ther 11:1–14

Heaton NS, Randall G (2011) Dengue virus and autophagy. Viruses 3(8):1332–1341

Heras-Sandoval D, Pérez-Rojas JM, Hernández-Damián J, Pedraza-Chaverri J (2014) The role of PI3K/AKT/mTOR pathway in the modulation of autophagy and the clearance of protein aggregates in neurodegeneration. Cell Signal 26(12):2694–2701

Herbein G (2018) The human cytomegalovirus, from oncomodulation to oncogenesis. Viruses 10(8):408

Hessvik NP, Øverbye A, Brech A, Torgersen ML, Jakobsen IS, Sandvig K et al (2016) PIKfyve inhibition increases exosome release and induces secretory autophagy. Cell Mol Life Sci 73(24):4717–4737

Hou C, Zhu M, Sun M, Lin Y (2014a) MicroRNA let-7i induced autophagy to protect T cell from apoptosis by targeting IGF1R. Biochem Biophys Res Commun 453(4):728–734

Hou N, Han J, Li J, Liu Y, Qin Y, Ni L et al (2014b) MicroRNA profiling in human colon cancer cells during 5-fluorouracil-induced autophagy. PLoS One 9(12):e114779

Hou L, Ge X, Xin L, Zhou L, Guo X, Yang H (2014c) Nonstructural proteins 2C and 3D are involved in autophagy as induced by the encephalomyocarditis virus. Virol J 11(1):156

Hu B, Zhang Y, Jia L, Wu H, Fan C, Sun Y et al (2015) Binding of the pathogen receptor HSP90AA1 to avibirnavirus VP2 induces autophagy by inactivating the AKT-MTOR pathway. Autophagy 11(3):503–515

Hu S, Liu X, Gao Y, Zhou R, Yan H, Zhao Y (2018) Hepatitis B virus inhibits neutrophil extracellular traps release by modulating reactive oxygen species production and autophagy. bioRxiv:334227

Hu Z, Cai M, Zhang Y, Tao L, Guo R (2020) miR-29c-3p inhibits autophagy and cisplatin resistance in ovarian cancer by regulating FOXP1/ATG14 pathway. Cell Cycle 19(2):193–206

Hua L, Zhu G, Wei J (2018) MicroRNA-1 overexpression increases chemosensitivity of non-small cell lung cancer cells by inhibiting autophagy related 3-mediated autophagy. Cell Biol Int 42(9):1240–1249

Huang SC, Chang CL, Wang PS, Tsai Y, Liu HS (2009) Enterovirus 71-induced autophagy detected in vitro and in vivo promotes viral replication. J Med Virol 81(7):1241–1252

Huang H, Kang R, Wang J, Luo G, Yang W, Zhao Z (2013) Hepatitis C virus inhibits AKT-tuberous sclerosis complex (TSC), the mechanistic target of rapamycin (MTOR) pathway, through endoplasmic reticulum stress to induce autophagy. Autophagy 9(2):175–195

Huang Z, Zhou L, Chen Z, Nice EC, Huang C (2016) Stress management by autophagy: implications for chemoresistance. Int J Cancer 139(1):23–32

Huang K-T, Kuo I-Y, Tsai M-C, Wu C-H, Hsu L-W, Chen L-Y et al (2017) Factor VII-induced microRNA-135a inhibits autophagy and is associated with poor prognosis in hepatocellular carcinoma. Mol Ther Nucleic Acids 9:274–283

Huang T, Wan X, Alvarez AA, James CD, Song X, Yang Y et al (2019a) MIR93 (microRNA-93) regulates tumorigenicity and therapy response of glioblastoma by targeting autophagy. Autophagy 15(6):1100–1111

Huang S, Qi P, Zhang T, Li F, He X (2019b) The HIF-1α/miR-224-3p/ATG5 axis affects cell mobility and chemosensitivity by regulating hypoxia-induced protective autophagy in glioblastoma and astrocytoma. Oncol Rep 41(3):1759–1768

Huang Y, Liu W, He B, Wang L, Zhang F, Shu H et al (2020) Exosomes derived from bone marrow mesenchymal stem cells promote osteosarcoma development by activating oncogenic autophagy. J Bone Oncol 21: 100280

Huangfu L, Liang H, Wang G, Su X, Li L, Du Z et al (2016) miR-183 regulates autophagy and apoptosis in colorectal cancer through targeting of UVRAG. Oncotarget 7(4):4735

Hung C-H, Chen L-W, Wang W-H, Chang P-J, Chiu Y-F, Hung C-C et al (2014) Regulation of autophagic activation by Rta of Epstein-Barr virus via ERK-kinase pathway. J Virol 88(20):12133–12145. https://doi.org/10.1128/JVI.02033-14

Irmler M, Thome M, Hahne M, Schneider P, Hofmann K, Steiner V et al (1997) Inhibition of death receptor signals by cellular FLIP. Nature 388(6638):190

Jackson WT (2015) Viruses and the autophagy pathway. Virology 479:450–456

Jackson WT, Giddings TH Jr, Taylor MP, Mulinyawe S, Rabinovitch M, Kopito RR et al (2005) Subversion of cellular autophagosomal machinery by RNA viruses. PLoS Biol 3(5):e156

Jafari SH, Saadatpour Z, Salmaninejad A, Momeni F, Mokhtari M, Nahand JS et al (2018) Breast cancer diagnosis: imaging techniques and biochemical markers. J Cell Physiol 233(7):5200–5213

Jamali Z, Taheri-Anganeh M, Shabaninejad Z, Keshavarzi A, Taghizadeh H, Razavi ZS et al (2020) Autophagy regulation by microRNAs: novel insights into osteosarcoma therapy. IUBMB Life 72(7): 1306–1321

Jankelson L, Karam G, Becker ML, Chinitz LA, Tsai M-C (2020) QT prolongation, torsades de pointes, and sudden death with short courses of chloroquine or hydroxychloroquine as used in COVID-19: a systematic review. Heart Rhythm 17(9):1472–1479

Jia H, Liu W, Zhang B, Wang J, Wu P, Tandra N et al (2018) HucMSC exosomes-delivered 14-3-3ζ enhanced autophagy via modulation of ATG16L in preventing cisplatin-induced acute kidney injury. Am J Transl Res 10(1):101

Jia Y, Lin R, Jin H, Si L, Jian W, Yu Q et al (2019) MicroRNA-34 suppresses proliferation of human ovarian cancer cells by triggering autophagy and apoptosis and inhibits cell invasion by targeting Notch 1. Biochimie 160:193–199

Jiang H, White EJ, Rios-Vicil CI, Xu J, Gomez-Manzano C, Fueyo J (2011) Human adenovirus type 5 induces cell lysis through autophagy and autophagy-triggered caspase activity. J Virol 85(10):4720–4729

Jiang M, Wang H, Jin M, Yang X, Ji H, Jiang Y et al (2018) Exosomes from MiR-30d-5p-ADSCs reverse acute ischemic stroke-induced, autophagy-mediated brain injury by promoting M2 microglial/macrophage polarization. Cell Physiol Biochem 47(2):864–878

Jin R, Zhu W, Cao S, Chen R, Jin H, Liu Y et al (2013) Japanese encephalitis virus activates autophagy as a viral immune evasion strategy. PLoS One 8(1):e52909

Jin W, Liang Y, Li S, Lin G, Liang H, Zhang Z et al (2021) MiR-513b-5p represses autophagy during the malignant progression of hepatocellular carcinoma by targeting PIK3R3. Aging (Albany NY) 13(12):16072

Joseph GP, McDermott R, Baryshnikova MA, Cobbs CS, Ulasov IV (2017) Cytomegalovirus as an oncomodulatory agent in the progression of glioma. Cancer Lett 384:79–85

Ju J-A, Huang C-T, Lan S-H, Wang T-H, Lin P-C, Lee J-C et al (2013) Characterization of a colorectal cancer migration and autophagy-related microRNA miR-338-5p and its target gene PIK3C3. Biomark Genom Med 5(3):74–78

Kang Y, Yuan R, Xiang B, Zhao X, Gao P, Dai X et al (2017) Newcastle disease virus-induced autophagy mediates antiapoptotic signaling responses in vitro and in vivo. Oncotarget 8(43):73981

Ke P-Y, Chen SS-L (2011a) Activation of the unfolded protein response and autophagy after hepatitis C virus infection suppresses innate antiviral immunity in vitro. J Clin Invest 121(1):37–56

Ke PY, Chen SS (2011b) Activation of the unfolded protein response and autophagy after hepatitis C virus infection suppresses innate antiviral immunity in vitro. J Clin Invest 121(1):37–56

Keller S, Ridinger J, Rupp A-K, Janssen JW, Altevogt P (2011) Body fluid derived exosomes as a novel template for clinical diagnostics. J Transl Med 9(1):86

Keshavarz M, Dianat-Moghadam H, Sofiani VH, Karimzadeh M, Zargar M, Moghoofei M et al (2018) miRNA-based strategy for modulation of influenza A virus infection. Epigenomics 10(6):829–844

Khalil H (2012) Influenza A virus stimulates autophagy to undermine host cell IFN-β production. Egypt J Biochem Mol Biol 30:283–299

Khaminets A, Heinrich T, Mari M, Grumati P, Huebner AK, Akutsu M et al (2015) Regulation of endoplasmic reticulum turnover by selective autophagy. Nature 522(7556):354–358

Kharaziha P, Ceder S, Li Q, Panaretakis T (2012) Tumor cell-derived exosomes: a message in a bottle. Biochim Biophys Acta 1826(1):103–111

Khatami F, Saatchi M, Zadeh SST, Aghamir ZS, Shabestari AN, Reis LO et al (2020) A meta-analysis of accuracy and sensitivity of chest CT and RT-PCR in COVID-19 diagnosis. Sci Rep 10(1):1–12

Kihara A, Kabeya Y, Ohsumi Y, Yoshimori T (2001) Beclin–phosphatidylinositol 3-kinase complex functions at the trans-Golgi network. EMBO Rep 2(4):330–335

Kim YS, Seo HW, Jung G (2015) Reactive oxygen species promote heat shock protein 90-mediated HBV capsid assembly. Biochem Biophys Res Commun 457(3):328–333

Koepke L, Hirschenberger M, Hayn M, Kirchhoff F, Sparrer KM (2021) Manipulation of autophagy by SARS-CoV-2 proteins. Autophagy 17(9):2659–2661

Kondo Y, Kanzawa T, Sawaya R, Kondo S (2005) The role of autophagy in cancer development and response to therapy. Nat Rev Cancer 5(9):726

Kong P, Zhu X, Geng Q, Xia L, Sun X, Chen Y et al (2017) The microRNA-423-3p-Bim axis promotes cancer progression and activates oncogenic autophagy in gastric cancer. Mol Ther 25(4):1027–1037

Kovaleva V, Mora R, Park YJ, Plass C, Chiramel A, Bartenschlager R et al (2012) MicroRNA-130a targets ATG2B and DICER1 to inhibit autophagy and trigger killing of chronic lymphocytic leukemia cells. Cancer Res 72(7):1763–1772. canres.3671.2011

Kroemer G, Marino G, Levine B (2010) Autophagy and the integrated stress response. Mol Cell 40(2):280–293

Kumar SH, Rangarajan A (2009) Simian virus 40 small T antigen activates AMPK and triggers autophagy to protect cancer cells from nutrient deprivation. J Virol 83(17):8565–8574

Kvansakul M, Caria S, Hinds MG (2017) The Bcl-2 family in host-virus interactions. Viruses 9(10):290

Kwon JJ, Willy JA, Quirin KA, Wek RC, Korc M, Yin X-M et al (2016) Novel role of miR-29a in pancreatic cancer autophagy and its therapeutic potential. Oncotarget 7(44):71635

Kyei GB, Dinkins C, Davis AS, Roberts E, Singh SB, Dong C et al (2009a) Autophagy pathway intersects with HIV-1 biosynthesis and regulates viral yields in macrophages. J Cell Biol 186(2):255–268

Kyei GB, Dinkins C, Davis AS, Roberts E, Singh SB, Dong C et al (2009b) Autophagy pathway intersects with HIV-1 biosynthesis and regulates viral yields in macrophages. J Cell Biol 186(2):255–268

Lakkaraju A, Rodriguez-Boulan E (2008) Itinerant exosomes: emerging roles in cell and tissue polarity. Trends Cell Biol 18(5):199–209

Lan SH, Wu SY, Zuchini R, Lin XZ, Su IJ, Tsai TF et al (2014) Autophagy suppresses tumorigenesis of hepatitis B virus-associated hepatocellular carcinoma through degradation of microRNA-224. Hepatology 59(2):505–517

Lan T, Shen Z, Yan B, Chen J (2020) New insights into the interplay between miRNA and autophagy in the ageing of intervertebral disc. Ageing Res Rev 65:101227

Lässer C (2015) Exosomes in diagnostic and therapeutic applications: biomarker, vaccine and RNA interference delivery vehicle. Expert Opin Biol Ther 15(1):103–117

Lazar C, Macovei A, Petrescu S, Branza-Nichita N (2012) Activation of ERAD pathway by human hepatitis B virus modulates viral and subviral particle production. PLoS One 7(3):e34169

Lee D, Sugden B (2008) The latent membrane protein 1 oncogene modifies B-cell physiology by regulating autophagy. Oncogene 27(20):2833

Lee J-S, Li Q, Lee J-Y, Lee S-H, Jeong JH, Lee H-R et al (2009) FLIP-mediated autophagy regulation in cell death control. Nat Cell Biol 11(11):1355

Lee Y-R, Hu H-Y, Kuo S-H, Lei H-Y, Lin Y-S, Yeh T-M et al (2013) Dengue virus infection induces autophagy: an in vivo study. J Biomed Sci 20(1):65

Lee Y-R, Wang P-S, Wang J-R, Liu H-S (2014) Enterovirus 71-induced autophagy increases viral replication and pathogenesis in a suckling mouse model. J Biomed Sci 21(1):80

Lennemann NJ, Coyne CB (2017) Dengue and Zika viruses subvert reticulophagy by NS2B3-mediated cleavage of FAM134B. Autophagy 13(2):322–332

Letafati A, Najafi S, Mottahedi M, Karimzadeh M, Shahini A, Garousi S et al (2022) MicroRNA let-7 and viral infections: focus on mechanisms of action. Cell Mol Biol Lett 27(1):14

Levine B (2005) Eating oneself and uninvited guests: autophagy-related pathways in cellular defense. Cell 120(2):159–162

Levine B, Liu R, Dong X, Zhong Q (2015) Beclin orthologs: integrative hubs of cell signaling, membrane trafficking, and physiology. Trends Cell Biol 25(9): 533–544

Li J, Liu Y, Wang Z, Liu K, Wang Y, Liu J et al (2011a) Subversion of cellular autophagy machinery by hepatitis B virus for viral envelopment. J Virol 85(13): 6319–6333. https://doi.org/10.1128/JVI.02627-10

Li JC, Au K-y, Fang J-w, Yim HC, Chow K-h, Ho P-l et al (2011b) HIV-1 trans-activator protein dysregulates IFN-γ signaling and contributes to the suppression of autophagy induction. AIDS 25(1):15–25

Li J, Liu Y, Wang Z, Liu K, Wang Y, Liu J et al (2011c) Subversion of cellular autophagy machinery by hepatitis B virus for viral envelopment. J Virol 85(13): 6319–6333

Li X, Wang S, Chen Y, Liu G, Yang X (2014a) miR-22 targets the 3′ UTR of HMGB1 and inhibits the HMGB1-associated autophagy in osteosarcoma cells during chemotherapy. Tumor Biol 35(6):6021–6028

Li M, Zeringer E, Barta T, Schageman J, Cheng A, Vlassov AV (2014b) Analysis of the RNA content of the exosomes derived from blood serum and urine and its potential as biomarkers. Phil Trans R Soc B 369(1652):20130502

Li Z, Han N, Tian Y (2016a) MicroRNA-130a promotes apoptosis of alveolar epithelia in COPD patients by inhibiting autophagy via the down-regulation of ATG16L expression. Int J Clin Exp Med 9:23039–23047

Li S-P, He J-D, Wang Z, Yu Y, Fu S-Y, Zhang H-M et al (2016b) miR-30b inhibits autophagy to alleviate hepatic ischemia-reperfusion injury via decreasing the Atg12-Atg5 conjugate. World J Gastroenterol 22(18):4501

Li S, Qiang Q, Shan H, Shi M, Gan G, Ma F et al (2016c) MiR-20a and miR-20b negatively regulate autophagy by targeting RB1CC1/FIP200 in breast cancer cells. Life Sci 147:143–152

Li Y, Jiang W, Hu Y, Da Z, Zeng C, Tu M et al (2016d) MicroRNA-199a-5p inhibits cisplatin-induced drug resistance via inhibition of autophagy in osteosarcoma cells. Oncol Lett 12(5):4203–4208

Li L, Wang Z, Hu X, Wan T, Wu H, Jiang W et al (2016e) Human aortic smooth muscle cell-derived exosomal miR-221/222 inhibits autophagy via a PTEN/Akt signaling pathway in human umbilical vein endothelial cells. Biochem Biophys Res Commun 479(2):343–350

Li X-Q, Liu J-T, Fan L-L, Liu Y, Cheng L, Wang F et al (2016f) Exosomes derived from gefitinib-treated EGFR-mutant lung cancer cells alter cisplatin sensitivity via up-regulating autophagy. Oncotarget 7(17): 24585

Li W, Jiang Y, Wang Y, Yang S, Bi X, Pan X et al (2018a) MiR-181b regulates autophagy in a model of Parkinson's disease by targeting the PTEN/Akt/mTOR signaling pathway. Neurosci Lett 675:83–88

Li H, Chen L, Li J-j, Zhou Q, Huang A, Liu W-w et al (2018b) miR-519a enhances chemosensitivity and promotes autophagy in glioblastoma by targeting STAT3/Bcl2 signaling pathway. J Hematol Oncol 11(1):70

Li H, He C, Wang X, Wang H, Nan G, Fang L (2019a) MicroRNA-183 affects the development of gastric cancer by regulating autophagy via MALAT1-miR-183-

SIRT1 axis and PI3K/AKT/mTOR signals. Artif Cells Nanomed Biotechnol 47(1):3163–3171

Li Q, Wang Y, Peng W, Jia Y, Tang J, Li W et al (2019b) microRNA-101a regulates autophagy phenomenon via the MAPK pathway to modulate Alzheimer's-associated pathogenesis. Cell Transplant 28(8): 1076–1084

Li Y, Zhou D, Ren Y, Zhang Z, Guo X, Ma M et al (2019c) Mir223 restrains autophagy and promotes CNS inflammation by targeting ATG16L1. Autophagy 15(3): 478–492

Li Y, Zhang G, Wu B, Yang W, Liu Z (2019d) miR-199a-5p represses protective autophagy and overcomes chemoresistance by directly targeting DRAM1 in acute myeloid leukemia. J Oncol 2019:5613417

Li M, Meng X, Li M (2020a) MiR-126 promotes esophageal squamous cell carcinoma via inhibition of apoptosis and autophagy. Aging (Albany NY) 12(12):12107

Li JP, Zhang HM, Liu MJ, Xiang Y, Li H, Huang F et al (2020b) miR-133a-3p/FOXP3 axis regulates cell proliferation and autophagy in gastric cancer. J Cell Biochem 121(5–6):3392–3405

Li X, He S, Ma B (2020c) Autophagy and autophagy-related proteins in cancer. Mol Cancer 19(1):1–16

Li T, Gu J, Yang O, Wang J, Wang Y, Kong J (2020d) Bone marrow mesenchymal stem cell-derived exosomal miRNA-29c decreases cardiac ischemia/reperfusion injury through inhibition of excessive autophagy via the PTEN/Akt/mTOR signaling pathway. Circ J 84(8):1304–1311

Li P, Cao G, Huang Y, Wu W, Chen B, Wang Z et al (2020e) siMTA1-loaded exosomes enhanced chemotherapeutic effect of gemcitabine in luminal-b type breast cancer by inhibition of EMT/HIF-α and autophagy pathways. Front Oncol 10:541262

Li M, Ball CB, Collins G, Hu Q, Luse DS, Price DH et al (2020f) Human cytomegalovirus IE2 drives transcription initiation from a select subset of late infection viral promoters by host RNA polymerase II. PLoS Pathog 16(4):e1008402

Li Y, Lin S, Xie X, Zhu H, Fan T, Wang S (2021) Highly enriched exosomal lncRNA OIP5-AS1 regulates osteosarcoma tumor angiogenesis and autophagy through miR-153 and ATG5. Am J Transl Res 13(5):4211

Liang XH, Kleeman LK, Jiang HH, Gordon G, Goldman JE, Berry G et al (1998) Protection against fatal Sindbis virus encephalitis by beclin, a novel Bcl-2-interacting protein. J Virol 72(11):8586–8596

Liang XH, Jackson S, Seaman M, Brown K, Kempkes B, Hibshoosh H et al (1999) Induction of autophagy and inhibition of tumorigenesis by beclin 1. Nature 402(6762):672–676

Liang C, Feng P, Ku B, Dotan I, Canaani D, Oh B-H et al (2006) Autophagic and tumour suppressor activity of a novel Beclin1-binding protein UVRAG. Nat Cell Biol 8(7):688

Liang C, Xiaofei E, Jung JU (2008) Downregulation of autophagy by herpesvirus Bcl-2 homologs. Autophagy 4(3):268–272

Liang Q, Luo Z, Zeng J, Chen W, Foo S-S, Lee S-A et al (2016) Zika virus NS4A and NS4B proteins deregulate

Akt-mTOR signaling in human fetal neural stem cells to inhibit neurogenesis and induce autophagy. Cell Stem Cell 19(5):663–671

Liang Y, Chen X, Liang Z (2017) MicroRNA-320 regulates autophagy in retinoblastoma by targeting hypoxia inducible factor-1α. Exp Ther Med 14(3): 2367–2372

Liao W, Zhang Y (2020) MicroRNA-381 facilitates autophagy and apoptosis in prostate cancer cells via inhibiting the RELN-mediated PI3K/AKT/mTOR signaling pathway. Life Sci 254:117672

Liao H, Xiao Y, Hu Y, Xiao Y, Yin Z, Liu L (2015) microRNA-32 induces radioresistance by targeting DAB2IP and regulating autophagy in prostate cancer cells. Oncol Lett 10(4):2055–2062

Liao D, Li T, Ye C, Zeng L, Li H, Pu X et al (2018) miR-221 inhibits autophagy and targets TP53INP1 in colorectal cancer cells. Exp Ther Med 15(2): 1712–1717

Lin J, Li J, Huang B, Liu J, Chen X, Chen X-M et al (2015) Exosomes: novel biomarkers for clinical diagnosis. Sci World J 2015:657086

Lin Y, Zhao J, Wang H, Cao J, Nie Y (2017) miR-181a modulates proliferation, migration and autophagy in AGS gastric cancer cells and downregulates MTMR3. Mol Med Rep 15(5):2451–2456

Lin X-T, Zheng X-B, Fan D-J, Yao Q-Q, Hu J-C, Lian L et al (2018) MicroRNA-143 targets ATG2B to inhibit autophagy and increase inflammatory responses in Crohn's disease. Inflamm Bowel Dis 24(4):781–791

Lin B, Feng D, Xu J (2019) Cardioprotective effects of microRNA-18a on acute myocardial infarction by promoting cardiomyocyte autophagy and suppressing cellular senescence via brain derived neurotrophic factor. Cell Biosci 9(1):38

Liu Q, Qin Y, Zhou L, Kou Q, Guo X, Ge X et al (2012) Autophagy sustains the replication of porcine reproductive and respiratory virus in host cells. Virology 429(2):136–147

Liu H, Cao W, Li Y, Feng M, Wu X, Yu K et al (2013) Subgroup J avian leukosis virus infection inhibits autophagy in DF-1 cells. Virol J 10(1):196

Liu B, Fang M, Hu Y, Huang B, Li N, Chang C et al (2014) Hepatitis B virus X protein inhibits autophagic degradation by impairing lysosomal maturation. Autophagy 10(3):416–430

Liu X, Hong Q, Wang Z, Yu Y, Zou X, Xu L (2015a) MiR-21 inhibits autophagy by targeting Rab11a in renal ischemia/reperfusion. Exp Cell Res 338(1):64–69

Liu C, Qu A, Han X, Wang Y (2015b) HCV core protein represses the apoptosis and improves the autophagy of human hepatocytes. Int J Clin Exp Med 8(9):15787

Liu L, He J, Wei X, Wan G, Lao Y, Xu W et al (2017a) MicroRNA-20a-mediated loss of autophagy contributes to breast tumorigenesis by promoting genomic damage and instability. Oncogene 36(42):5874–5884

Liu L, Ren W, Chen K (2017b) MiR-34a promotes apoptosis and inhibits autophagy by targeting HMGB1 in acute myeloid leukemia cells. Cell Physiol Biochem 41(5):1981–1992

Liu Y, Song Y, Zhu X (2017c) MicroRNA-181a regulates apoptosis and autophagy process in Parkinson's disease by inhibiting p38 Mitogen-Activated Protein Kinase (MAPK)/c-Jun N-Terminal Kinases (JNK) signaling pathways. Med Sci Monit 23:1597

Liu L, Jin X, Hu C-F, Li R, Shen C-X (2017d) Exosomes derived from mesenchymal stem cells rescue myocardial ischaemia/reperfusion injury by inducing cardiomyocyte autophagy via AMPK and Akt pathways. Cell Physiol Biochem 43(1):52–68

Liu Y, Pan J, Liu L, Li W, Tao R, Chen Y et al (2017e) The influence of HCMV infection on autophagy in THP-1 cells. Medicine 96(44):e8298

Liu F, Zhang Z, Xin G, Guo L, Jiang Q, Wang Z (2018a) miR-192 prevents renal tubulointerstitial fibrosis in diabetic nephropathy by targeting Egr1. Eur Rev Med Pharmacol Sci 22(13):4252–4260

Liu J, Jiang M, Deng S, Lu J, Huang H, Zhang Y et al (2018b) miR-93-5p-containing exosomes treatment attenuates acute myocardial infarction-induced myocardial damage. Mol Ther Nucleic Acids 11: 103–115

Liu W, Jiang D, Gong F, Huang Y, Luo Y, Rong Y et al (2020a) miR-210-5p promotes epithelial–mesenchymal transition by inhibiting PIK3R5 thereby activating oncogenic autophagy in osteosarcoma cells. Cell Death Dis 11(2):1–15

Liu S, Wang H, Mu J, Wang H, Peng Y, Li Q et al (2020b) MiRNA-211 triggers an autophagy-dependent apoptosis in cervical cancer cells: regulation of Bcl-2. Naunyn Schmiedeberg's Arch Pharmacol 393(3):359–370

Liu X, Zhou Z, Wang Y, Zhu K, Deng W, Li Y et al (2020c) Corrigendum: downregulation of HMGA1 mediates autophagy and inhibits migration and invasion in bladder cancer via miRNA-221/TP53INP1/p-ERK axis. Front Oncol 10:1735

Liu G, Kang X, Guo P, Shang Y, Du R, Wang X et al (2020d) miR-25-3p promotes proliferation and inhibits autophagy of renal cells in polycystic kidney mice by regulating ATG14-Beclin 1. Ren Fail 42(1):333–342

Liu J-J, Li Y, Yang M-S, Chen R, Cen C-Q (2020e) SP1-induced ZFAS1 aggravates sepsis-induced cardiac dysfunction via miR-590–3p/NLRP3-mediated autophagy and pyroptosis. Arch Biochem Biophys 695:108611

Liu F, Ai FY, Zhang DC, Tian L, Yang ZY, Liu SJ (2020f) LncRNA NEAT1 knockdown attenuates autophagy to elevate 5-FU sensitivity in colorectal cancer via targeting miR-34a. Cancer Med 9(3):1079–1091

Loi M, Müller A, Steinbach K, Niven J, da Silva RB, Paul P et al (2016) Macroautophagy proteins control MHC class I levels on dendritic cells and shape anti-viral CD8+ T cell responses. Cell Rep 15(5):1076–1087

Long J, He Q, Yin Y, Lei X, Li Z, Zhu W (2020) The effect of miRNA and autophagy on colorectal cancer. Cell Prolif 53(10):e12900

Lu W, Lin J, Zheng D, Hong C, Ke L, Wu X et al (2020) Overexpression of microRNA-133a inhibits apoptosis and autophagy in a cell model of Parkinson's disease by downregulating Ras-related C3 botulinum toxin substrate 1 (RAC1). Med Sci Monit 26:e922032–e922031

Lu X, Zhang Y, Zheng Y, Chen B (2021) The miRNA-15b/USP7/KDM6B axis engages in the initiation of osteoporosis by modulating osteoblast differentiation and autophagy. J Cell Mol Med 25(4):2069–2081

Luo H-C, Yi T-Z, Huang F-G, Wei Y, Luo X-P, Luo Q-S (2020) Role of long noncoding RNA MEG3/miR-378/GRB2 axis in neuronal autophagy and neurological functional impairment in ischemic stroke. J Biol Chem 295(41):14125–14139

Luo L, Jian X, Sun H, Qin J, Wang Y, Zhang J et al (2021) Cartilage endplate stem cells inhibit intervertebral disc degeneration by releasing exosomes to nucleus pulposus cells to activate Akt/autophagy. Stem Cells 39(4):467–481

Lussignol M, Queval C, Bernet-Camard M-F, Cotte-Laffitte J, Beau I, Codogno P et al (2013) The herpes simplex virus type 1 Us11 protein inhibits autophagy through its interaction with the protein kinase PKR. J Virol 87(2):859–871. https://doi.org/10.1128/JVI.01158-12

Lv S, Xu Q, Sun E, Zhang J, Wu D (2016) Impaired cellular energy metabolism contributes to bluetongue-virus-induced autophagy. Arch Virol 161(10):2807–2811

Lv X, Wang K, Tang W, Yu L, Cao H, Chi W et al (2019) miR-34a-5p was involved in chronic intermittent hypoxia-induced autophagy of human coronary artery endothelial cells via Bcl-2/beclin 1 signal transduction pathway. J Cell Biochem 120(11):18871–18882

Ma J, Sun Q, Mi R, Zhang H (2011a) Avian influenza A virus H5N1 causes autophagy-mediated cell death through suppression of mTOR signaling. J Genet Genomics 38(11):533–537

Ma J, Sun Q, Mi R, Zhang H (2011b) Avian influenza A virus H5N1 causes autophagy-mediated cell death through suppression of mTOR signaling. J Genet Genomics 38(11):533–537

Ma L, Li Z, Li W, Ai J, Chen X (2019) MicroRNA-142-3p suppresses endometriosis by regulating KLF9-mediated autophagy in vitro and in vivo. RNA Biol 16(12):1733–1748

Ma Z, Li L, Livingston MJ, Zhang D, Mi Q, Zhang M et al (2020) p53/microRNA-214/ULK1 axis impairs renal tubular autophagy in diabetic kidney disease. J Clin Invest 130(9):5011–5026

Ma W, Zhou Y, Liu M, Qin Q, Cui Y (2021) Long non-coding RNA LINC00470 in serum derived exosome: a critical regulator for proliferation and autophagy in glioma cells. Cancer Cell Int 21(1):1–16

Mack HI, Munger K (2012) Modulation of autophagy-like processes by tumor viruses. Cells 1(3):204–247

Manning BD, Cantley LC (2007) AKT/PKB signaling: navigating downstream. Cell 129(7):1261–1274

Martin JL, Maldonado JO, Mueller JD, Zhang W, Mansky LM (2016) Molecular studies of HTLV-1 replication: an update. Viruses 8(2):31

McLean JE, Wudzinska A, Datan E, Quaglino D, Zakeri Z (2011) Flavivirus NS4A-induced autophagy protects cells against death and enhances virus replication. J

Biol Chem 286(25):22147–22159. https://doi.org/10.1074/jbc.M110.192500

Meng C, Zhou Z, Jiang K, Yu S, Jia L, Wu Y et al (2012a) Newcastle disease virus triggers autophagy in U251 glioma cells to enhance virus replication. Arch Virol 157(6):1011–1018

Meng S, Jiang K, Zhang X, Zhang M, Zhou Z, Hu M et al (2012b) Avian reovirus triggers autophagy in primary chicken fibroblast cells and Vero cells to promote virus production. Arch Virol 157(4):661–668

Meng F, Zhang Y, Li X, Wang J, Wang Z (2017) Clinical significance of miR-138 in patients with malignant melanoma through targeting of PDK1 in the PI3K/AKT autophagy signaling pathway. Oncol Rep 38(3):1655–1662

Meng C, Liu Y, Shen Y, Liu S, Wang Z, Ye Q et al (2018) MicroRNA-26b suppresses autophagy in breast cancer cells by targeting DRAM1 mRNA, and is downregulated by irradiation. Oncol Lett 15(2):1435–1440

Meng CY, Zhao ZQ, Bai R, Zhao W, Wang YX, Sun L et al (2020) MicroRNA-22 regulates autophagy and apoptosis in cisplatin resistance of osteosarcoma. Mol Med Rep 22(5):3911–3921

Meo S, Klonoff D, Akram J (2020) Efficacy of chloroquine and hydroxychloroquine in the treatment of COVID-19. Eur Rev Med Pharmacol Sci 24(8):4539–4547

Miller S, Kastner S, Krijnse-Locker J, Bühler S, Bartenschlager R (2007) The non-structural protein 4A of dengue virus is an integral membrane protein inducing membrane alterations in a 2K-regulated manner. J Biol Chem 282(12):8873–8882

Mirzaei H, Hamblin MR (2020) Regulation of glycolysis by non-coding RNAs in cancer: switching on the Warburg effect. Mol Ther Oncol 19:218–239

Mizushima N (2007) Autophagy: process and function. Genes Dev 21(22):2861–2873

Mizushima N, Levine B (2010) Autophagy in mammalian development and differentiation. Nat Cell Biol 12(9):823–830

Mlera L, Moy M, Maness K, Tran LN, Goodrum FD (2020) The role of the human cytomegalovirus UL133-UL138 gene locus in latency and reactivation. Viruses 12(7):714

Molina JM, Delaugerre C, Le Goff J, Mela-Lima B, Ponscarme D, Goldwirt L et al (2020) No evidence of rapid antiviral clearance or clinical benefit with the combination of hydroxychloroquine and azithromycin in patients with severe COVID-19 infection. Med Mal Infect 50(4):384

Mollazadeh S, Bazzaz BSF, Neshati V, de Vries AA, Naderi-Meshkin H, Mojarad M et al (2019) Overexpression of MicroRNA-148b-3p stimulates osteogenesis of human bone marrow-derived mesenchymal stem cells: the role of MicroRNA-148b-3p in osteogenesis. BMC Med Genet 20(1):1–10

Møller R, Schwarz TM, Noriega VM, Panis M, Sachs D, Tortorella D (2018) miRNA-mediated targeting of human cytomegalovirus reveals biological host and viral targets of IE2. Proc Natl Acad Sci 115(5):1069–1074

Moloughney JG, Monken CE, Tao H, Zhang H, Thomas JD, Lattime EC et al (2011) Vaccinia virus leads to ATG12-ATG3 conjugation and deficiency in autophagosome formation. Autophagy 7(12): 1434–1447

Monaco DC, Ende Z, Hunter E (2017) Virus-host gene interactions define HIV-1 disease progression. Viruses Genes Cancer 407:31–63

Mouna L, Hernandez E, Bonte D, Brost R, Amazit L, Delgui LR et al (2016) Analysis of the role of autophagy inhibition by two complementary human cytomegalovirus BECN1/Beclin 1-binding proteins. Autophagy 12(2):327–342

Mousavi SM, Derakhshan M, Baharloii F, Dashti F, Mirazimi SMA, Mahjoubin-Tehran M et al (2022) Non-coding RNAs and glioblastoma: insight into their roles in metastasis. Mol Ther Oncol 24:262–287

Mui UN, Haley CT, Tyring SK (2017) Viral oncology: molecular biology and pathogenesis. J Clin Med 6(12):111

Mukai R, Ohshima T (2014) HTLV-1 HBZ positively regulates the mTOR signaling pathway via inhibition of GADD34 activity in the cytoplasm. Oncogene 33(18):2317–2328

Mukhopadhyay U, Chanda S, Patra U, Mukherjee A, Rana S, Mukherjee A et al (2019) Synchronized orchestration of miR-99b and let-7g positively regulates rotavirus infection by modulating autophagy. Sci Rep 9(1):1–13

Nahand JS, Rabiei N, Fathazam R, Taghizadieh M, Ebrahimi MS, Mahjoubin-Tehran M et al (2021) Oncogenic viruses and chemoresistance: what do we know? Pharmacol Res 170:105730

Nakahata S, Chilmi S, Nakatake A, Sakamoto K, Yoshihama M, Nishikata I et al (2021) Clinical significance of soluble CADM1 as a novel marker for adult T-cell leukemia/lymphoma. Haematologica 106(2):532

Nakashima A, Tanaka N, Tamai K, Kyuuma M, Ishikawa Y, Sato H et al (2006) Survival of parvovirus B19-infected cells by cellular autophagy. Virology 349(2):254–263

Neshati V, Mollazadeh S, Bazzaz BSF, De Vries AA, Mojarrad M, Naderi-Meshkin H et al (2018) MicroRNA-499a-5p promotes differentiation of human bone marrow-derived mesenchymal stem cells to cardiomyocytes. Appl Biochem Biotechnol 186(1): 245–255

Neumann S, El Maadidi S, Faletti L, Haun F, Labib S, Schejtman A et al (2015) How do viruses control mitochondria-mediated apoptosis? Virus Res 209:45–55

Nyhan MJ, O'Donovan TR, Boersma AW, Wiemer EA, McKenna SL (2016) MiR-193b promotes autophagy and non-apoptotic cell death in oesophageal cancer cells. BMC Cancer 16(1):101

O'Donnell TB, Hyde JL, Mintern JD, Mackenzie JM (2016) Mouse Norovirus infection promotes autophagy induction to facilitate replication but prevents final autophagosome maturation. Virology 492:130–139

Oberstein A, Jeffrey PD, Shi Y (2007) Crystal structure of the Bcl-XL-Beclin 1 peptide complex Beclin 1 is a novel BH3-only protein. J Biol Chem 282(17): 13123–13132

Offerdahl DK, Dorward DW, Hansen BT, Bloom ME (2017) Cytoarchitecture of Zika virus infection in human neuroblastoma and Aedes albopictus cell lines. Virology 501:54–62

Orvedahl A, Alexander D, Tallóczy Z, Sun Q, Wei Y, Zhang W et al (2007) HSV-1 ICP34. 5 confers neurovirulence by targeting the Beclin 1 autophagy protein. Cell Host Microbe 1(1):23–35

Ou Y, He J, Liu Y (2018) MiR-490-3p inhibits autophagy via targeting ATG7 in hepatocellular carcinoma. IUBMB Life 70(6):468–478

Ouimet M, Ediriweera H, Afonso MS, Ramkhelawon B, Singaravelu R, Liao X et al (2017) microRNA-33 regulates macrophage autophagy in atherosclerosis. Arterioscler Thromb Vasc Biol 37(6):1058–1067. https://doi.org/10.1161/ATVBAHA.116.308916

Paludan C, Schmid D, Landthaler M, Vockerodt M, Kube D, Tuschl T et al (2005) Endogenous MHC class II processing of a viral nuclear antigen after autophagy. Science 307(5709):593–596

Pan J-A, Tang Y, Yu J-Y, Zhang H, Zhang J-F, Wang C-Q et al (2019) miR-146a attenuates apoptosis and modulates autophagy by targeting TAF9b/P53 pathway in doxorubicin-induced cardiotoxicity. Cell Death Dis 10(9):1–15

Pei X, Li Y, Zhu L, Zhou Z (2020) Astrocyte-derived exosomes transfer miR-190b to inhibit oxygen and glucose deprivation-induced autophagy and neuronal apoptosis. Cell Cycle 19(8):906–917

Peng J, Zhu S, Hu L, Ye P, Wang Y, Tian Q et al (2016) Wild-type rabies virus induces autophagy in human and mouse neuroblastoma cell lines. Autophagy 12(10):1704–1720

Peng H, Liu B, Yves TD, He Y, Wang S, Tang H et al (2018) Zika virus induces autophagy in human umbilical vein endothelial cells. Viruses 10(5):259

Petrovski G, Pásztor K, Orosz L, Albert R, Mencel E, Moe MC et al (2014) Herpes simplex virus types 1 and 2 modulate autophagy in SIRC corneal cells. J Biosci 39(4):683–692

Phatak P, Noe M, Asrani K, Chesnick IE, Greenwald BD, Donahue JM (2021) MicroRNA-141-3p regulates cellular proliferation, migration, and invasion in esophageal cancer by targeting tuberous sclerosis complex 1. Mol Carcinog 60(2):125–137

Plotkin SA, Boppana SB (2019) Vaccination against the human cytomegalovirus. Vaccine 37(50):7437–7442

Pourhanifeh MH, Vosough M, Mahjoubin-Tehran M, Hashemipour M, Nejati M, Abbasi-Kolli M et al (2020a) Autophagy-related microRNAs: possible regulatory roles and therapeutic potential in and gastrointestinal cancers. Pharmacol Res 161:105133

Pourhanifeh MH, Mahjoubin-Tehran M, Karimzadeh MR, Mirzaei HR, Razavi ZS, Sahebkar A et al (2020b) Autophagy in cancers including brain tumors: role of MicroRNAs. Cell Commun Signal 18(1):88

Qased AB, Yi H, Liang N, Ma S, Qiao S, Liu X (2013) MicroRNA-18a upregulates autophagy and ataxia telangiectasia mutated gene expression in HCT116 colon cancer cells. Mol Med Rep 7(2):559–564

Qu Y, Zhang Q, Cai X, Li F, Ma Z, Xu M et al (2017) Exosomes derived from miR-181-5p-modified adipose-derived mesenchymal stem cells prevent liver fibrosis via autophagy activation. J Cell Mol Med 21(10):2491–2502

Qu Y, Wang X, Zhu Y, Wang W, Wang Y, Hu G et al (2021) ORF3a-mediated incomplete autophagy facilitates severe acute respiratory syndrome coronavirus-2 replication. Front Cell Dev Biol 9: 716208

Ramalinga M, Roy A, Srivastava A, Bhattarai A, Harish V, Suy S et al (2015) MicroRNA-212 negatively regulates starvation induced autophagy in prostate cancer cells by inhibiting SIRT1 and is a modulator of angiogenesis and cellular senescence. Oncotarget 6(33):34446

Rautou PE, Mansouri A, Lebrec D, Durand F, Valla D, Moreau R (2010) Autophagy in liver diseases. J Hepatol 53(6):1123–1134

Razavi ZS, Tajiknia V, Majidi S, Ghandali M, Mirzaei HR, Rahimian N et al (2021) Gynecologic cancers and non-coding RNAs: epigenetic regulators with emerging roles. Crit Rev Oncol Hematol 157:103192

Reddehase MJ, Lemmermann NA (2019) Cellular reservoirs of latent cytomegaloviruses. Med Microbiol Immunol 208(3):391–403

Reggiori F, Monastyrska I, Verheije MH, Calì T, Ulasli M, Bianchi S et al (2010) Coronaviruses Hijack the LC3-I-positive EDEMosomes, ER-derived vesicles exporting short-lived ERAD regulators, for replication. Cell Host Microbe 7(6):500–508

Ren T, Takahashi Y, Liu X, Loughran TP, Sun SC, Wang HG et al (2015) HTLV-1 Tax deregulates autophagy by recruiting autophagic molecules into lipid raft microdomains. Oncogene 34(3):334–345

Ren W-W, Li D-D, Li X-L, He Y-P, Guo L-H, Liu L-N et al (2018) MicroRNA-125b reverses oxaliplatin resistance in hepatocellular carcinoma by negatively regulating EVA1A mediated autophagy. Cell Death Dis 9(5):547

Rezaei S, Mahjoubin-Tehran M, Aghaee-Bakhtiari SH, Jalili A, Movahedpour A, Khan H et al (2020) Autophagy-related MicroRNAs in chronic lung diseases and lung cancer. Crit Rev Oncol Hematol 153:103063

Robinson SM, Tsueng G, Sin J, Mangale V, Rahawi S, McIntyre LL et al (2014) Coxsackievirus B exits the host cell in shed microvesicles displaying autophagosomal markers. PLoS Pathog 10(4):e1004045

Rodriguez-Rocha H, Gomez-Gutierrez JG, Garcia-Garcia-A, Rao X-M, Chen L, McMasters KM et al (2011) Adenoviruses induce autophagy to promote virus replication and oncolysis. Virology 416(1–2):9–15

Romao S, Gannage M, Münz C (eds) (2013) Checking the garbage bin for problems in the house, or how autophagy assists in antigen presentation to the immune system. Semin Cancer Biol 23(5):391–396; Elsevier

Rosati A, Graziano V, De Laurenzi V, Pascale M, Turco M (2011) BAG3: a multifaceted protein that regulates major cell pathways. Cell Death Dis 2(4):e141

Rubinstein AD, Kimchi A (2012) Life in the balance–a mechanistic view of the crosstalk between autophagy and apoptosis. J Cell Sci 125(22):5259–5268

Sadri Nahand J, Shojaie L, Akhlagh SA, Ebrahimi MS, Mirzaei HR, Bannazadeh Baghi H et al (2021) Cell death pathways and viruses: role of microRNAs. Mol Ther Nucleic Acids 24:487–511

Saha S, Panigrahi DP, Patil S, Bhutia SK (2018) Autophagy in health and disease: a comprehensive review. Biomed Pharmacother 104:485–495

Salimi L, Akbari A, Jabbari N, Mojarad B, Vahhabi A, Szafert S et al (2020) Synergies in exosomes and autophagy pathways for cellular homeostasis and metastasis of tumor cells. Cell Biosci 10:1–18

Sanche S, Lin YT, Xu C, Romero-Severson E, Hengartner N, Ke R (2020) High contagiousness and rapid spread of severe acute respiratory syndrome coronavirus 2. Emerg Infect Dis 26(7):1470

Santoso MR, Ikeda G, Tada Y, Jung JH, Vaskova E, Sierra RG et al (2020) Exosomes from induced pluripotent stem cell–derived cardiomyocytes promote autophagy for myocardial repair. J Am Heart Assoc 9(6):e014345

Sarkar B, Nishikata I, Nakahata S, Ichikawa T, Shiraga T, Saha HR et al (2019) Degradation of p47 by autophagy contributes to CADM1 overexpression in ATLL cells through the activation of NF-κB. Sci Rep 9(1): 1–14

Schierhout G, McGregor S, Gessain A, Einsiedel L, Martinello M, Kaldor J (2020) Association between HTLV-1 infection and adverse health outcomes: a systematic review and meta-analysis of epidemiological studies. Lancet Infect Dis 20(1):133–143

Seca H, Lima RT, Lopes-Rodrigues V, Guimaraes JE, Gabriela GM, Vasconcelos MH (2013) Targeting miR-21 induces autophagy and chemosensitivity of leukemia cells. Curr Drug Targets 14(10):1135–1143

Senft D, Ze'ev AR (2015) UPR, autophagy, and mitochondria crosstalk underlies the ER stress response. Trends Biochem Sci 40(3):141–148

Shafabakhsh R, Arianfar F, Vosough M, Mirzaei HR, Mahjoubin-Tehran M, Khanbabaei H et al (2021) Autophagy and gastrointestinal cancers: the behind the scenes role of long non-coding RNAs in initiation, progression, and treatment resistance. Cancer Gene Ther 28(12):1229–1255

Shao Y, Liu X, Meng J, Zhang X, Ma Z, Yang G (2019) MicroRNA-1251-5p promotes carcinogenesis and autophagy via targeting the tumor suppressor TBCC in ovarian cancer cells. Mol Ther 27(9):1653–1664

Sharma T, Radosevich JA, Mandal CC (2021) Dual role of microRNAs in autophagy of colorectal cancer. Endocr Metab Immune Disord Drug Targets 21(1):56–66

Shi Y, He X, Zhu G, Tu H, Liu Z, Li W et al (2015) Coxsackievirus A16 elicits incomplete autophagy involving the mTOR and ERK pathways. PLoS One 10(4):e0122109

Shi JY, Chen C, Xu X, Lu Q (2019) miR-29a promotes pathological cardiac hypertrophy by targeting the PTEN/AKT/mTOR signalling pathway and suppressing autophagy. Acta Physiol 227(2):e13323

Shi C, Pan L, Peng Z, Li J (2020) MiR-126 regulated myocardial autophagy on myocardial infarction. Eur Rev Med Pharmacol Sci 24(12):6971–6979

Shrivastava S, Raychoudhuri A, Steele R, Ray R, Ray RB (2011) Knockdown of autophagy enhances the innate immune response in hepatitis C virus–infected hepatocytes. Hepatology 53(2):406–414

Shrivastava S, Chowdhury JB, Steele R, Ray R, Ray RB (2012) Hepatitis C virus upregulates Beclin1 for induction of autophagy and activates mTOR signaling. J Virol 86(16):8705–8712. https://doi.org/10.1128/JVI.00616-12

Singh SB, Davis AS, Taylor GA, Deretic V (2006) Human IRGM induces autophagy to eliminate intracellular mycobacteria. Science 313(5792):1438–1441

Singh SV, Dakhole AN, Deogharkar A, Kazi S, Kshirsagar R, Goel A et al (2017) Restoration of miR-30a expression inhibits growth, tumorigenicity of medulloblastoma cells accompanied by autophagy inhibition. Biochem Biophys Res Commun 491(4):946–952

Singh AK, Singh A, Singh R, Misra A (2020) Hydroxychloroquine in patients with COVID-19: a systematic review and meta-analysis. Diabetes Metab Syndr Clin Res Rev 14(4):589–596

Singletary K, Milner J (2008) Diet, autophagy, and cancer: a review. Cancer Epidemiol Biomarkers Prev 17(7):1596–1610

Sinha SC, Colbert CL, Becker N, Wei Y, Levine B (2008) Molecular basis of the regulation of Beclin 1-dependent autophagy by the γ-herpesvirus 68 Bcl-2 homolog M11. Autophagy 4(8):989–997

Sir D, Tian Y (2010) Chen W-l, Ann DK, Yen T-SB, Ou J-hJ. The early autophagic pathway is activated by hepatitis B virus and required for viral DNA replication. Proc Natl Acad Sci 107(9):4383–4388

Sir D, Liang C, Chen W-l, Jung JU, James Ou J-H (2008a) Perturbation of autophagic pathway by hepatitis C virus. Autophagy 4(6):830–831

Sir D, Chen W, Choi J, Wakita T, Yen TB, Ou JHJ (2008b) Induction of incomplete autophagic response by hepatitis C virus via the unfolded protein response. Hepatology 48(4):1054–1061

Sir D, Ann DK, Ou JH (2010a) Autophagy by hepatitis B virus and for hepatitis B virus. Autophagy 6(4):548–549

Sir D, Tian Y, Chen WL, Ann DK, Yen TS, Ou JH (2010b) The early autophagic pathway is activated by hepatitis B virus and required for viral DNA replication. Proc Natl Acad Sci U S A 107(9):4383–4388

Sir D, Kuo C-F, Tian Y, Liu HM, Huang EJ, Jung JU et al (2012) Replication of hepatitis C virus RNA on autophagosomal membranes. J Biol Chem 287(22):18036–18043. https://doi.org/10.1074/jbc.M111.320085

Song L, Zhou F, Cheng L, Hu M, He Y, Zhang B et al (2017) MicroRNA-34a suppresses autophagy in alveolar type II epithelial cells in acute lung injury by inhibiting FoxO3 expression. Inflammation 40(3):927–936

Song H, Du C, Wang X, Zhang J, Shen Z (2019) MicroRNA-101 inhibits autophagy to alleviate liver ischemia/reperfusion injury via regulating the mTOR signaling pathway erratum in/10.3892/ijmm.2019.4160. Int J Mol Med 43(3):1331–1342

Soni M, Patel Y, Markoutsa E, Jie C, Liu S, Xu P et al (2018a) Autophagy, cell viability, and chemoresistance are regulated by miR-489 in breast cancer. Mol Cancer Res 16(9):1348–1360

Soni M, Patel Y, Markoutsa E, Jie C, Liu S, Xu P et al (2018b) Autophagy, cell viability, and chemoresistance are regulated by miR-489 in breast cancer. Mol Cancer Res 16(9):1348–1360

Stiuso P, Potenza N, Lombardi A, Ferrandino I, Monaco A, Zappavigna S et al (2015) MicroRNA-423-5p promotes autophagy in cancer cells and is increased in serum from hepatocarcinoma patients treated with sorafenib. Mol Ther Nucleic Acids 4:E233

Su W-C, Chao T-C, Huang Y-L, Weng S-C, Jeng K-S, Lai MM (2011) Rab5 and class III PI-3-kinase Vps34 are involved in hepatitis C virus NS4B-induced autophagy. J Virol 85(20):10561–10571. https://doi.org/10.1128/JVI.00173-11

Su Z, Yang Z, Xu Y, Chen Y, Yu Q (2015) MicroRNAs in apoptosis, autophagy and necroptosis. Oncotarget 6(11):8474

Su B, Wang X, Sun Y, Long M, Zheng J, Wu W et al (2020) miR-30e-3p promotes cardiomyocyte autophagy and inhibits apoptosis via regulating Egr-1 during ischemia/hypoxia. Biomed Res Int 2020:1–10

Suares A, Medina MV, Coso O (2021) Autophagy in viral development and progression of cancer. Front Oncol 11:147

Sun Y, Li C, Shu Y, Ju X, Zou Z, Wang H et al (2012) Inhibition of autophagy ameliorates acute lung injury caused by avian influenza A H5N1 infection. Sci Signal 5(212):ra16

Sun AG, Meng FG, Wang MG (2017a) CISD2 promotes the proliferation of glioma cells via suppressing beclin-1-mediated autophagy and is targeted by microRNA-449a. Mol Med Rep 16(6):7939–7948

Sun L, Zhao M, Wang Y, Liu A, Lv M, Li Y et al (2017b) Neuroprotective effects of miR-27a against traumatic brain injury via suppressing FoxO3a-mediated neuronal autophagy. Biochem Biophys Res Commun 482(4):1141–1147

Sun M, Hou L, Tang Y-d, Liu Y, Wang S, Wang J et al (2017c) Pseudorabies virus infection inhibits autophagy in permissive cells in vitro. Sci Rep 7:39964

Surviladze Z, Sterk RT, DeHaro SA, Ozbun MA (2013) Cellular entry of human papillomavirus type 16 involves activation of the PI3K/Akt/mTOR pathway and inhibition of autophagy. J Virol 87(5): 2508–2517. https://doi.org/10.1128/JVI.02319-12

Taguwa S, Kambara H, Fujita N, Noda T, Yoshimori T, Koike K et al (2011) Dysfunction of autophagy participates in vacuole formation and cell death in cells replicating hepatitis C virus. J Virol 85(24):13185–13194

Takahashi M-N, Jackson W, Laird DT, Culp TD, Grose C, Haynes JI et al (2009) Varicella-zoster virus infection induces autophagy in both cultured cells and human skin vesicles. J Virol 83(11):5466–5476

Tan S, Shi H, Ba M, Lin S, Tang H, Zeng X et al (2016) miR-409-3p sensitizes colon cancer cells to oxaliplatin by inhibiting Beclin-1-mediated autophagy. Int J Mol Med 37(4):1030–1038

Tan D, Zhou C, Han S, Hou X, Kang S, Zhang Y (2018) MicroRNA-378 enhances migration and invasion in cervical cancer by directly targeting autophagy-related protein 12. Mol Med Rep 17(5):6319–6326

Tang H, Da L, Mao Y, Li Y, Li D, Xu Z et al (2009a) Hepatitis B virus X protein sensitizes cells to starvation-induced autophagy via up-regulation of beclin 1 expression. Hepatology 49(1):60–71

Tang YC, Thoman M, Linton PJ, Deisseroth A (2009b) Use of CD40L immunoconjugates to overcome the defective immune response to vaccines for infections and cancer in the aged. Cancer Immunol Immunother 58(12):1949–1957

Tang S-W, Chen C-Y, Klase Z, Zane L, Jeang K-T (2013) The cellular autophagy pathway modulates human T-cell leukemia virus type 1 replication. J Virol 87(3):1699–1707

Tang H, Xu X, Xiao W, Liao Y, Xiao X, Li L et al (2019) Silencing of microRNA-27a facilitates autophagy and apoptosis of melanoma cells through the activation of the SYK-dependent mTOR signaling pathway. J Cell Biochem 120(8):13262–13274

Tavakolizadeh J, Roshanaei K, Salmaninejad A, Yari R, Nahand JS, Sarkarizi HK et al (2018) MicroRNAs and exosomes in depression: potential diagnostic biomarkers. J Cell Biochem 119(5):3783–3797

Taylor DD, Gercel-Taylor C (2008) MicroRNA signatures of tumor-derived exosomes as diagnostic biomarkers of ovarian cancer. Gynecol Oncol 110(1):13–21

Tey S-K, Khanna R (2012) Autophagy mediates transporter associated with antigen processing-independent presentation of viral epitopes through MHC class I pathway. Blood 120(5):994–1004

Théry C, Zitvogel L, Amigorena S (2002) Exosomes: composition, biogenesis and function. Nat Rev Immunol 2(8):569

Thome M, Schneider P, Hofmann K, Fickenscher H, Meinl E, Neipel F et al (1997) Viral FLICE-inhibitory proteins (FLIPs) prevent apoptosis induced by death receptors. Nature 386(6624):517

To KK-W, Tsang OT-Y, Yip CC-Y, Chan K-H, Wu T-C, Chan JM-C et al (2020) Consistent detection of 2019

novel coronavirus in saliva. Clin Infect Dis 71(15): 841–843

Tomlinson CC, Damania B (2004) The K1 protein of Kaposi's sarcoma-associated herpesvirus activates the Akt signaling pathway. J Virol 78(4):1918–1927

Torresilla C, Larocque É, Landry S, Halin M, Coulombe Y, Masson J-Y et al (2013) Detection of the HIV-1 minus strand-encoded Antisense Protein and its association with autophagy. J Virol 87(9): 5089–5105. https://doi.org/10.1128/JVI.00225-13

Utama A, Siburian MD, Purwantomo S, Intan MDB, Kurniasih TS, Gani RA et al (2011) Association of core promoter mutations of hepatitis B virus and viral load is different in HBeAg (+) and HBeAg (−) patients. World J Gastroenterol 17(6):708

Van Grol J, Subauste C, Andrade RM, Fujinaga K, Nelson J, Subauste CS (2010) HIV-1 inhibits autophagy in bystander macrophage/monocytic cells through Src-Akt and STAT3. PLoS One 5(7):e11733

Van Niel G, Porto-Carreiro I, Simoes S, Raposo G (2006) Exosomes: a common pathway for a specialized function. J Biochem 140(1):13–21

Vescarelli E, Gerini G, Megiorni F, Anastasiadou E, Pontecorvi P, Solito L et al (2020) MiR-200c sensitizes Olaparib-resistant ovarian cancer cells by targeting Neuropilin 1. J Exp Clin Cancer Res 39(1):1–15

Vojtechova Z, Tachezy R (2018) The role of miRNAs in virus-mediated oncogenesis. Int J Mol Sci 19(4):1217

Wang L, Damania B (2008) Kaposi's sarcoma–associated herpesvirus confers a survival advantage to endothelial cells. Cancer Res 68(12):4640–4648

Wang X, Gao Y, Tan J, Devadas K, Ragupathy V, Takeda K et al (2012) HIV-1 and HIV-2 infections induce autophagy in Jurkat and CD4+ T cells. Cell Signal 24(7):1414–1419

Wang P, Guo Q-s, Wang Z-w, Qian H-x (2013a) HBx induces HepG-2 cells autophagy through PI3K/Akt–mTOR pathway. Mol Cell Biochem 372(1–2):161–168

Wang J, Niu Z, Shi Y, Gao C, Wang X, Han J et al (2013c) Bcl-3, induced by Tax and HTLV-1, inhibits NF-κB activation and promotes autophagy. Cell Signal 25(12):2797–2804

Wang W, Zhou J, Shi J, Zhang Y, Liu S, Liu Y et al (2014) HTLV-1 Tax-deregulated both autophagy pathway and c-FLIP expression contribute to the resistance against death receptor-mediated apoptosis. J Virol 88(5): 2786–2798. https://doi.org/10.1128/JVI.03025-13

Wang J, Kang R, Huang H, Xi X, Wang B, Wang J et al (2014a) Hepatitis C virus core protein activates autophagy through EIF2AK3 and ATF6 UPR pathway-mediated MAP1LC3B and ATG12 expression. Autophagy 10(5):766–784

Wang HY, Yang GF, Huang YH, Huang QW, Gao J, Zhao XD et al (2014b) Reduced expression of autophagy markers correlates with high-risk human papillomavirus infection in human cervical squamous cell carcinoma. Oncol Lett 8(4):1492–1498

Wang I-K, Sun K-T, Tsai T-H, Chen C-W, Chang S-S, Yu T-M et al (2015a) MiR-20a-5p mediates hypoxia-induced autophagy by targeting ATG16L1 in ischemic kidney injury. Life Sci 136:133–141

Wang G, Yu Y, Tu Y, Tong J, Liu Y, Zhang C et al (2015b) Highly pathogenic porcine reproductive and respiratory syndrome virus infection induced apoptosis and autophagy in thymi of infected piglets. PLoS One 10(6):e0128292

Wang L, Tian Y, Ou JH (2015c) HCV induces the expression of Rubicon and UVRAG to temporally regulate the maturation of autophagosomes and viral replication. PLoS Pathog 11(3):e1004764

Wang H, Ye Y, Zhu Z, Mo L, Lin C, Wang Q et al (2016) MiR-124 regulates apoptosis and autophagy process in MPTP model of P arkinson's disease by targeting to B im. Brain Pathol 26(2):167–176

Wang Y, Luo J, Wang X, Yang B, Cui L (2017a) MicroRNA-199a-5p induced autophagy and inhibits the pathogenesis of ankylosing spondylitis by modulating the mTOR signaling via directly targeting Ras homolog enriched in brain (Rheb). Cell Physiol Biochem 42(6):2481–2491

Wang Y, Wang Q, Song J (2017b) Inhibition of autophagy potentiates the proliferation inhibition activity of microRNA-7 in human hepatocellular carcinoma cells. Oncol Lett 14(3):3566–3572

Wang B, Jia H, Zhang B, Wang J, Ji C, Zhu X et al (2017c) Pre-incubation with hucMSC-exosomes prevents cisplatin-induced nephrotoxicity by activating autophagy. Stem Cell Res Ther 8(1):75

Wang Z, Hu J, Pan Y, Shan Y, Jiang L, Qi X et al (2018a) miR-140-5p/miR-149 affects chondrocyte proliferation, apoptosis, and autophagy by targeting FUT1 in osteoarthritis. Inflammation 41(3):959–971

Wang B, Huang J, Li J, Zhong Y (2018b) Control of macrophage autophagy by miR-384-5p in the development of diabetic encephalopathy. Am J Transl Res 10(2):511

Wang Y, Zhang S, Dang S, Fang X, Liu M (2019a) Overexpression of microRNA-216a inhibits autophagy by targeting regulated MAP1S in colorectal cancer. Onco Targets Ther 12:4621

Wang D, Bao F, Teng Y, Li Q, Li J (2019b) MicroRNA-506-3p initiates mesenchymal-to-epithelial transition and suppresses autophagy in osteosarcoma cells by directly targeting SPHK1. Biosci Biotechnol Biochem 83(5):836–844

Wang P, Zhao ZQ, Guo SB, Yang TY, Chang ZQ, Li DH et al (2019c) Roles of microRNA-22 in suppressing proliferation and promoting sensitivity of osteosarcoma cells via metadherin-mediated autophagy. Orthop Surg 11(2):285–293

Wang Z-C, Huang F-Z, Xu H-B, Sun J-C, Wang C-F (2019d) MicroRNA-137 inhibits autophagy and chemosensitizes pancreatic cancer cells by targeting ATG5. Int J Biochem Cell Biol 111:63–71

Wang J, Chen J, Liu Y, Zeng X, Wei M, Wu S et al (2019e) Hepatitis B virus induces autophagy to promote its replication by the axis of miR-192-3p-XIAP

through NF kappa B signaling. Hepatology 69(3):974–992

Wang R, Zhu Y, Lin X, Ren C, Zhao J, Wang F et al (2019f) Influenza M2 protein regulates MAVS-mediated signaling pathway through interacting with MAVS and increasing ROS production. Autophagy 15(7):1163–1181

Wang L, Xu P, Xie X, Hu F, Jiang L, Hu R et al (2020a) Down regulation of SIRT2 reduced ASS induced NSCLC apoptosis through the release of autophagy components via exosomes. Front Cell Dev Biol 8:1495

Wang B, Mao J-h, Wang B-Y, Wang L-X, Wen H-Y, Xu L-J et al (2020b) Exosomal miR-1910-3p promotes proliferation, metastasis, and autophagy of breast cancer cells by targeting MTMR3 and activating the NF-κB signaling pathway. Cancer Lett 489:87–99

Wang L, Wang Y, Quan J (2020c) Exosomal miR-223 derived from natural killer cells inhibits hepatic stellate cell activation by suppressing autophagy. Mol Med 26(1):1–9

Wang Y, Wang P, Zhao L, Chen X, Lin Z, Zhang L et al (2021a) miR-224-5p carried by human umbilical cord mesenchymal stem cells-derived exosomes regulates autophagy in breast cancer cells via HOXA5. Front Cell Dev Biol 9:1308

Wang Y, He SH, Liang X, Zhang XX, Li SS, Li TF (2021b) ATF4-modified serum exosomes derived from osteoarthritic mice inhibit osteoarthritis by inducing autophagy. IUBMB Life 73(1):146–158

Wei J, Ma Z, Li Y, Zhao B, Wang D, Jin Y et al (2015) miR-143 inhibits cell proliferation by targeting autophagy-related 2B in non-small cell lung cancer H1299 cells. Mol Med Rep 11(1):571–576

Wei R, Cao G, Deng Z, Su J, Cai L (2016) miR-140-5p attenuates chemotherapeutic drug induced cell death by regulating autophagy through IP3k2 in human osteosarcoma cells. Biosci Rep 36(5):e00392. https://doi.org/10.1042/BSR20160238

Wei X, Yi X, Lv H, Sui X, Lu P, Li L et al (2020) MicroRNA-377-3p released by mesenchymal stem cell exosomes ameliorates lipopolysaccharide-induced acute lung injury by targeting RPTOR to induce autophagy. Cell Death Dis 11(8):1–14

Wei-Wei R, Dan-Dan L, Chen X, Xiao-Long L, He Y-P, Le-Hang G et al (2018) MicroRNA-125b reverses oxaliplatin resistance in hepatocellular carcinoma by negatively regulating EVA1A mediated autophagy. Cell Death Dis 9:1–15

Wen H-J, Yang Z, Zhou Y, Wood C (2010) Enhancement of autophagy during lytic replication by the Kaposi's sarcoma-associated herpesvirus replication and transcription activator. J Virol 84(15):7448–7458

Wen Z, Zhang J, Tang P, Tu N, Wang K, Wu G (2018a) Overexpression of miR-185 inhibits autophagy and apoptosis of dopaminergic neurons by regulating the AMPK/mTOR signaling pathway in Parkinson's disease. Mol Med Rep 17(1):131–137

Wen Z, Zhang J, Tang P, Tu N, Wang K, Wu G (2018b) Overexpression of miR185 inhibits autophagy and apoptosis of dopaminergic neurons by regulating the AMPK/mTOR signaling pathway in Parkinson's disease. Mol Med Rep 17(1):131–137

Wen D, Liu W-l, Lu Z-W, Cao Y-M, Ji Q-H, Wei W-J (2021) SNHG9, a papillary thyroid cancer cell exosome-enriched lncRNA, inhibits cell autophagy and promotes cell apoptosis of normal thyroid epithelial cell Nthy-ori-3 through YBOX3/P21 pathway. Front Oncol 11:1538

Wirawan E, Lippens S, Vanden Berghe T, Romagnoli A, Fimia GM, Piacentini M et al (2012) Beclin1: a role in membrane dynamics and beyond. Autophagy 8(1): 6–17

Worldometer (2020) Covid-19

Wu X-Y, Yao X-Q, Wu Z-F, Chen C, Liu J-Y, Wu G-N et al (2016a) MiR-32 induces radio-resistance by targeting DOC-2/DAB2 interactive protein and regulating autophagy in gastric carcinoma. Int J Clin Exp Pathol 9(9):8933–8942

Wu L, Liu T, Xiao Y, Li X, Zhu Y, Zhao Y et al (2016b) Polygonatum odoratum lectin induces apoptosis and autophagy by regulation of microRNA-1290 and microRNA-15a-3p in human lung adenocarcinoma A549 cells. Int J Biol Macromol 85:217–226

Wu H, Zhai X, Chen Y, Wang R, Lin L, Chen S et al (2016c) Protein 2B of coxsackievirus B3 induces autophagy relying on its transmembrane hydrophobic sequences. Viruses 8(5):131

Wu S-Y, Lan S-H, Liu H-S (2016d) Autophagy and microRNA in hepatitis B virus-related hepatocellular carcinoma. World J Gastroenterol 22(1):176

Wu J, Gao F, Xu T, Deng X, Wang C, Yang X et al (2018) miR-503 suppresses the proliferation and metastasis of esophageal squamous cell carcinoma by triggering autophagy via PKA/mTOR signaling. Int J Oncol 52(5):1427–1442

Wu K, Huang J, Xu T, Ye Z, Jin F, Li N et al (2019) MicroRNA-181b blocks gensenoside Rg3-mediated tumor suppression of gallbladder carcinoma by promoting autophagy flux via CREBRF/CREB3 pathway. Am J Transl Res 11(9):5776

Wu H, Liu C, Yang Q, Xin C, Du J, Sun F et al (2020) MIR145-3p promotes autophagy and enhances bortezomib sensitivity in multiple myeloma by targeting HDAC4. Autophagy 16(4):683–697

Xi Z, Si J, Nan J (2019) LncRNA MALAT1 potentiates autophagy-associated cisplatin resistance by regulating the microRNA-30b/autophagy-related gene 5 axis in gastric cancer. Int J Oncol 54(1):239–248

Xiao W, Dai B, Zhu Y, Ye D (2015) Norcantharidin induces autophagy-related prostate cancer cell death through Beclin-1 upregulation by miR-129-5p suppression. Tumour Biol. https://doi.org/10.1007/s13277-015-4488-6. Online ahead of print

Xiao W, Dai B, Zhu Y, Ye D (2016) Norcantharidin induces autophagy-related prostate cancer cell death through Beclin-1 upregulation by miR-129-5p suppression. Tumor Biol 37(12):15643–15648

Xie N, Yuan K, Zhou L, Wang K, Chen HN, Lei Y et al (2016) PRKAA/AMPK restricts HBV replication through promotion of autophagic degradation. Autophagy 12(9):1507–1520

Xin L, Ma X, Xiao Z, Yao H, Liu Z (2015) Coxsackievirus B3 induces autophagy in HeLa cells via the AMPK/MEK/ERK and Ras/Raf/MEK/ERK signaling pathways. Infect Genet Evol 36:46–54

Xing H, Tan J, Miao Y, Lv Y, Zhang Q (2021) Crosstalk between exosomes and autophagy: a review of molecular mechanisms and therapies. J Cell Mol Med 25(5): 2297–2308

Xiong J (2015) Atg7 in development and disease: panacea or Pandora's box? Protein Cell 6(10):722–734

Xiong J, Wang D, Wei A, Ke N, Wang Y, Tang J et al (2017) MicroRNA-410-3p attenuates gemcitabine resistance in pancreatic ductal adenocarcinoma by inhibiting HMGB1-mediated autophagy. Oncotarget 8(64):107500

Xu N, Zhang J, Shen C, Luo Y, Xia L, Xue F et al (2012) Cisplatin-induced downregulation of miR-199a-5p increases drug resistance by activating autophagy in HCC cell. Biochem Biophys Res Commun 423(4): 826–831

Xu Y, An Y, Wang Y, Zhang C, Zhang H, Huang C et al (2013) miR-101 inhibits autophagy and enhances cisplatin-induced apoptosis in hepatocellular carcinoma cells. Oncol Rep 29(5):2019–2024

Xu L, Beckebaum S, Iacob S, Wu G, Kaiser GM, Radtke A et al (2014) MicroRNA-101 inhibits human hepatocellular carcinoma progression through EZH2 downregulation and increased cytostatic drug sensitivity. J Hepatol 60(3):590–598

Xu R, Liu S, Chen H, Lao L (2016) MicroRNA-30a downregulation contributes to chemoresistance of osteosarcoma cells through activating Beclin-1-mediated autophagy. Oncol Rep 35(3):1757–1763

Xu J, Huang H, Peng R, Ding X, Jiang B, Yuan X et al (2018a) MicroRNA-30a increases the chemosensitivity of U251 glioblastoma cells to temozolomide by directly targeting beclin 1 and inhibiting autophagy. Exp Ther Med 15(6):4798–4804

Xu J, Su Y, Xu A, Fan F, Huang H, Hu Y et al (2018b) MiR-221/222 promote dexamethasone resistance of multiple myeloma through inhibition of autophagy by targeting ATG12. Blood 132:4469

Xu TH, Qiu XB, Sheng ZT, Han YR, Wang J, Tian BY et al (2019a) Restoration of microRNA-30b expression alleviates vascular calcification through the mTOR signaling pathway and autophagy. J Cell Physiol 234(8):14306–14318

Xu J, Su Y, Xu A, Fan F, Mu S, Chen L et al (2019b) miR-221/222-mediated inhibition of autophagy promotes dexamethasone resistance in multiple myeloma. Mol Ther 27(3):559–570

Xu Y, Xu Y, Wang S (2019c) Effect of exosome-carried miR-30a on myocardial apoptosis in myocardial ischemia-reperfusion injury rats through regulating autophagy. Eur Rev Med Pharmacol Sci 23(16): 7066–7072

Xu W-P, Liu J-P, Feng J-F, Zhu C-P, Yang Y, Zhou W-P et al (2020a) miR-541 potentiates the response of human hepatocellular carcinoma to sorafenib treatment by inhibiting autophagy. Gut 69(7):1309–1321

Xu Z, Shi L, Wang Y, Zhang J, Huang L, Zhang C et al (2020b) Pathological findings of COVID-19 associated with acute respiratory distress syndrome. Lancet Respir Med 8(4):420–422

Xu JX, Yang Y, Zhang X, Luan XP (2021) Micro-RNA29b enhances the sensitivity of glioblastoma multiforme cells to temozolomide by promoting autophagy. Anat Rec 304(2):342–352

Xue K, Li J, Nan S, Zhao X, Xu C (2019) Downregulation of LINC00460 decreases STC2 and promotes autophagy of head and neck squamous cell carcinoma by up-regulating microRNA-206. Life Sci 231:116459

Xue J, Hu B, Xing W, Li F, Huang Z, Zheng W et al (2021) Low expression of miR-142-3p promotes intervertebral disk degeneration. J Orthop Surg Res 16(1): 1–10

Yang Z, Klionsky DJ (2010) Mammalian autophagy: core molecular machinery and signaling regulation. Curr Opin Cell Biol 22(2):124–131

Yang ZJ, Chee CE, Huang S, Sinicrope FA (2011) The role of autophagy in cancer: therapeutic implications. Mol Cancer Ther 10(9):1533–1541

Yang X, Xu X, Zhu J, Zhang S, Wu Y, Wu Y et al (2016a) miR-31 affects colorectal cancer cells by inhibiting autophagy in cancer-associated fibroblasts. Oncotarget 7(48):79617

Yang Y, Li Y, Chen X, Cheng X, Liao Y, Yu X (2016b) Exosomal transfer of miR-30a between cardiomyocytes regulates autophagy after hypoxia. J Mol Med 94(6):711–724

Yang J, He Y, Zhai N, Ding S, Li J, Peng Z (2018) MicroRNA-181a inhibits autophagy by targeting Atg5 in hepatocellular carcinoma. Front Biosci (Landmark Ed) 23:388–396

Yang L, Peng X, Jin H, Liu J (2019a) Long non-coding RNA PVT1 promotes autophagy as ceRNA to target ATG3 by sponging microRNA-365 in hepatocellular carcinoma. Gene 697:94–102

Yang CL, Zheng XL, Ye K, Sun YN, Lu YF, Ge H et al (2019b) Effects of microRNA-217 on proliferation, apoptosis, and autophagy of hepatocytes in rat models of CCL4-induced liver injury by targeting NAT2. J Cell Physiol 234(4):3410–3424

Yang B, Zang L-E, Cui J-W, Zhang M-Y, Ma X, Wei L-L (2020) Melatonin plays a protective role by regulating miR-26a-5p-NRSF and JAK2-STAT3 pathway to improve autophagy, inflammation and oxidative stress of cerebral ischemia-reperfusion injury. Drug Des Devel Ther 14:3177

Yang B, Zang J, Yuan W, Jiang X, Zhang F (2021) The miR-136-5p/ROCK1 axis suppresses invasion and migration, and enhances cisplatin sensitivity in head and neck cancer cells. Exp Ther Med 21(4):317

Yao L, Zhu Z, Wu J, Zhang Y, Zhang H, Sun X et al (2019) MicroRNA-124 regulates the expression of

p62/p38 and promotes autophagy in the inflammatory pathogenesis of Parkinson's disease. FASEB J 33(7): 8648–8665

Yao W, Guo P, Mu Q, Wang Y (2021) Exosome-derived Circ-PVT1 contributes to cisplatin resistance by regulating autophagy, invasion, and apoptosis via miR-30a-5p/YAP1 Axis in gastric cancer cells. Cancer Biother Radiopharm 36(4):347–359

Ye Z, Fang B, Pan J, Zhang N, Huang J, Xie C et al (2017) miR-138 suppresses the proliferation, metastasis and autophagy of non-small cell lung cancer by targeting Sirt1. Oncol Rep 37(6):3244–3252

Yin G, Yu B, Liu C, Lin Y, Xie Z, Hu Y et al (2021) Exosomes produced by adipose-derived stem cells inhibit schwann cells autophagy and promote the regeneration of the myelin sheath. Int J Biochem Cell Biol 132:105921

YiRen H, YingCong Y, Sunwu Y, Keqin L, Xiaochun T, Senrui C et al (2017) Long noncoding RNA MALAT1 regulates autophagy associated chemoresistance via miR-23b-3p sequestration in gastric cancer. Mol Cancer 16(1):174

Yoon J-H, Ahn S-G, Lee B-H, Jung S-H, Oh S-H (2012) Role of autophagy in chemoresistance: regulation of the ATM-mediated DNA-damage signaling pathway through activation of DNA–PKcs and PARP-1. Biochem Pharmacol 83(6):747–757

Yu X, Luo A, Liu Y, Wang S, Li Y, Shi W et al (2015a) MiR-214 increases the sensitivity of breast cancer cells to tamoxifen and fulvestrant through inhibition of autophagy. Mol Cancer 14(1):208

Yu J, Bao C, Dong Y, Liu X (2015b) Activation of autophagy in rat brain cells following focal cerebral ischemia reperfusion through enhanced expression of Atg1/pULK and LC3. Mol Med Rep 12(3):3339–3344

Yu G, Jia Z, Dou Z (2017a) miR-24-3p regulates bladder cancer cell proliferation, migration, invasion and autophagy by targeting DEDD. Oncol Rep 37(2): 1123–1131

Yu X, Shi W, Zhang Y, Wang X, Sun S, Song Z et al (2017b) CXCL12/CXCR4 axis induced miR-125b promotes invasion and confers 5-fluorouracil resistance through enhancing autophagy in colorectal cancer. Sci Rep 7:42226

Yu K, Li N, Cheng Q, Zheng J, Zhu M, Bao S et al (2018a) miR-96-5p prevents hepatic stellate cell activation by inhibiting autophagy via ATG7. J Mol Med (Berl) 96(1):65–74

Yu Y, Zhang J, Jin Y, Yang Y, Shi J, Chen F et al (2018b) MiR-20a-5p suppresses tumor proliferation by targeting autophagy-related gene 7 in neuroblastoma. Cancer Cell Int 18(1):5

Yu K, Li N, Cheng Q, Zheng J, Zhu M, Bao S et al (2018c) miR-96-5p prevents hepatic stellate cell activation by inhibiting autophagy via ATG7. J Mol Med 96(1): 65–74

Yu Q, Zhao B, He Q, Zhang Y, Peng XB (2019) microRNA-206 is required for osteoarthritis development through its effect on apoptosis and autophagy of articular chondrocytes via modulating the phosphoinositide 3-kinase/protein kinase B-mTOR pathway by targeting insulin-like growth factor-1. J Cell Biochem 120(4):5287–5303

Yu B, Li C, Chen P, Zhou N, Wang L, Li J et al (2020) Low dose of hydroxychloroquine reduces fatality of critically ill patients with COVID-19. Sci China Life Sci 63(10):1515–1521

Yun Z, Wang Y, Feng W, Zang J, Zhang D, Gao Y (2020) Overexpression of microRNA-185 alleviates intervertebral disc degeneration through inactivation of the Wnt/β-catenin signaling pathway and downregulation of Galectin-3. Mol Pain 16:1744806920902559

Yuwen D, Sheng B, Liu J, Wenyu W, Shu Y (2017) MiR-146a-5p level in serum exosomes predicts therapeutic effect of cisplatin in non-small cell lung cancer. Eur Rev Med Pharmacol Sci 21(11):2650–2658

Zeng LP, Hu ZM, Li K, Xia K (2016) miR-222 attenuates cisplatin-induced cell death by targeting the PPP 2R2A/Akt/mTOR Axis in bladder cancer cells. J Cell Mol Med 20(3):559–567

Zeng R, Song X-J, Liu C-W, Ye W (2019) LncRNA ANRIL promotes angiogenesis and thrombosis by modulating microRNA-99a and microRNA-449a in the autophagy pathway. Am J Transl Res 11(12):7441

Zhai H, Song B, Xu X, Zhu W, Ju J (2013) Inhibition of autophagy and tumor growth in colon cancer by miR-502. Oncogene 32(12):1570

Zhai H, Fesler A, Ba Y, Wu S, Ju J (2015) Inhibition of colorectal cancer stem cell survival and invasive potential by hsa-miR-140-5p mediated suppression of Smad2 and autophagy. Oncotarget 6(23):19735

Zhang H, Monken CE, Zhang Y, Lenard J, Mizushima N, Lattime EC et al (2006) Cellular autophagy machinery is not required for vaccinia virus replication and maturation. Autophagy 2(2):91–95

Zhang HT, Chen G, Hu BG, Zhang ZY, Yun JP, He ML et al (2014) Hepatitis B virus x protein induces autophagy via activating death-associated protein kinase. J Viral Hepat 21(9):642–649

Zhang X, Shi H, Lin S, Ba M, Cui S (2015a) MicroRNA-216a enhances the radiosensitivity of pancreatic cancer cells by inhibiting beclin-1-mediated autophagy. Oncol Rep 34(3):1557–1564

Zhang Q, Zhu H, Xu X, Li L, Tan H, Cai X (2015b) Inactivated Sendai virus induces apoptosis and autophagy via the PI3K/Akt/mTOR/p70S6K pathway in human non-small cell lung cancer cells. Biochem Biophys Res Commun 465(1):64–70

Zhang Y, Liu Y, Xu X (2017a) Upregulation of miR-142-3p improves drug sensitivity of acute myelogenous leukemia through reducing P-glycoprotein and repressing autophagy by targeting HMGB1. Transl Oncol 10(3):410–418

Zhang L, Cheng R, Huang Y (2017b) MiR-30a inhibits BECN1-mediated autophagy in diabetic cataract. Oncotarget 8(44):77360

Zhang Y, Zhao S, Wu D, Liu X, Shi M, Wang Y et al (2018a) MicroRNA-22 promotes renal tubulointerstitial fibrosis by targeting PTEN and suppressing

autophagy in diabetic nephropathy. J Diabetes Res 2018:4728645

Zhang K, Chen J, Zhou H, Chen Y, Zhi Y, Zhang B et al (2018b) PU. 1/microRNA-142-3p targets ATG5/ATG16L1 to inactivate autophagy and sensitize hepatocellular carcinoma cells to sorafenib. Cell Death Dis 9(3):1–16

Zhang K, Chen J, Zhou H, Chen Y, Zhi Y, Zhang B et al (2018c) PU. 1/microRNA-142-3p targets ATG5/ATG16L1 to inactivate autophagy and sensitize hepatocellular carcinoma cells to sorafenib. Cell Death Dis 9(3):312

Zhang HH, Huang ZX, Zhong SQ, Fei KL, Cao YH (2020a) miR-21 inhibits autophagy and promotes malignant development in the bladder cancer T24 cell line. Int J Oncol 56(4):986–998

Zhang C, Gan X, Liang R, Jian J (2020b) Exosomes derived from epigallocatechin gallate-treated cardiomyocytes attenuated acute myocardial infarction by modulating microRNA-30a. Front Pharmacol 11:126

Zhang L, Song Y, Chen L, Li D, Feng H, Lu Z et al (2020c) MiR-20a-containing exosomes from umbilical cord mesenchymal stem cells alleviates liver ischemia/reperfusion injury. J Cell Physiol 235(4): 3698–3710

Zhang H, Liang H, Wu S, Zhang Y, Yu Z (2021a) MicroRNA-638 induces apoptosis and autophagy in human liver cancer cells by targeting enhancer of zeste homolog 2 (EZH2). Environ Toxicol Pharmacol 82:103559

Zhang B, Lin F, Dong J, Liu J, Ding Z, Xu J (2021b) Peripheral macrophage-derived exosomes promote repair after spinal cord injury by inducing local anti-inflammatory type microglial polarization via increasing autophagy. Int J Biol Sci 17(5):1339

Zhang X, Xi T, Zhang L, Bi Y, Huang Y, Lu Y et al (2021c) The role of autophagy in human cytomegalovirus IE2 expression. J Med Virol 93(6):3795–3803

Zhang Y, Sun H, Pei R, Mao B, Zhao Z, Li H et al (2021d) The SARS-CoV-2 protein ORF3a inhibits fusion of autophagosomes with lysosomes. Cell Discov 7(1): 1–12

Zhao S, Yao D, Chen J, Ding N, Ren F (2015a) MiR-20a promotes cervical cancer proliferation and metastasis in vitro and in vivo. PLoS One 10(3):e0120905

Zhao W, Zheng X-L, Zhao S-P (2015b) Exosome and its roles in cardiovascular diseases. Heart Fail Rev 20(3): 337–348

Zhao X, Li H, Wang L (2019a) MicroRNA-107 regulates autophagy and apoptosis of osteoarthritis chondrocytes by targeting TRAF3. Int Immunopharmacol 71:181–187

Zhao XH, Wang YB, Yang J, Liu HQ, Wang LL (2019b) MicroRNA-326 suppresses iNOS expression and promotes autophagy of dopaminergic neurons through the JNK signaling by targeting XBP1 in a mouse model of Parkinson's disease. J Cell Biochem 120(9): 14995–15006

Zhao Y, Wang P, Wu Q (2020) miR-1278 sensitizes nasopharyngeal carcinoma cells to cisplatin and suppresses autophagy via targeting ATG2B. Mol Cell Probes 53:101597

Zheng D, Wang W, Zhou J, Shi J, Liu Y (2014) HTLV-1 Tax protein induces autophagy via IKK in human astroglioma cells: a protective mechanism against death receptor-mediated apoptosis (610.1). FASEB J 28(1_supplement):610.1

Zheng B, Zhu H, Gu D, Pan X, Qian L, Xue B et al (2015) MiRNA-30a-mediated autophagy inhibition sensitizes renal cell carcinoma cells to sorafenib. Biochem Biophys Res Commun 459(2):234–239

Zheng Y, Liu L, Wang Y, Xiao S, Mai R, Zhu Z et al (2021) Glioblastoma stem cell (GSC)-derived PD-L1-containing exosomes activates AMPK/ULK1 pathway mediated autophagy to increase temozolomide-resistance in glioblastoma. Cell Biosci 11(1):1–12

Zhirnov O, Klenk H (2013a) Influenza A virus proteins NS1 and HA along with M2 are involved in stimulation of autophagy in infected cells. J Virol 87(24): 13107–13114. https://doi.org/10.1128/JVI.02148-13

Zhirnov OP, Klenk HD (2013b) Influenza A virus proteins NS1 and hemagglutinin along with M2 are involved in stimulation of autophagy in infected cells. J Virol 87(24):13107–13114

Zhong L, Shu W, Dai W, Gao B, Xiong S (2017) Reactive oxygen species-mediated c-Jun NH(2)-terminal kinase activation contributes to hepatitis B virus X protein-induced autophagy via regulation of the Beclin-1/Bcl-2 interaction. J Virol 91(15):e00001–e00017

Zhou D, Spector SA (2008) Human immunodeficiency virus type-1 infection inhibits autophagy. AIDS (London, England) 22(6):695

Zhou Z, Jiang X, Liu D, Fan Z, Hu X, Yan J et al (2009) Autophagy is involved in influenza A virus replication. Autophagy 5(3):321–328

Zhou L, Liu S, Han M, Feng S, Liang J, Li Z et al (2017) MicroRNA-185 induces potent autophagy via AKT signaling in hepatocellular carcinoma. Tumor Biol 39(2):1010428317694313

Zhou Y, Geng P, Liu Y, Wu J, Qiao H, Xie Y et al (2018) Rotavirus-encoded virus-like small RNA triggers autophagy by targeting IGF1R via the PI3K/Akt/mTOR pathway. Biochim Biophys Acta 1864(1):60–68

Zhou S, Lei D, Bu F, Han H, Zhao S, Wang Y (2019) MicroRNA-29b-3p targets SPARC gene to protect cardiocytes against autophagy and apoptosis in hypoxic-induced H9c2 cells. J Cardiovasc Transl Res 12(4):358–365

Zhu H, Wu H, Liu X, Li B, Chen Y, Ren X et al (2009) Regulation of autophagy by a beclin 1-targeted microRNA, miR-30a, in cancer cells. Autophagy 5(6):816–823

Zhu B, Zhou Y, Xu F, Shuai J, Li X, Fang W (2012) Porcine circovirus type 2 induces autophagy via AMPK/ERK/TSC2/mTOR signaling pathway in PK-15 cells. J Virol 86(22):12003–12012. https://doi.org/10.1128/JVI.01434-12

Zhu M, Liu X, Li W, Wang L (2020) Exosomes derived from mmu_circ_0000623-modified ADSCs prevent liver fibrosis via activating autophagy. Hum Exp Toxicol 39(12):1619–1627

Adv Exp Med Biol - Cell Biology and Translational Medicine (2022) 17: 163–171
https://doi.org/10.1007/5584_2022_721
© Springer Nature Switzerland AG 2022
Published online: 4 July 2022

Epitranscriptomics Changes the Play: m⁶A RNA Modifications in Apoptosis

Azime Akçaöz and Bünyamin Akgül

Abstract

Apoptosis is a form of programmed cell death that is essential for cellular and organismal homeostasis. Any irregularities that disturb the balance between apoptosis and cell survival have severe implications, such as improper development or life-threatening diseases. Thus, it is highly critical to maintain a proper rate of apoptosis throughout development. In fact, several complex transcriptional and posttranscriptional mechanisms exist in eukaryotes to critically regulate the rate of apoptotic processes. Recent studies suggest that not only RNA sequences but also their modifications, such as m^6A methylation, play a fundamental role in these transcriptional and posttranscriptional processes. A specific set of proteins, called writer, eraser, and reader of m^6A marks, modulate the rate of apoptosis by determining the m^6A repertoire and the fate of certain transcripts associated with apoptosis. In this *Review*, we will cover the dynamic m^6A RNA modifications and their impact on modulation of apoptosis.

Keywords

Apoptosis · Epitranscriptomics · m^6A RNA modification

Abbreviations

ac^4C	N4-acetylcytidine
circRNA	Circular RNA
CP	Cisplatin
DISC	Death including signaling complex
Fas-L	Fas ligand
hm5C	Hydroxymethylcytosine
lncRNA	Long noncoding RNA
m^1A	N1-methyladenosine
m^5C	5-methylcytosine
m^6A	N6-Methyladenosine
m^6Am	N6,2′-O-dimethyladenosine
m^7G	7-methylguanosine
miRNA	microRNA
ncRNA	Noncoding RNA
TGCT	Testicular germ cell tumors
TNF	Tumor necrosis factor
TRAIL	TNF-related apoptosis-inducing ligand
Y	Pseudouridine

A. Akçaöz and B. Akgül (✉)
Noncoding RNA Laboratory, Department of Molecular Biology and Genetics, İzmir Institute of Technology, Urla, İzmir, Turkey
e-mail: bunyaminakgul@iyte.edu.tr

1 Introduction

Deoxyribonucleic acids (DNA) and ribonucleic acids (RNA) were thought to harbor solely canonical nucleotides until the detection of the first distinct chemical moiety, deoxy-5-methylcytosine, on DNA in 1948 (Hotchkiss 1948). Subsequently, similar modifications were also reported to exist on RNAs, such as 5-ribosyluracil (Davis and Allen 1957), pseudouridine (Y) (Cohn 1960), 2′-O-methylribose (Rabczenko and Shugar 1971), 5-methylribouridine and 5-methylribocytosine (Grosjean 2005). This novel and exciting area of RNA modification, called epitranscriptomics, refers to altered chemical structure of RNA without any change in the ribonucleotide sequence and implies extensive regulatory effects on various layers of gene expression (Saletore et al. 2012). RNA modifications were first reported to exist on ribosomal RNAs (rRNAs) and transfer RNAs (tRNAs). Ensuing advances in detection methods have led to the discovery of other types of RNA modifications on mRNAs and noncoding RNAs (ncRNAs) (Motorin and Helm 2011).

Currently, almost 170 modifications have emerged in all RNA types (Wiener and Schwartz 2021). At least, thirteen different chemical modifications have been reported to exist on mRNAs, which include 7-methylguanosine (m^7G), 2′-O-methylated at the ribose (cOMe), N6,2′-O-dimethyladenosine (m^6Am), N6-methyladenosine (m^6A), 5-methylcytosine (m^5C), pseudouridine (Ψ), N1-methyladenosine (m^1A), N4-acetylcytidine (ac^4C), hydroxymethylcytosine (hm5C), 3-methylcytidine (m^3C), cytosine to uridine (C to U) editing, m^7G, Nm, and 7,8-dihydro-8-oxoguanosine (Nachtergaele and He 2018; Anreiter et al. 2021). The type and location of modification appear to modulate different aspects of RNA processing, such as mRNA abundance (Jia et al. 2011), splicing (Xiao et al. 2016), export (Zheng et al. 2013), stability (Huang et al. 2018), and translation (Zhou et al. 2015). Elucidation of epitranscriptomics processes should have implications in furthering our understanding of biological processes such as cell cycle, proliferation, development, and cell death (Zhu et al. 2020). In fact, the existing evidence clearly suggests that RNA modifications modulate various types of cell death, including apoptosis (Lin et al. 2019; Vu et al. 2017; Wang et al. 2020). In this *Review*, we will cover a succinct description of apoptosis followed by a detailed discussion on

RNA m^6A marks and their impact on apoptotic processes.

2 Apoptosis

Apoptosis is described as a mode of programmed cell death characterized by a series of biochemical and morphological features resulting in elimination of excess cells (Elmore 2007). Initial morphological hallmarks of apoptosis include condensation of the chromatin as well as shrinkage of the cell followed by fragmentation of the condensed nucleus. The process is then followed by detachment of the cell from the surrounding environment and formation of cytoplasmic blebs (Kerr and Wyllie 1972). Throughout the apoptotic process, cellular organelles maintain their compact structures. Subsequently, the cell fragmentation results in the formation of apoptotic bodies, which are bound to the plasma membrane and include fragmented nuclear materials, packed organelles, and condensed cytoplasm (Kurosaka et al. 2003). Phagocytosis of apoptotic bodies is carried out by neighboring cells, macrophages, and parenchymal cells. The maintenance of membrane integrity ensures that all of these events are completed without causing any inflammation (Saraste and Pulkki 2000; Kurosaka et al. 2003). Characteristic biochemical features of apoptosis also include the fragmentation of DNA and proteins by caspases, cysteine proteases (Saraste and Pulkki 2000).

There are two canonical pathways of apoptosis: (1) intrinsic and (2) extrinsic (Julien and Wells 2017). The intrinsic pathway can be triggered by many extra- or intracellular stimuli such as toxins, radiation, oxidative stress, or treatment with chemotherapeutic agents (Sivamani and Kar 2015; Pistritto et al. 2016). It is mediated by mitochondria and results in activation of caspase -9, -3, -6 and -7 (Jan and Chaudhry 2019; Xu and Shi 2007). However, death receptors and their ligands are involved in activation of the extrinsic apoptotic pathway. The binding of extracellular ligands such as tumor necrosis factor (TNF) alpha, Fas ligand (Fas-L) or TNF-related apoptosis-inducing ligand (TRAIL) to their receptors activates death including signaling complex (DISC) and activates caspase -8, -10, -3, -6 and -7 (Sivamani and Kar 2015; Carneiro and El-Deiry 2020). Both pathways terminate with the activation of effector caspases, which degrade all nuclear materials by stimulating endonucleases and proteases (Elmore 2007). Other than the intrinsic and extrinsic pathways, a perforin/granzyme pathway is triggered by cytotoxic T lymphocytes and natural killer cells to eliminate infected cells (Igney and Krammer 2002; Nirmala and Lopus 2020). Perforin released by immune cells disrupts the membrane of the target cells while granzymes facilitate DNA fragmentation in a caspase-independent manner.

Apoptosis plays a critical role in the maintenance of organismal homeostasis. Consequently, apoptosis must be tightly regulated as part of cellular hemostasis, development, and elimination of pathogens or diseases. The existing evidence clearly documents the significance of transcriptional regulatory mechanisms that target proapoptotic and antiapoptotic proteins (Budhidarmo and Day 2015; Hotchkiss et al. 2009). Different stages of apoptosis can also be regulated by posttranscriptional mechanisms (Guttman and Rinn 2012). We have reported the contribution to this process of microRNAs (miRNAs) and circular RNAs (circRNAs) (Erdoğan et al. 2018; Tuncel et al. 2021; Yaylak et al. 2019). Recent studies suggest that both transcriptional and posttranscriptional regulatory mechanisms may be modulated by RNA modifications (Zhao et al. 2016). Especially, m⁶A modifications have been documented to regulate apoptotic mechanisms through a diverse array of regulatory factors (Huang et al. 2019; Liu et al. 2018a, b, 2019; Vu et al. 2017; Wei et al. 2019; Xu et al. 2019).

3 m⁶A RNA Modification

m⁶A RNA modification is a highly dynamic and reversible process that forms the basis of gene regulation by m⁶A RNA methylation. The first example of m⁶A RNA modification on eukaryotic mRNAs was reported in 1970s (Desrosiers et al.

1974; Perry and Kelley 1974). Of numerous RNA chemical marks, m⁶A methylation constitutes 80% of all mRNA modifications. Adenosine residing in a highly conserved consensus sequence, called the DRACH motif ([G / A / U] [G> A] m⁶AC [U> A> C], undergoes m⁶A methylation upon the reception of cellular signals that activate a special set of methylation enzymes (Niu et al. 2013). m⁶A marks may be distributed throughout the body of the transcript. The abundance of m⁶A-methylated residues is attained by a combinatorial effect of methyltransferases (writers) and demethyltransferases (erasers) (Fig. 1). The impact of m⁶A modification is then dictated by reader proteins that recognize the m⁶A site and probably recruit other proteins that dictate the fate of the transcript (Huang and Yin 2018).

3.1 Deposition of m⁶A Modification

A multiprotein writer complex carries out the m⁶A methylation of the transcriptome in a highly specific manner. The core complex consists of METTL3, METTL14, and WTAP, and other writer proteins (Table 1) have been reported to interact with the core complex. METTL3 has a consensus methylation motif-I that contains an Adomet binding site and a consensus motif-II that harbors the catalytic domain (Bokar et al. 1996). In fact, METTL3 is the only enzyme with a catalytic site in the core complex. Although METTL14 possesses a methyltransferase domain, it contributes to m⁶A methylation merely by enhancing the catalysis in the presence of METTL3. METTL14 also helps change the local RNA structure to increase its binding efficiency with the writer complex (Liu et al. 2014). WTAP, on the other hand, stabilizes the core complex by interacting with METTL3-METTL14 and dictates the localization of the core complex (Ping et al. 2014). RBM15, VIRMA, and HAKAI are responsible for recruiting the core writer complex to specific RNA sites for selective methylation (Bawankar et al. 2021; Liu et al. 2018a, b; Ortega et al. 2003; Patil et al. 2016; Wen et al. 2018).

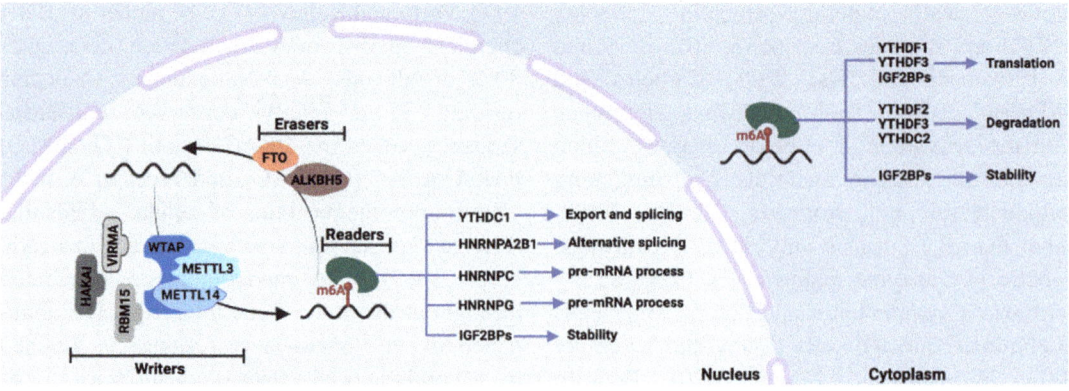

Fig. 1 Mechanism of m^{6}A modification

Table 1 The function of writers, erasers, and readers

	Name	Function	References
Writers	METTL3	Catalysis of m^{6}A methylation on adenosine	Bokar et al. (1996)
	METTL14	Boosts the catalytic activity by stabilizing the interaction between substrates and the writer complex	Liu et al. (2014)
	WTAP	Interacts with and stabilizes METTL3-METTL14	Ping et al. (2014)
	RBM15	Recruits the writer complex via its RNA-binding domains and methylates target mRNAs	Patil et al. (2016)
	VIRMA	Guides the core complex to specific RNA sites	Ortega et al. (2003)
	HAKAI	Interacts with WTAP for its target mRNAs	Bawankar et al. (2021)
Erasers	FTO	Removes m^{6}A marks in two steps via oxidation	Jia et al. (2011)
	ALKBH5	Directly removes the methyl group without oxidative demethylation	Zheng et al. (2013)
Readers	YTHDF1	Promotes translation	Wang et al. (2015)
	YTHDF2	Regulates RNA stability	Du et al. (2016)
	YTHDF3	Role in translation and mRNA decay	Shi et al. (2017)
	YTHDC1	Responsible for mRNA export and splicing	Roundtree et al. (2017)
	YTHDC2	Reduces mRNA stability	Mao et al. (2019)
	HNRNPA2B1	Modulates alternative splicing	Alarcón et al. (2015)
	HNRNPC	Role in pre-mRNA processing	Liu et al. (2015)
	HNRNPG	Role in pre-mRNA processing	Liu et al. (2015)
	IGF2BP-1/2/3	Enhances mRNA stability	Huang et al. (2018)

Erasers are RNA demethyltransferases that remove m^{6}A residues from RNAs. FTO, which is localized both in the nucleus and the cytoplasm, has been identified as the first demethylase that eliminates the methyl group in a two-step reaction that involves oxidation (Jia et al. 2011). Subsequently, ALKBH5 was reported as an eraser protein that directly discards the m^{6}A residue without oxidative demethylation (Zheng et al. 2013). Interestingly, ALKBH5 is enriched in the nuclear speckles. Although ALKBH5 plays a role in gene expression primarily by regulating splicing and the nucleocytoplasmic export of RNAs (Zheng et al. 2013), FTO appears to function in mRNA splicing, cell differentiation, and other gene regulatory processes (Jia et al. 2011).

3.2 Fates of m^{6}A Methylated RNAs

A special set of proteins, called reader proteins, dictate the fate of m^{6}A-methylated RNAs by directly or indirectly interacting with m^{6}A marks (Fig. 1). YT521-B-homology-(YTH)-domain-containing proteins (YTHDF1, YTHDF2, YTHDF3, YTHDC1, and YTHDC2) recognize and bind preferentially to the m^{6}A residues (Wang et al. 2014). There appears to be a specificity between the type of reader protein and its effect on the mRNA fate. For example, YTHDF1 and YTHDF3 enhance the translational efficiency of m^{6}A-methylated RNAs (Li et al. 2017; Shi et al. 2017; Wang et al. 2014). On the other hand, YTHDF2 and YTHDC2 appear to modulate the stability of m^{6}A-methylated RNAs by recruiting the RNA degradation machinery (Du et al. 2016; Mao et al. 2019). YTHDF3 has a dual function in that it cooperates with YTHDF1 and YTHDF2 to modulate translation and mRNA degradation, respectively (Li et al. 2017; Shi et al. 2017). YTHDC1 is a reader protein primarily involved in the splicing and nucleocytoplasmic transport of mRNAs (Xiao et al. 2016). The reader proteins of the heterogenous nuclear ribonucleoprotein (HNRNP) family recognize m^{6}A-methylated mRNAs via alteration of the mRNA structure. HNRNPA2B1 modulates alternative splicing and processing of primary miRNAs (Alarcón et al. 2015; Wu et al. 2018). HNRNPC and HNRNPG have a role in pre-mRNA processing (Liu et al. 2015). In addition, insulin-like growth factor 2 binding proteins (IGF2BPs) strengthen the stability and enhance the translation efficiency of transcripts with m^{6}A marks (Huang et al. 2018). In brief, m^{6}A residues possess the capability of dictating multiple fates of mRNAs, such as stability, translation, alternative splicing, and degradation. By doing so, m^{6}A methylation has the potential to modulate a variety of biological processes, such as apoptosis.

4 m^{6}A Modifications in Apoptosis

Recent years have witnessed a significant progress in the regulation of apoptosis by the m^{6}A methylation machinery. Initial efforts have been geared toward understanding the relationship between apoptosis and m^{6}A modifications

particularly from the perspective of cancer (An and Duan 2022). As expectedly, apoptotic stimuli could activate or inactivate methyltransferases or demethylases targeting either pro-apoptotic or anti-apoptotic transcripts, resulting in the spatial or temporal regulation of apoptotic pathways. In turn, transcript-specific recognition of m^{6}A marks by readers modulates a series of molecular effects such as enhanced stability, degradation, or translation efficiency of pro- and anti-apoptotic transcripts (Fig. 2).

The importance of m^{6}A RNA methylation was first documented in HepG2 cells, in which *METTL3* knockdown led to apoptosis (Dominissini et al. 2012). Congruently, *METTL3* overexpression was reported to reduce the rate of apoptosis through the elevated m^{6}A methylation of *PTEN*, *BCL2,* and *MYC*, whose translation efficiency is enhanced upon methylation in human hematopoietic stem/progenitor cells (HSPCs) (Vu et al. 2017). *METTL3*-mediated modulation of apoptosis was shown in testicular germ cell tumors (TGCT) in which *METTL3* upregulation decreased cisplatin (CP) sensitivity by targeting ATG5 (Chen et al. 2021). ATG5 promotes autophagy by inhibiting apoptosis, leading to chemoresistance. Similarly, *TFAP2C* m^{6}A methylation enhances the stability of the transcript, resulting in enhanced cell viability under CP treatment conditions in seminoma (Wei et al. 2020). m^{6}A modification of *BCL2* under CP and TNF-alpha treatment conditions positively regulates its mRNA stability. Increased BCL2 in turn blocks apoptosis and enhances invasion in TGCT and temporomandibular joint osteoarthritis (Peng et al. 2021; He et al. 2022). Interestingly, METTL3 and ALKBH5 inversely modulate the m^{6}A RNA methylation of the *TFEB* transcript in hypoxia/reoxygenation-treated mouse cardiomyocytes (Song et al. 2019). *TFEB* transcripts m^{6}A-methylated in their 3′ untranslated regions (UTRs) exhibit a lower stability, leading to an increase in the rate of apoptosis in cardiomyocytes. Modulation of apoptosis is not limited to the METTL3 component of the core writer complex. For example, METTL14 silencing reduces the apoptotic rate via m^{6}A RNA methylation of miR-375 that in turn inactivates the mTOR

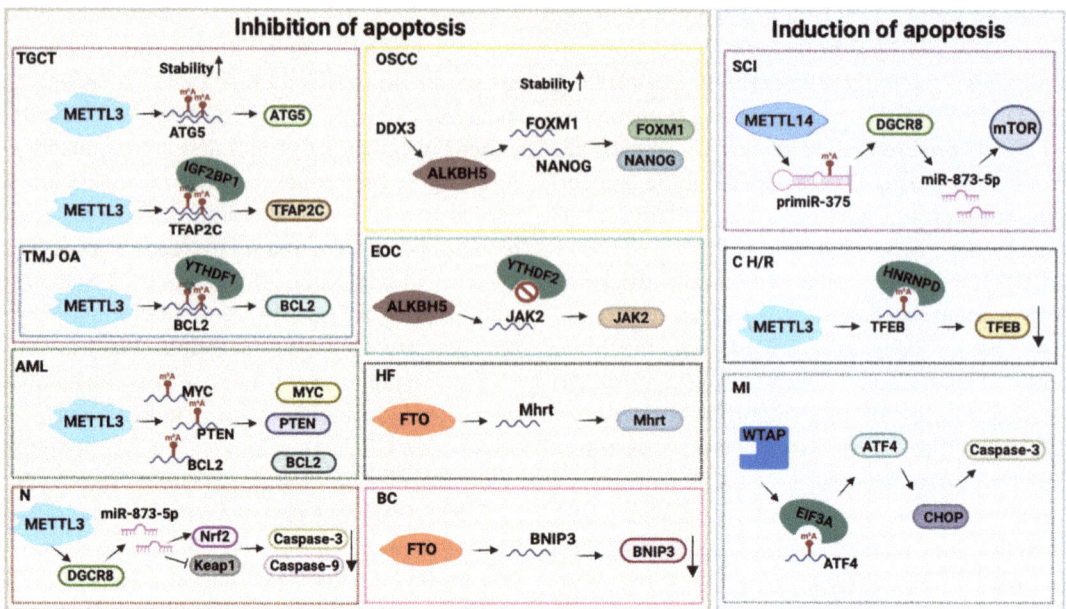

Fig. 2 m⁶A-mediated regulatory mechanisms in apoptosis. *TGCT* testicular germ cell tumor, *TMJ OA* temporomandibular joint osteoarthritis, *AML* acute myeloid leukemia, *N* nephrotoxicity, *OSCC* oral squamous cell carcinoma, *EOC* epithelial ovarian cancer, *HF* heart failure, *BC* breast cancer, *SCI* spinal cord injury, *C H/R* cardiomyocyte treated with hypoxia/reoxygenation, *MI* myocardial infarction

pathway in the spinal cord injury (Wang et al. 2021a, b). Additionally, WTAP was reported to facilitate m⁶A modification on *ATF4* in myocardial infarction (Wang et al. 2021a, b). Consequently, enhanced translation of the *ATF4* transcript promotes the ER stress and apoptosis in myocardial infarction. microRNAs can be targeted by the m⁶A RNA methylation machinery as well as mRNAs. For example, METL3-mediated modification of miR-873-5p leads to downregulation of caspase -3 and -9 via Keap1 and Nrf2 in mouse renal tubular epithelial cells (Wang et al. 2019).

The dynamic nature of m⁶A marks on pro- or antiapoptotic RNAs may be modulated by erasers through demethylation of these transcripts and thereby controlling the rate of apoptosis. For example, the DDX3-ALKBH5 axis removes the m⁶A residues from *FOXM1* and *NANOG* and reduces the rate of apoptosis in head and oral squamous cell carcinoma by enhancing their translation rates (Shriwas et al. 2020). ALKBH5 was also reported to function in a loop with HOXA10 to facilitate chemoresistance in epithelial ovarian

cancer cells (Nie et al. 2021). ALKBH5-mediated demethylation of *JAK2* stabilizes the transcript and inhibits apoptosis, leading to chemoresistance to cisplatin. FTO is another eraser whose overexpression downregulates apoptosis by inducing the m⁶A modification of *MHRT* in myocardial cells (Shen et al. 2021). FTO also eliminates the m⁶A marks from *BNIP3* in breast cancer cells, and downregulation of BNIP3 results in the blockage of apoptosis (Niu et al. 2019).

There exists strong evidence for the critical role of writers, erasers, and readers on regulation of apoptosis through the differential m⁶A methylation of specific transcripts. However, the extent of m⁶A methylation under apoptotic conditions is unknown. Recently, we employed cisplatin as a universal inducer of apoptosis to examine the scope of m⁶A methylation through m⁶A miCLIP-seq analysis in HeLa cells (Akcaoz et al. 2022). Our analyses revealed that a total of 972 transcripts are subjected to differentially m⁶A methylation in HeLa cells. Interestingly, 132 mRNAs associated with apoptosis are targeted for m^6A methylation under cisplatin-induced apoptosis. Further analyses have uncovered a METTL3-p53-NOXA axis that might be important in regulating the intrinsic apoptotic pathway in HeLa cells.

In conclusion, the m^6A methylation machinery, including writers, erasers, and readers play an essential role in orchestrating the apoptotic processes through m^6A methylation of various transcripts (Fig. 2). Writers and erasers determine the abundance and location of m^6A residues on transcripts while readers recognize these m^6A marks and dictate the fate of transcripts. One of the pressing questions in the field is what determines the pathway-specific induction of apoptosis and whether different mechanisms function in a pathway-specific manner. Yet another interesting question is how the m^6A methylation machinery communicates with the apoptotic pathways that results in a pathway- or cell-specific response under apoptotic conditions. The existing information suggests that readers might play a pivotal role in serving as a link between the m^6A machinery and apoptotic pathways. Uncovering these mechanisms shall certainly pave the way for discovering novel targets that can be exploited both in basic and translational research.

Acknowledgments The authors would like to thank BIOMER (IZTECH, Turkey) for the instrumental help during m⁶A miCLIP-seq analyses. This study was funded by the Scientific and Technological Research Council of Turkey (TÜBİTAK) (Project No: 217Z234 to BA).

Conflict of Interest The authors declare that they have no conflict of interest.

References

Akcaoz A, Tuncel O, Saglam B et al (2022) Genomewide m6A mapping uncovers dynamic changes in the m6A epitranscriptome of cisplatin-treated apoptotic HeLa cells. BioRxiv:481057

Alarcón CR, Goodarzi H, Lee H et al (2015) HNRNPA2B1 is a mediator of m6A-dependent nuclear RNA processing events. Cell 162:1299–1308

An Y, Duan H (2022) The role of m6A RNA methylation in cancer metabolism. Mol Cancer 21:1–24

Anreiter I, Mir Q, Simpson JT et al (2021) New twists in detecting mRNA modification dynamics. Trends Biotechnol 39:72–89

Bawankar P, Lence T, Paolantoni C et al (2021) Hakai is required for stabilization of core components of the m6A mRNA methylation machinery. Nat Commun 12:1–15

Bokar JA, Shambaugh ME, Polayes D et al (1996) Purification and cDNA cloning of the AdoMet-binding subunit of the human mRNA (N6-adenosine)-methyltransferase. RNA 2:1033–1045

Budhidarmo R, Day CL (2015) IAPs: modular regulators of cell signalling. Semin Cell Dev Biol 39:80–90

Carneiro BA, El-Deiry WS (2020) Targeting apoptosis in cancer therapy. Nat Rev Clin Oncol 17:395–417

Chen H, Xiang Y, Yin Y et al (2021) The m6A methyltransferase METTL3 regulates autophagy and sensitivity to cisplatin by targeting ATG5 in seminoma. Transl Androl Urol 10:1711–1722

Cohn WE (1960) Pseudouridine, a carbon-carbon linked ribonucleoside in ribonucleic acids: isolation, structure, and chemical characteristics. J Biol Chem 235:1488–1498

Davis FF, Allen FW (1957) Ribonucleic acids from yeast which contain a fifth nucleotide. J Biol Chem 227:907–915

Desrosiers R, Friderici K, Rottman F (1974) Identification of methylated nucleosides in messenger RNA from Novikoff hepatoma cells. Proc Natl Acad Sci U S A 71:3971–3975

Dominissini D, Moshitch-Moshkovitz S, Schwartz S et al (2012) Topology of the human and mouse m6A RNA methylomes revealed by m6A-seq. Nature 485:201–206

Du H, Zhao Y, He J et al (2016) YTHDF2 destabilizes m 6 A-containing RNA through direct recruitment of the CCR4-NOT deadenylase complex. Nat Commun 7:12626

Elmore S (2007) Apoptosis: a review of programmed cell death. Toxicol Pathol 35:495–516

Erdoğan İ, Coşacak Mİ, Nalbant A, Akgül B (2018) Deep sequencing reveals two Jurkat subpopulations with distinct miRNA profiles during camptothecin-induced apoptosis. Turk J Biol 42:113–122

Grosjean H (2005) Modification and editing of RNA: historical overview and important facts to remember. Top Curr Genet 12:1–22

Guttman M, Rinn JL (2012) Modular regulatory principles of large non–coding RNAs. Nature 482:339–346

He Y, Wang W, Xu X et al (2022) Mettl3 inhibits the apoptosis and autophagy of chondrocytes in inflammation through mediating Bcl2 stability via Ythdf1-mediated m6A modification. Bone 154:116182

Hotchkiss D (1948) The quantitative separation of purines, snd nucleosides by the separation of amino acid mixtures by migration with organic solvents in filter paper has been successfully accomplished by many workers since it was first described by Consden, Gordon, and. J Biol Chem 175:315–332

Hotchkiss RS, Strasser A, McDunn JE, Swanson PE (2009) Cell death. N Engl J Med 361(16):1570–1583

Huang J, Yin P (2018) Structural insights into N6-methyladenosine (m6A) modification in the transcriptome. Genomic Proteomics Bioinforma 16: 85–98

Huang H, Weng H, Sun W et al (2018) Recognition of RNA N 6 -methyladenosine by IGF2BP proteins enhances mRNA stability and translation. Nat Cell Biol 20:285–295

Huang Y, Su R, Sheng Y et al (2019) Small-molecule targeting of oncogenic FTO demethylase in acute myeloid leukemia. Cancer Cell 35:677–691.e10

Igney FH, Krammer PH (2002) Death and anti-death: tumour resistance to apoptosis. Nat Rev Cancer 2: 277–288

Jan R, Chaudhry G-S (2019) Understanding apoptosis and apoptotic pathways targeted cancer therapeutics. Adv Pharm Bull 9:203–218

Jia G, Fu Y, Zhao X et al (2011) N6-Methyladenosine in nuclear RNA is a major substrate of the obesity-associated FTO. Nat Chem Biol 7:885–887

Julien O, Wells JA (2017) Caspases and their substrates. Cell Death Differ 24:1380–1389

Kerr JFR, Wyllie AH, Currie AR (1972) Apoptosis: a basic biological phenomenon with wide-ranging implications in tissue kinetics. Br J Cancer 26:239–257

Kurosaka K, Takahashi M, Watanabe N, Kobayashi Y (2003) Silent cleanup of very early apoptotic cells by macrophages. J Immunol 171:4672–4679

Li A, Chen YS, Ping XL et al (2017) Cytoplasmic m 6 a reader YTHDF3 promotes mRNA translation. Cell Res 27:444–447

Lin S, Liu J, Jiang W et al (2019) METTL3 promotes the proliferation and mobility of gastric cancer cells. Open Med 14:25–31

Liu J, Yue Y, Han D et al (2014) A METTL3-METTL14 complex mediates mammalian nuclear RNA N6-adenosine methylation. Nat Chem Biol 10:93–95

Liu N, Dai Q, Zheng G et al (2015) N6 -methyladenosine-dependent RNA structural switches regulate RNA-protein interactions. Nature 518:560–564

Liu J, Ren D, Du Z et al (2018a) m6A demethylase FTO facilitates tumor progression in lung squamous cell carcinoma by regulating MZF1 expression. Biochem Biophys Res Commun 502:456–464

Liu J, Yue Y, Liu J et al (2018b) VIRMA mediates preferential m6A mRNA methylation in 3′UTR and near stop codon and associates with alternative polyadenylation. Cell Discov 4. https://doi.org/10.1038/s41421-018-0019-0

Mao Y, Dong L, Liu XM et al (2019) m6A in mRNA coding regions promotes translation via the RNA helicase-containing YTHDC2. Nat Commun 10:1–11

Motorin Y, Helm M (2011) RNA nucleotide methylation. Wiley Interdiscip Rev RNA 2:611–631

Nachtergaele S, He C (2018) Chemical modifications in the life of an mRNA transcript. Annu Rev Genet 52: 349–372

Nie S, Zhang L, Liu J et al (2021) ALKBH5-HOXA10 loop-mediated JAK2 m6A demethylation and cisplatin resistance in epithelial ovarian cancer. J Exp Clin Cancer Res 40:1–18

Nirmala JG, Lopus M (2020) Cell death mechanisms in eukaryotes. Cell Biol Toxicol 36:145–164

Niu Y, Zhao X, Wu YS et al (2013) N6-methyl-adenosine (m6A) in RNA: an old modification with a novel epigenetic function. Genomics Proteomics Bioinforma 11:8–17

Niu Y, Lin Z, Wan A et al (2019) RNA N6-methyladenosine demethylase FTO promotes breast tumor progression through inhibiting BNIP3. Mol Cancer 18:1–16

Ortega A, Niksic M, Bachi A et al (2003) Biochemical function of female-lethal (2)D/Wilms' tumor suppressor-1-associated proteins in alternative pre-mRNA splicing. J Biol Chem 278:3040–3047

Patil DP, Chen CK, Pickering BF et al (2016) M6 A RNA methylation promotes XIST-mediated transcriptional repression. Nature 537:369–373

Peng J, Xiang Y, Lian W et al (2021) The m6A methyltransferase METTL3 promotes cisplatin resistance and invasion in testicular seminoma via BCL2. Research Square PPR: PPR408693 p.s. https://europepmc.org/article/ppr/ppr408693

Perry RP, Kelley DE (1974) Existence of methylated messenger RNA in mouse L cells. Cell 1:37–42

Ping XL, Sun BF, Wang L et al (2014) Mammalian WTAP is a regulatory subunit of the RNA N6-methyladenosine methyltransferase. Cell Res 24:177–189

Pistritto G, Trisciuoglio D, Ceci C et al (2016) Apoptosis as anticancer mechanism: function and dysfunction of its modulators and targeted therapeutic strategies. Aging (Albany NY) 8:603–619

Rabczenko A, Shugar D (1971) Studies on the conformation of nucleosides, dinucleoside monophosphates and homopolynucleotides containing uracil or thymine base residues, and ribose, deoxyribose or 2'-O-methylribose. Acta Biochim Pol 18(4):387–402

Roundtree IA, Luo GZ, Zhang Z et al (2017) YTHDC1 mediates nuclear export of N6-methyladenosine methylated mRNAs. Elife 6:1–28

Saletore Y, Meyer K, Korlach J et al (2012) The birth of the Epitranscriptome: deciphering the function of RNA modifications Yogesh. Genome Biol 13:8–12

Saraste A, Pulkki K (2000) Morphologic and biochemical hallmarks of apoptosis. Cardiovasc Res 45:528–537

Shen W, Li H, Su H et al (2021) FTO overexpression inhibits apoptosis of hypoxia/reoxygenation-treated myocardial cells by regulating m6A modification of Mhrt. Mol Cell Biochem 476:2171–2179

Shi H, Wang X, Lu Z et al (2017) YTHDF3 facilitates translation and decay of N 6-methyladenosine-modified RNA. Cell Res 27:315–328

Shriwas O, Priyadarshini M, Samal SK et al (2020) DDX3 modulates cisplatin resistance in OSCC through ALKBH5-mediated m6A-demethylation of FOXM1 and NANOG. Apoptosis 25:233–246

Sivamani B, Kar B (2015) Apoptosis: basic concepts, mechanisms and clinical implications. Int J Pharm Sci Res 6:940–950

Song H, Feng X, Zhang H et al (2019) METTL3 and ALKBH5 oppositely regulate m6A modification of TFEB mRNA, which dictates the fate of hypoxia/reoxygenation-treated cardiomyocytes. Autophagy 15:1419–1437

Tuncel O, Kara M, Yaylak B et al (2021) Noncoding RNAs in apoptosis: identification and function. Turk J Biol 46:1–40

Vu LP, Pickering BF, Cheng Y et al (2017) The N 6 -methyladenosine (m 6 A)-forming enzyme METTL3 controls myeloid differentiation of normal hematopoietic and leukemia cells. Nat Med 23:1369–1376

Wang X, Lu Z, Gomez A et al (2014) N 6-methyladenosine-dependent regulation of messenger RNA stability. Nature 505:117–120

Wang X, Zhao BS, Roundtree IA et al (2015) N6-methyladenosine modulates messenger RNA translation efficiency. Cell 161:1388–1399

Wang J, Ishfaq M, Xu L et al (2019) METTL3/m6A/miRNA-873-5p attenuated oxidative stress and apoptosis in colistin-induced kidney injury by modulating Keap1/Nrf2 pathway. Front Pharmacol 10:1–14

Wang H, Xu B, Shi J (2020) N6-methyladenosine METTL3 promotes the breast cancer progression via targeting Bcl-2. Gene 722:144076

Wang H, Yuan J, Dang X et al (2021a) Mettl14-mediated m6A modification modulates neuron apoptosis during the repair of spinal cord injury by regulating the transformation from pri-mir-375 to miR-375. Cell Biosci 11:1–14

Wang J, Zhang J, Ma Y et al (2021b) WTAP promotes myocardial ischemia/reperfusion injury by increasing endoplasmic reticulum stress via regulating m6A modification of ATF4 mRNA. Aging (Albany NY) 13:11135–11149

Wei W, Huo B, Shi X (2019) miR-600 inhibits lung cancer via downregulating the expression of METTL3. Cancer Manag Res 11:1177–1187

Wei J, Yin Y, Zhou J et al (2020) METTL3 potentiates resistance to cisplatin through m6A modification of TFAP2C in seminoma. J Cell Mol Med 24:11366–11380

Wen J, Lv R, Ma H et al (2018) Zc3h13 regulates nuclear RNA m6A methylation and mouse embryonic stem cell self-renewal. Mol Cell 69:1028–1038.e6

Wiener D, Schwartz S (2021) The epitranscriptome beyond m6A. Nat Rev Genet 22:119–131

Wu B, Su S, Patil DP et al (2018) Molecular basis for the specific and multivariant recognitions of RNA substrates by human hnRNP A2/B1. Nat Commun 9

Xiao W, Adhikari S, Dahal U et al (2016) Nuclear m6A reader YTHDC1 regulates mRNA splicing. Mol Cell 61:507–519

Xu G, Shi Y (2007) Apoptosis signaling pathways and lymphocyte homeostasis. Cell Res 17:759–771

Xu F, Li CH, Wong CH et al (2019) Genome-wide screening and functional analysis identifies tumor suppressor long noncoding RNAs epigenetically silenced in hepatocellular carcinoma. Cancer Res 79:1305–1317

Yaylak B, Erdogan I, Akgul B (2019) Transcriptomics analysis of circular RNAs differentially expressed in apoptotic HeLa cells. Front Genet 10:1–10

Zhao BS, Roundtree IA, He C (2016) Post-transcriptional gene regulation by mRNA modifications. Nat Rev Mol Cell Biol 18:31–42

Zheng G, Dahl JA, Niu Y et al (2013) ALKBH5 is a mammalian RNA demethylase that impacts RNA metabolism and mouse fertility. Mol Cell 49:18–29

Zhou J, Wan J, Gao X et al (2015) Dynamic m6 A mRNA methylation directs translational control of heat shock response. Nature 526:591–594

Zhu ZM, Huo FC, Pei DS (2020) Function and evolution of rna n6-methyladenosine modification. Int J Biol Sci 16:1929–1940

Adv Exp Med Biol - Cell Biology and Translational Medicine (2022) 17: 173–189
https://doi.org/10.1007/5584_2022_729
© Springer Nature Switzerland AG 2022
Published online: 19 July 2022

The Fingerprints of Biomedical Science in Internal Medicine

Babak Arjmand ⓘ, Sepideh Alavi-Moghadam ⓘ,
Masoumeh Sarvari, Akram Tayanloo-Beik ⓘ,
Hamid Reza Aghayan ⓘ, Neda Mehrdad, Hossein Adibi,
Mostafa Rezaei-Tavirani ⓘ, and Bagher Larijani ⓘ

Abstract

With the development of numerous advances in science and technologies, medical science has also been updated. Internal medicine is one of the most valuable specialized fields of medical sciences that review a broad range of diseases. Herein, the internal medicine specialist (internist) is obliged to do diagnostic measures to evaluate disease signs and symptoms. In recent times, biomedical sciences as the new emergence science (including cellular and molecular biology, genetics, nanobiotechnology, bioinformatics, biochemistry, etc.) have been capable of providing more specific diagnostic methods together with techniques for better understanding the mechanism of the disease and the best diseases modeling and offering proper therapies. Accordingly, the authors have tried to review the link between biomedical sciences and medicine, particularly internal medicine.

B. Arjmand (✉), S. Alavi-Moghadam, M. Sarvari,
A. Tayanloo-Beik, and H. R. Aghayan
Cell Therapy and Regenerative Medicine Research Center,
Endocrinology and Metabolism Molecular-Cellular
Sciences Institute, Tehran University of Medical Sciences,
Tehran, Iran
e-mail: barjmand@sina.tums.ac.ir; sepidalavi@gmail.
com; maasoomehsarvari@yahoo.com; a.tayanloo@gmail.
com; hr.aghayan@gmail.com

N. Mehrdad
Elderly Health Research Center, Endocrinology and
Metabolism Population Sciences Institute, Tehran
University of Medical Sciences, Tehran, Iran
e-mail: emri-research@tums.ac.ir

H. Adibi
Diabetes Research Center, Endocrinology and Metabolism
Clinical Sciences Institute, Tehran University of Medical
Sciences, Tehran, Iran
e-mail: adibi@tums.ac.ir

M. Rezaei-Tavirani
Proteomics Research Center, Shahid Beheshti University
of Medical Sciences, Tehran, Iran
e-mail: Tavirany@yahoo.com

B. Larijani (✉)
Endocrinology and Metabolism Research Center,
Endocrinology and Metabolism Clinical Sciences
Institute, Tehran University of Medical Sciences,
Tehran, Iran
e-mail: emrc@tums.ac.ir

Keywords

Advanced technology · Biomedical research ·
Internal medicine · Medical informatics ·
Molecular diagnostic techniques ·
Regenerative medicine

Abbreviations

CAR T cell	Chimeric antigen receptor T cells
CFTR	Cystic fibrosis transmembrane conductance regulator

CNB	Core needle biopsy
CT	Computed tomography
DNA	Deoxyribonucleic acid
EBUS	Endobronchial ultrasound
ECG	Electrocardiography
ELISA	Enzyme-linked immunosorbent assay
EMBL	The European Bioinformatics Institute
EMG	Electromyography
FNA	Fine needle aspiration
GGBN	Global Genome Biodiversity Network
GI	Gastrointestinal
GWAS	Genome-wide association studies
HRT	Hormone replacement therapy
IBD	Inflammatory bowel disease
IL	Interleukin
KNB	Knowledge Network for Biocomplexity
MR	Magnetic resonance
MRA	Magnetic resonance angiography
MRI	Magnetic resonance imaging
NCBI	National Centre for Biotechnology Information
NGS	Next-generation sequencing
PCR	Polymerase chain reaction
PET	Positron-emission tomography
PTH	Parathyroid hormone
RNA	Ribonucleic acid
SCID	Severe combined immunodeficiency
SPECT	Single-photon emission computed tomography

1   Introduction

Before the advent of medical science, people believed that the cause of many diseases was supernatural and should resort to magic for treatment (Major 1954, Ackerknecht and Haushofer 2016). Gradually, the use of medical sciences with the approach of herbal medicine was formed among ancient civilizations such as Greece, Babylon, Egypt, China, and India (Organization 2002; Weatherall et al. 2006; Hajar 2012;

Jamshidi-Kia et al. 2018). With the passage of time and the creation of numerous advances in science and technologies such as pharmacology and drug production, medical science has also been upgraded (Weatherall et al. 2006). In this respect, medical specialists are trained for the treatment of special disorders in various organs of the body (Weiland et al. 2015). Herein, internal medicine as one of the most efficient specialized fields of medical sciences examines and treats a wide range of diseases, disorders, and syndromes such as rheumatologic, immunologic, allergic, endocrine and metabolic, infectious, pulmonary, etc. On the other hand, during facing an unknown cause disease, an internal medicine specialist can usually give the best advice, because they are known as specialists with comprehensive information on a wide range of diseases (West and Dupras 2012). In all specialized medical disciplines, the physician is required to use methods to diagnose the signs and symptoms of the disease and ultimately prescribe appropriate treatment. In this context, choosing the proper diagnostic methods is often a challenging issue, because many of the signs and symptoms are nonspecific or common to many diseases (Crombie 1963; Organization 2018). Hereupon, in recent centuries, the emergence and development of biomedical sciences with different approaches including cellular and molecular biology, genetics, nanobiotechnology, bioinformatics, biochemistry, etc. have been able to prepare more specific diagnostic methods (e.g., molecular imaging) along with methods to study the mechanism of the disease and provide the best practices for diseases modeling. Moreover, it can also offer modern therapies (e.g., regenerative medicine). In other words, biomedical sciences have played a significant role in the advancement of medical science around the world (Cambrosio and Keating 2001; Wade and Halligan 2004; Quirke and Gaudillière 2008; Gwee et al. 2010; Blann and Ahmed 2014; Fuller 2017). Here, the authors have sought to review the connection between biomedical sciences and medicine, especially internal medicine, as one of the most general medical disciplines.

2 Internal Medicine: Background and Present Status

Internal medicine as a branch of medicine deals with the prevention, diagnosis, and medical treatment of diseases in adults through understanding the basic pathological causes of symptoms and signs of patients. Throughout the history of medicine, during the nineteenth and twentieth century, the combination of three way of medical thinking including the anatomoclinic, the physiopathologic, and the ethiopatogenic mentality led to the emergence of new conception, holistic medicine or medicine of the person, in Europe. In this regard, the "internal medicine" expression originated from a German term *Innere Medizin*. Additionally, owing to the first written book about the internal diseases and the first convened world congress of internal medicine during that period, the twentieth century has been mentioned the golden century of internal medicine (Fordtran et al. 2004; Amatriain 2007). Nowadays, internal medicine has diverse subspecialties including rheumatology, pulmonary disease, hematology, endocrinology, nephrology, gastroenterology, etc. that the internists can follow specialty training in internal medicine if they wish.

2.1 Common Diagnostic Methods

In addition to history taking and accurate physical examination, how to reach differential diagnosis and final diagnosis is an important step in managing the patients. Nowadays, novel achievements in technology and basic sciences had shed light on the future of different fields of medicine and improve the diagnosis and management of the patients. In this regard, internal medicine has also benefited greatly in clinical approaches by applying para clinical investigations including laboratory and imaging investigations. The common diagnostic methods including imaging, laboratory data, pathology, etc. have their own pros and cons. Different imaging technologies assist the internists a lot in different areas and subspecialties. Laboratory testing plays a pivotal role in screening, diagnosis, treatment planning, and follow-up not only in internal medicine but also in every field of medicine. In recent years, new immunologic-based tests such as enzyme-linked immunosorbent assay (ELISA) and Western blot analysis are routinely applied in medicine (Swanson et al. 2018). Imaging is known as one of the noninvasive diagnostic method in medicine. Different imaging techniques like X-ray, computed tomography (CT) scan, magnetic resonance imaging (MRI), ultrasound, positron-emission tomography (PET), etc. through providing images from internal tissues and organs help internists in the diagnosis and management of diseases. The major advantage of MRI and ultrasound over the previous modalities is lack of X-ray and reduced the exposure to ionizing radiation. Due to the ability of MRI in showing the soft tissues and vasculature with high resolution, it is widely applied in many fields of medicine. Different methods of MRI are available: functional MRI (brain mapping) and magnetic resonance (MR) spectroscopy (measuring the chemical components of tissues, e.g., the brain tumors). One of the best modalities with high sensitivity for diagnosing the acute ischemic stroke is diffusion-weighted magnetic resonance imaging. PET as a functional imaging technique is broadly performed to evaluate the malignancies and their spread. Fluoroscopy, angiography, magnetic resonance angiography (MRA), and single-photon emission computed tomography (SPECT) are some the other imaging techniques (Ahn et al. 2002; Hansell et al. 2009; Kang et al. 2009; Goodarzi et al. 2019d). Pathological study is another diagnostic method which needs bio-fluids and tissue samples. Various methods are applied for obtaining tissue samples including fine needle aspiration (FNA), core needle biopsy (CNB), and open incisional/excisional biopsies. Hence, pathological study unlike the other methods often is an invasive diagnostic method (Tayanloo-Beik et al. 2020). Furthermore, there are other modalities applied in subspecialties of internal medicine. Accordingly, endoscopy, colonoscopy, rectosigmoidoscopy, endobronchial ultrasound (EBUS), electrocardiography (ECG), echocardiography, and electromyography (EMG) are some of these modalities. Although all of the above-discussed methods most of the time assist the internists, sometimes these common diagnostic techniques are not sufficient enough, and their sensitivity and specificity are different. Hence, novel tests and methods are required to progress the diagnosis and management of patients.

2.2   Common Treatment Options

Internists, unlike the surgeons, commonly deal with medical treatments. Prevention, diagnosis, and treatment are the basis of an internist career. Prevention has some levels that internal medicine can concern with various levels in different ways. Accordingly, recommendations for changing lifestyle, screening, and treatment for preventing/limiting the progress of diseases and their complications are examples of internal medicine role in prevention. Besides the wide range of treatments applied in internal medicine, the medication has an essential role among other options. Hormone therapy including hormone replacement therapy (HRT), insulin therapy, corticosteroid therapy, levothyroxine, and parathyroid hormone (PTH) replacement therapy is considered as one of the treatment bases in internal diseases (Forsblad d'Elia and Carlsten 2006; Cutolo 2010; Gluvic et al. 2015). Recently, interventional therapy by novel achievements finds a special place in various subspecialties of internal medicine. In this regard, coronary angioplasty with or without stenting in cardiology, endoscopic and colonoscopic intervention for sphincterotomy, stent placement, stone removal, polypectomy, clip application, submucosal injections in gastroenterology, radiology-guided intra-articular/periarticular/myofascial trigger point injection in rheumatology, ultrasound-guided renal biopsy, and insertion of peritoneal dialysis catheters in nephrology are some examples of interventional therapies (Khan 2005; Efstratiadis et al. 2007; Lee-Kong and Feingold 2017; Ramírez and Plasencia 2018; Tseng et al. 2019).

3   Biomedical Science: Subsets and Applications

Biomedical sciences (*biomedicine*) include a set of natural science disciplines which help advance the goals of medical science by using physiological and biological principles. The mentioned natural science disciplines include cellular and molecular biology, genetics, biochemistry, bionanotechnology, bioinformatics, bioengineering, microbiology, embryology, and physiology (Kirschner et al. 1994, Cambrosio and Keating 2001, Pal 2007, Quirke and Gaudillière 2008, Nass et al. 2009, Arjmand et al. 2020d, e).

3.1   Cellular and Molecular Biology

Cellular and molecular biology is a science which generally studies the function, evolution, and development of the cellular structures of living organisms and their molecular basis. The mentioned studies include investigating cells and molecules' interaction with each other and the environment (Weatherall 1998; Karp 2009; Wei and Huang 2013; Alberts et al. 2018). In other words, it evaluates cellular and molecular signaling and metabolic pathways as well as the cell cycle regulation. Furthermore, molecular biology strongly overlaps with some other biological sciences including genetics and biochemistry (Swanson 2018). Indeed, cellular and molecular biology studies can help physicians to understand the precise pathogenesis of the disease (Beenhouwer 2018; Williams and Silverman 2018). On the other hand, it can be important for the development of diagnostic methods. In other words, cellular and molecular diagnoses, including the analysis of different cell phenotypes and tissue derivatives along with the measurement of various macromolecules and metabolites, can indicate the presence of a disease and abnormal body function (DeBerardinis and Thompson 2012; Tan 2016; Raghavendra and Pullaiah 2018). Further, cellular- and molecular-based diagnoses can improve simulating the disease in appropriate preclinical models (including cell-based and animal models) and lead to select the most appropriate treatment options. Additionally, investigations based on cellular and molecular sciences have opened a new window of therapeutic approaches, including cell therapy and regenerative medicine (Weatherall 1998; Wang et al. 2013).

3.1.1 Biochemistry

Biochemistry (investigation of chemical compounds and essential chemical processes in living organisms) and medicine share a relationship of mutual collaboration. In other words, biochemical experiments have shed light on multiple phases of the disease (Kogut 1977; Baynes and Dominiczak 2009). Herein, different diseases are classified as being associated with the main types of biochemical molecules (including proteins, carbohydrates, lipids, and nucleic acids) and their related signaling pathways (Stryer et al. 2002; Blanco and Blanco 2017).

3.1.2 Genetics

Genetics as a branch of biology investigates the genetic material in an organism (including deoxyribonucleic acid (DNA) or ribonucleic acid (RNA)), genes (as sequences of nucleotides in DNA or RNA), genetic variations, and heredity (Griffiths et al. 2000). Also, genetic studies (by high-throughput methods, i.e., genome-wide association studies (GWAS) and next-generation sequencing (NGS)) can play a pivotal part to develop understanding of the several disease (e.g., diabetes, obesity, arthritis rheumatoid, Alzheimer's disease, Parkinson's disease, etc.) mechanisms via evaluation the specific involved biological pathways in pathogenesis and implementing accurate diagnostic methods (Edwards 1963; Claussnitzer et al. 2020; Jackson et al. 2020) (Fig. 1). Accordingly, understanding the mechanism of disease can lead to ameliorating therapeutic tactics and finding novel biomarkers and drug targets. Moreover, advances in genetic studies have led to the emergence of a new and effective treatment called gene therapy as approach for improving mutant genes (altered genes) or site-specific modifications (Abati et al. 2019).

3.2 Bioinformatics

Bioinformatics is the use of computing, statistics, and research techniques to collect, analyze, and handle data in recent biology and medicine. In this context, physicians and biologists can detect the structure of biological molecules, e.g., nucleic acids and proteins, via accessing the Internet and bioinformatics-related websites (Table 1), together with simple bioinformatics methods (Lesk 2019; Azodi et al. 2020; Baxevanis et al. 2020). Additionally, bioinformatics has become an important component of omics (genomics, transcriptomics, proteomics, and metabolomics) (Fig. 2) investigations (Mayer 2011; Schneider and Orchard 2011; Yadav 2015). The goal of omics investigations is to identify and quantify the biological molecules on which the structure, dynamics, and function of organisms depend (Horgan and Kenny 2011; Agharezaee et al. 2018; Arjmand 2019; Gilany et al. 2019a, b c; Goodarzi et al. 2019a; Khatami et al. 2019; Larijani et al. 2019a, b; Mehrparavar et al. 2019; Mehrparvar et al. 2020; Tayanloo-Beik et al. 2020). Moreover, the broad omics information achievement can lead to biology development and contribute to the emergence of system biology (research area which focuses on the understanding of whole biological processes, i.e., metabolic pathways and gene regulation) (Chen and Snyder 2012; Yan et al. 2018). On the other hand, individual omics evaluation is expected to lead to substantial improvement in personalized medicine (Chen and Snyder 2013; Ibrahim et al. 2016).

3.3 Bioengineering

Bioengineering uses a range of sciences such as mathematics, biomechanics, tissue engineering, and polymer science to design and develop some areas (including medical devices, diagnostic instruments, biocompatible products, ecological engineering, agricultural engineering, etc.) in order to improve living a healthy lifestyle in this modern world (Valentinuzzi et al. 2017; Sharma and Khurana 2018).

Genetics Diagnostics Methods

PCR Test

This method is usually applied to copy DNA and promote the study of genetic variants identified as certain causes associated with diseases.

DNA Sequencing

DNA sequencing (sanger sequencing, next-generation sequencing) refers to identifying the order of bases that make up DNA, which essentially helps clinicians to determine if there are variations or variants related to a disease in the regulatory region of DNA.

Cytogenetics

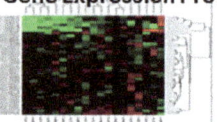

This method (including Karyotyping and Fluorescence in situ hybridization) evaluates the number, shape, and staining pattern of chromosomes to detect genetic disorders by special technologies.

Microarray Test

This method can be used to assess if there is duplication, deletion, or large stretches of identical DNA in an individual.

Gene Expression Profiling

To assess which genes are actively producing proteins, gene expression tests analyze the RNA in an individual's tissue sample.

Fig. 1 Genetics diagnostic methods. Genetics diagnostic methods are including polymerase chain reaction (PCR), DNA sequencing, cytogenetics, microarray testing, and gene expression profiling which help look for disease mechanisms (Dwivedi et al. 2017)

3.4 Bionanotechnology

Bionanotechnology, which involves many scientific fields such as cellular and molecular biology, physical sciences, bioengineering, chemistry, nanotechnology, and medicine, can incorporate biological molecules into nanotechnological applications. In other words, it uses knowledge of the characteristics acquired by living organisms on the evolutionary path for technological purposes. Hereupon, the production and design of multifunctional nanoparticles focuses on improving diagnostic techniques, drug delivery system, and therapeutic approaches (Kumar et al. 2013; Ramsden 2016; Zhang et al. 2017; Rauta et al. 2019).

3.5 Microbiology

Microbiology that investigates microscopic organisms (viruses, bacteria, fungi, protozoa, and archaea) includes fundamental evaluation of microorganisms' physiology, cell biology, biochemistry, and ecology. In this respect, it offers services to help diagnose and manage infectious diseases (Glazer and Nikaido 2007; Brooks 2013; Murray et al. 2020).

Table 1 Some of the useful bioinformatics websites

Websites	Application and services
Allen brain atlas	It can provide a unique online public source of broad gene expression, connectivity, and neuroanatomical data about the brain in mice, humans, and nonhuman primates
BLAST	It can be applied to understand functional and evolutionary connections between sequences and recognize gene family members
ChemSpider	It can provide instant access to more than 67 million chemical structures from hundreds of data sources
The European bioinformatics institute (EMBL)	It provides a freely accessible and up-to-date comprehensive collection of molecular data resources
ExPASy	It can provide access to over 160 databases and software resources for the study of genomics, proteomics, structural biology, evolution and phylogeny, system biology, and medical chemistry as an extensible and integrative portal.
Global genome biodiversity network (GGBN)	It can provide a set of vocabulary designed to describe samples of tissue, DNA, or RNA linked to voucher specimens and samples of tissue
Knowledge network for biocomplexity (KNB)	It is an international repository designed to promote environmental and ecological studies around biocomplexity
National Centre for biotechnology information (NCBI)	It can provide access to biomedical and genomic data

Altschul et al. 1997; Andelman et al. 2004; Kanz et al. 2005; Dong 2008; Pence and Williams 2010; Artimo et al. 2012; Barrett et al. 2012; Droege et al. 2014

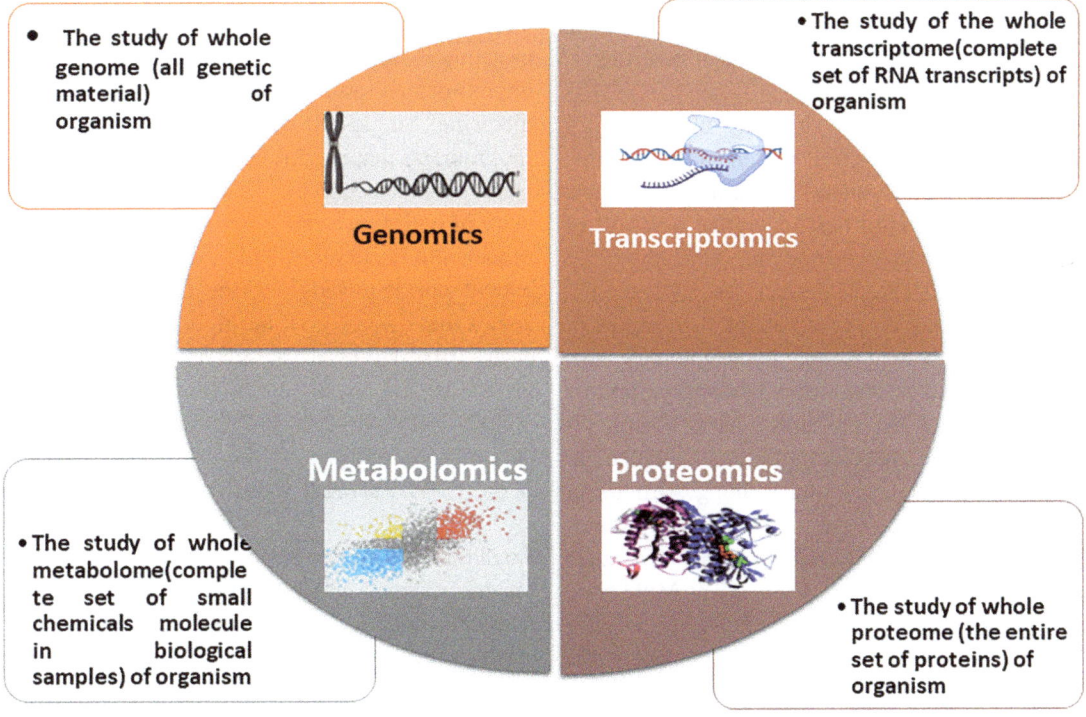

Fig. 2 Omics technologies. Omics technologies seek to the studying of whole genome (genomics), mRNAs (transcriptomics), proteome (proteomics), and metabolome (metabolomics) in specific biological sample (Arjmand 2019; b; Larijani et al. 2019a, b)

3.6   Embryology

Embryology studies the evolutionary and development procedure of various tissues of the living organism from the embryonic stage. In addition, embryological investigations can be effective in the treatment process of fertility-related disorders as well as advancing tissue engineering studies (Patten 1954; De Ferraris and Muñoz 2009; Appasani and Appasani 2010).

3.7   Physiology

Physiology is a branch of biology which concentrates on the biomolecules, cells, and organs' mechanisms of function in a living organism. Indeed, it evaluates the chemical and physical mechanisms (Withers 1992; Ganong 1995; Feder et al. 2000).

4   Molecular Diagnostics and Multi-Omics Approaches

Molecular diagnostics is a group of techniques used to examine biological markers in the genome, proteome, and metabolome. In the recent decades, molecular diagnostics has undergone a period of rapid growth and development (Chehab 1993; Buckingham 2019). Moreover, to advance the goal of achieving proper treatment, it is important to introduce new high-throughput technologies in a clinical molecular diagnostic laboratory. In this context, one of the promising technologies for accelerating the detection process is molecular detection by analytical omics along with using different nanotechnologies (application of numerous nano-devices and nano-systems) (Quezada et al. 2017; Chakraborty et al. 2018; Mukherjee et al. 2020).

4.1   Molecular Imaging

Molecular imaging as part of medical imaging techniques focuses on the use of specific imaging molecules (special probes, i.e., metal ion and radioactive isotope which is injected into a specific anatomical location of living organisms) and imaging modalities (i.e., MRI, CT scan, and PET scan) with the aim of noninvasively studying at the molecular and micromolecular level. The mentioned imaging technique is used to identify metabolic pathways and tissue structures and to evaluate small laboratory animals. Recently, it is applied specifically for infectious diseases, congenital abnormalities, and cancer subjects, from diagnosis to therapy (Aghayan et al. 2014b; Abou-Elkacem et al. 2015; Haris et al. 2015; Saadatpour et al. 2016; Saadatpour et al. 2017).

4.2   Single-Cell Multi-Omics Analysis

Recent technical advancements (including groundbreaking single-cell assays) are promising to overcome the limitation of genome-wide assays (which offers an average of a large number of cells). Accordingly, single-cell sequencing is now becoming available for genomes, transcriptomes, proteomes, and metabolomes, and it provides unprecedented insights into basic biology and biomedicine. Single-cell multi-omics profiling may fix problems that are difficult for other techniques. Hereupon, the genotypic and phenotypic heterogeneity of bulk tissue can be analyzed by single-cell sequencing technology. Indeed, it promises to extend our knowledge of the fundamental processes that control both health and disease (Bock et al. 2016; Hu et al. 2018; Packer and Trapnell 2018; Lee and Hwang 2020; Samir et al. 2020).

5   Advanced Preclinical Models

To simulate a human disease condition (e.g., psychiatric disease), using preclinical models in biomedical studies has become near-universal. Additionally, preclinical models can increase knowledge of cellular signaling pathways and recognizing possible drug targets and novel treatment options (Pan et al. 2020; Scearce-Levie et al. 2020). In this respect, since the past, the use of animal models (especially mammalians) has been popular (Goodarzi et al. 2019d; Larijani et al.

2019a, b; Arjmand et al. 2020b; Baradaran-Rafii et al. 2020). Herein, extrapolating outcomes from models to humans have become an important topic in the evaluation process of the novel treatments. Moreover, based on recent investigations, some biological conditions (such as mental development) have been described which are unique to the human and cannot be modeled in other organisms. In this context, to overcome the mentioned limitations, the advent of in vitro approaches to 3D cell culture systems (employing the genetic engineered stem cells derived from various tissues) or organoids as fast-emerging technology has drawn extensive attention. Organoids (Fig. 3) are able to regenerate, reorganize themselves, and display the function of organs (Li and Izpisua Belmonte 2019; Maximino and van der Staay 2019; Duque-Correa et al. 2020; Jimenez-Palomares et al. 2020; Kim et al. 2020).

6 The Next Generation of Treatments

New breakthroughs in science show great promise in the future of medicine through novel alternative treatments. In this respect, cell-based and gene-based therapies in recent decades greatly progress and become some light of hope for the treatment of incurable diseases. On the other hand, a modern mantra emerging in healthcare is personalized medicine which is powered by providing the clinical, genetic, and environmental knowledge of each individual.

6.1 Cell-Based Therapies

Recently, cell therapy and regenerative medicine are extensively considered in various area of

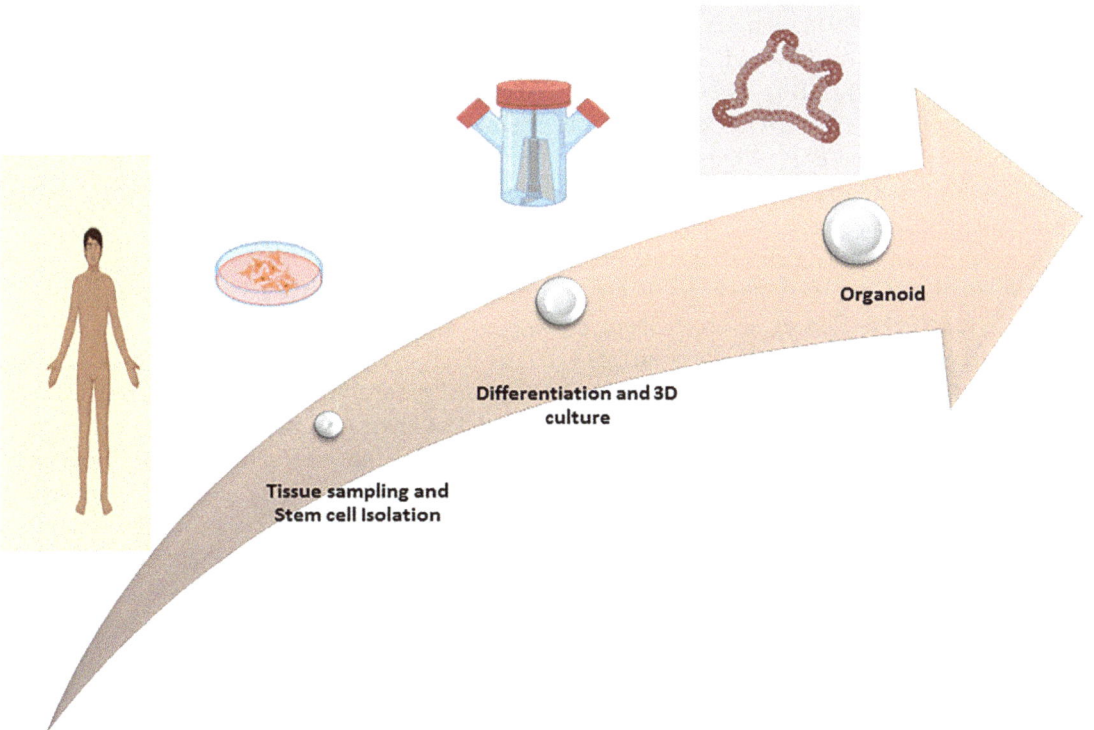

Fig. 3 Three-dimensional (3D) organoid culture. Organoids are 3D cell cultured models which include a population of self-renewing stem cells that differentiate into multiple forms of organ-specific cells and finally show a spatial organization similar to their origin organ and are capable of imitating a highly physiologically certain functions of that organ (Kaushik et al. 2018)

medicine, especially in internal medicine (Arjmand and Aghayan 2014; Arjmand et al. 2017a; Goodarzi et al. 2018b, 2019b, c; Aghayan et al. 2020; Arjmand et al. 2020c, 2021; Ebrahimi-Barough et al. 2020; Hosseini et al. 2020). Different types of stem cell are now available and can be used to generate the healthy and functioning specialized cells, which can then replace diseased or dysfunctional cells. Investigations of cell therapy for cirrhosis, perianal disease of inflammatory bowel disease (IBD), cystic fibrosis of the lung, different types of lymphoma, chronic angina, heart failure, myocardial infarction, ischemic stroke, critical limb ischemia, Parkinson' disease, Alzheimer' disease, spinal cord injuries, etc. have significant progress, but most of them are not yet ready for routine clinical application (Saberi et al. 2008; Aghayan et al. 2014a; Arjmand et al. 2014, 2019a; Goodarzi et al. 2014; Larijani et al. 2014; Derakhshanrad et al. 2015; Goodarzi et al. 2015; Larijani et al. 2015; Shirian et al. 2016; Soleimani et al. 2016; Arjmand et al. 2017c; Rahim and Arjmand 2017; Goodarzi et al. 2018a, c; Payab et al. 2018; Rahim et al. 2018a, b, c, d; Larijani et al. 2020; Payab et al. 2020; Roudsari et al. 2020a, b). Hence, although regenerative medicine and cell therapy as a new technique of treatment have received widespread attention in recent decades, there are a lot of pitfalls in this field (Cho et al. 2015; Nguyen et al. 2016; Shah et al. 2018; Yong et al. 2018; Abramson et al. 2019; Chen et al. 2019; Guo et al. 2019; Hayes Jr et al. 2019).

6.2 Gene-Based Therapies

Gene therapy assists in preventing/treating/or curing disorders through introducing genes into cells (Hashemi et al. 2016; Arjmand et al. 2019b, 2020; Hasanzad and Larijani 2019). This technique is used in treating several inherited disorders; however, it is more feasible for diseases with a single gene involved. There are two types of gene therapy: somatic gene therapy and germ cell gene therapy. Results of the somatic gene therapy unlike the germ cell gene therapy will not be continued by the patients' offspring. Somatic gene therapy is acceptable for diseases such as cystic fibrosis, muscular dystrophy, cancer, inherited blindness, Parkinson's disease, etc. (Verma and Weitzman 2005; Larijani et al. 2019a, b). The first approved gene therapy is introduced in 1990 for severe combined immunodeficiency (SCID) (Ferrua and Aiuti 2017). Thereafter, this technique is applied for other blood disorders like thalassemia, hemophilia, and sickle cell anemia. Also, investigations for gene therapy of cystic fibrosis through introducing normal cystic fibrosis transmembrane conductance regulator (CFTR) gene are ongoing (Yan et al. 2015). Application of gene therapy in the field of cancer treatment was a hot topic in recent years. Accordingly, introducing tumor suppressor genes like p53 or cytokine encoding genes (such as IL 2) and chimeric antigen receptor T cells (CAR T cells) is an example of this application (Ginn et al. 2018). However, gene therapy is not considered to be applied routinely because of some safety and ethical problems such as probability of gene therapy effects on descendants of the patient.

6.3 Regenerative Personalized Medicine

Personalized medicine in biomedical sciences is a special modern treatment method in the current century. The inherent heterogeneity of patients, which indicates different prevalence, different clinical symptoms, and different treatment responses in individuals, families, and ethnicities, is considered to be a very important element in the method described. Personalized medicine needs an examination of the genotype, physiology, and clinical and behavioral details of each individual in order to diagnose and create a personalized treatment plan rapidly. The use of bioinformatics and omics technologies is very helpful in this context (Arjmand et al. 2017a, b, 2020b; Arjmand and Larijani 2017).

7 Conclusion and Future Perspectives

Internal medicine focuses on both acute and chronic diseases. In recent years, combination of technology with basic and medical sciences ameliorates the disease managements. Common available treatment options can cure some of the diseases. On the other hand, due to diverse reasons such as lack of knowledge about the pathophysiology of some diseases and the absence of effective therapeutic options, there are no curative treatments for some of the other diseases. In such circumstances, the goal of management is palliative care and to prevent the progression of disease and its consequences. Further, better understanding the etiology of diseases, new discoveries in basic sciences, novel advances in technology, and proper combination of these elements are required. Here through the experiments on human tissue and fluid samples carried out in high-tech laboratories, biomedical scientists have a huge effect on the development of innovative therapies for human diseases. In other words, the emphasis of biomedical research in medicine has specific advantages for the future of internal medicine and offers satisfaction with the application of fundamental science to the resolution of clinical problems. Therein, in the coming years, medical practice is expected to continue to evolve toward improved and personalized healthcare, with new techniques to be implemented in order to recognize and improve existing clinical limitations, with an increased effort to tailor medical services to each patient's unique characteristics. In order to understand and direct tailored strategies for an effective clinical context, technological advances and clinical trials can help. Hereupon, also the innovation in stem cell-based technology would enable this pathway such mentioned therapies can be readily accessible in medical procedures.

Compliance with Ethical Statements

Conflict of interest: There is no conflict of interest.

Funding: This article received no specific grant from any funding agency.

Ethical approval: This article does not contain any studies with human participants or animals performed by any of the authors.

Authors' Contributions All authors contributed to the study conception and design. Speeder Alavi-Moghadam and Masoumeh Sarvari wrote the first draft. Akram Tayanloo-Beik, Mostafa Rezaei-Tavirani, and Hossein Adibi helped to study and gather information. Neda Mehrdad and Hamid Reza Aghayan extensively edited the manuscript. Bagher Larijani participated in a critical review. Babak Arjmand helped supervise the project and gave final approval of the version to be published.

References

Abati E, Bresolin N, Comi G, Corti S (2019) Advances, challenges, and perspectives in translational stem cell therapy for amyotrophic lateral sclerosis. Mol Neurobiol 56(10):6703–6715

Abou-Elkacem L, Bachawal SV, Willmann JK (2015) Ultrasound molecular imaging: moving toward clinical translation. Eur J Radiol 84(9):1685–1693

Abramson JS, Irwin KE, Frigault MJ, Dietrich J, McGree B, Jordan JT, Yee AJ, Chen YB, Raje NS, Barnes JA (2019) Successful anti-CD19 CAR T-cell therapy in HIV-infected patients with refractory high-grade B-cell lymphoma. Cancer 125(21):3692–3698

Ackerknecht EH, Haushofer L (2016) A short history of medicine. JHU Press, Baltimore

Agharezaee N, Marzbani R, Rezadoost H, Zamani Koukhaloo S, Arjmand B, Gilany K (2018) Metabolomics: a bird's eye view of infertile men. Tehran Univ Med J TUMS Publicat 75(12):860–868

Aghayan H-R, Goodarzi P, Arjmand B (2014a) GMPcompliant human adipose tissue-derived mesenchymal stem cells for cellular therapy. In: Stem cells and good manufacturing practices. Springer, pp 93–107. Publisher: (Humana Press an American academic publisher)

Aghayan HR, Soleimani M, Goodarzi P, Norouzi-Javidan-A, Emami-Razavi SH, Larijani B, Arjmand B (2014b) Magnetic resonance imaging of transplanted stem cell fate in stroke. J Res Med Sci 19(5):465

Aghayan HR, Payab M, Mohamadi-Jahani F, Aghayan SS, Larijani B, Arjmand B (2020) GMP-compliant production of human placenta-derived mesenchymal stem cells. Methods Mol Biol. 2286:213–225

Ahn SH, Mayo-Smith WW, Murphy BL, Reinert SE, Cronan JJ (2002) Acute nontraumatic abdominal pain in adult patients: abdominal radiography compared with CT evaluation. Radiology 225(1):159–164

Alberts B, Johnson A, Lewis J, Morgan D, Raff M, Keith Roberts PW (2018) Molecular biology of the cell. In: Annals of botany. Oxford Academic, Oxford

Altschul SF, Madden TL, Schäffer AA, Zhang J, Zhang Z, Miller W, Lipman DJ (1997) Gapped BLAST and PSI-BLAST: a new generation of protein database search programs. Nucleic Acids Res 25(17):3389–3402

Amatriain RC (2007) Internal medicine: past, present and future. Italian J Med 1:10–11

Andelman SJ, Bowles CM, Willig MR, Waide RB (2004) Understanding environmental complexity through a distributed knowledge network. Bioscience 54(3): 240–246

Appasani K, Appasani RK (2010) Stem cells & regenerative medicine: from molecular embryology to tissue engineering. Springer, Cham

Arjmand B (2019) Genomics, proteomics, and metabolomics. Springer, Cham

Arjmand B, Aghayan HR (2014) Cell manufacturing for clinical applications. Stem Cells (Dayton, Ohio) 32(9): 2557

Arjmand B, Larijani B (2017) Personalized medicine: a new era in endocrinology. Acta Med Iran 55(1):142–143

Arjmand B, Safavi M, Heydari R, Aghayan H, Bazargani S, Dehghani S, Goodarzi P, Mohammadi-Jahani F, Pourmand G (2014) 481 Concomittant transurethral and transvaginal-periurethral injection of autologous adipose stem cells for treatment of female stress urinary incontinence. Eur Urol Suppl 13(1):e481

Arjmand B, Goodarzi P, Mohamadi-Jahani F, Falahzadeh K, Larijani B (2017a) Personalized regenerative medicine. Acta Med Iran 55:144–149

Arjmand B, Abdollahi M, Larijani B (2017b) Precision medicine: a new revolution in healthcare system. Iran Biomed J 21(5):282–283

Arjmand B, Safavi M, Heidari R, Aghayan H, Bazargani ST, Dehghani S, Goodarzi P, Mohammadi-Jahani F, Heidari F, Payab M (2017c) Concomitant transurethral and transvaginal-periurethral injection of autologous adipose derived stem cells for treatment of female stress urinary incontinence: a phase one clinical trial. Acta Med Iran 199:368–374

Arjmand B, Goodarzi P, Aghayan H, Payab M, Rahim F, Alavi-Moghadam S, Mohamadi-Jahani F, Larijani B (2019a) Co-transplantation of human fetal mesenchymal and hematopoietic stem cells in type 1 diabetic mice model. Front Endocrinol 10:761

Arjmand B, Larijani B, Hosseini MS, Payab M, Gilany K, Goodarzi P, Roudsari PP, Baharvand MA (2019b) The Horizon of gene therapy in modern medicine: advances and challenges. Adv Exp Med Biol. 1247:33–64

Arjmand B, Alavi-Moghadam S, Payab M, Goodarzi P, Hosseini MS, Tayanloo-Beik A, Rezaei-Tavirani M, Larijani B (2020a) GMP-Compliant adenoviral vectors for gene therapy. Springer, Cham

Arjmand B, Tayanloo-Beik A, Heravani NF, Alaei S, Payab M, Alavi-Moghadam S, Goodarzi P, Gholami M, Larijani B (2020b) Zebrafish for personalized regenerative medicine; a more predictive humanized model of endocrine disease. Front Endocrinol 11:396

Arjmand B, Sarvari M, Alavi-Moghadam S, Payab M, Goodarzi P, Gilany K, Mehrdad N, Larijani B (2020c) Prospect of stem cell therapy and regenerative medicine in osteoporosis. Front Endocrinol 11

Arjmand B, Payab M, Goodarzi P (2020d) Biomedical product development: bench to bedside. Springer, Cham

Arjmand B, Payab M, Goodarzi P (2020e) Correction to: biomedical product development: bench to bedside, learning materials in biosciences. In: Biomedical product development: bench to bedside. Frontiers Media, Cham, pp C1–C1

Arjmand B, Goodarzi P, Alavi-Moghadam S, Payab M, Aghayan HR, Mohamadi-jahani F, Tayanloo-Beik A, Mehrdad N, Larijani B (2021) GMP-compliant human schwann cell manufacturing for clinical application. Methods Mol Biol 2286:227–235. https://doi.org/10.1007/7651_2020_283

Artimo P, Jonnalagedda M, Arnold K, Baratin D, Csardi G, De Castro E, Duvaud S, Flegel V, Fortier A, Gasteiger E (2012) ExPASy: SIB bioinformatics resource portal. Nucleic Acids Res 40(W1):W597–W603

Azodi MZ, Arjmand B, Zali A, Razzaghi M (2020) Introducing APOA1 as a key protein in COVID-19 infection: a bioinformatics approach. Gastroenterol Hepatol Bed Bench 13(4):367

Baradaran-Rafii A, Sarvari M, Alavi-Moghadam S, Payab M, Goodarzi P, Aghayan HR, Larijani B, Rezaei-Tavirani M, Biglar M, Arjmand B (2020) Cell-based approaches towards treating age-related macular degeneration. Cell Tissue Bank 21(3):339–347

Barrett T, Wilhite SE, Ledoux P, Evangelista C, Kim IF, Tomashevsky M, Marshall KA, Phillippy KH, Sherman PM, Holko M (2012) NCBI GEO: archive for functional genomics data sets—update. Nucleic Acids Res 41(D1):D991–D995

Baxevanis AD, Bader GD, Wishart DS (2020) Bioinformatics. Wiley

Baynes J, Dominiczak MH (2009) Medical biochemistry. Elsevier Health Sciences, Amsterdam

Beenhouwer DO (2018) Molecular basis of diseases of immunity. Mol Pathol: 329–345. Elsevier

Blanco G, Blanco A (2017) Medical biochemistry. Academic Press, Amsterdam

Blann A, Ahmed N (2014) Blood science: principles and pathology. Wiley, Chichester

Bock C, Farlik M, Sheffield NC (2016) Multi-omics of single cells: strategies and applications. Trends Biotechnol 34(8):605–608

Brooks F (2013) Medical microbiology. The McGraw-Hill Companies, Inc., London

Buckingham L (2019) Molecular diagnostics: fundamentals, methods, and clinical applications. FA Davis, Philadelphia

Cambrosio A, Keating P (2001) Biomedical sciences and technology: history and sociology. In: Smelser NJ, Baltes PB (eds) International encyclopedia of the social & behavioral sciences. Pergamon, Oxford, pp 1222–1226

Chakraborty S, Hosen M, Ahmed M, Shekhar HU (2018) Onco-multi-OMICS approach: a new frontier in cancer research. BioMed Res Int 2018:9836256

Chehab FF (1993) Molecular diagnostics: past, present, and future. Hum Mutat 2(5):331–337

Chen R, Snyder M (2012) Systems biology: personalized medicine for the future? Curr Opin Pharmacol 12: 623–628

Chen R, Snyder M (2013) Promise of personalized omics to precision medicine. Wiley Interdiscip Rev Syst Biol Med 5(1):73–82

Chen B, Pang L, Cao H, Wu D, Wang Y, Tao Y, Wang M, Chen E (2019) Autologous stem cell transplantation for patients with viral hepatitis-induced liver cirrhosis: a systematic review and meta-analysis. Eur J Gastroenterol Hepatol 31(10):1283–1291

Cho YB, Park KJ, Yoon SN, Song KH, Kim DS, Jung SH, Kim M, Jeong HY, Yu CS (2015) Long-term results of adipose-derived stem cell therapy for the treatment of Crohn's fistula. Stem Cells Transl Med 4(5): 532–537

Claussnitzer M, Cho JH, Collins R, Cox NJ, Dermitzakis ET, Hurles ME, Kathiresan S, Kenny EE, Lindgren CM, MacArthur DG (2020) A brief history of human disease genetics. Nature 577(7789):179–189

Crombie DL (1963) Diagnostic process. J Coll Gen Pract 6(4):579

Cutolo M (2010) Hormone therapy in rheumatic diseases. Curr Opin Rheumatol 22(3):257–263

DeBerardinis RJ, Thompson CB (2012) Cellular metabolism and disease: what do metabolic outliers teach us? Cell 148(6):1132–1144

De Ferraris MEG, Muñoz AC. (2009) Histologa, embriologa e ingeniera tisular bucodental/Histology, embryology and oral tissue engineering, Ed. Médica Panamericana, Mexico

Derakhshanrad N, Saberi H, Meybodi KT, Taghvaei M, Arjmand B, Aghayan HR, Kohan AH, Haghpanahi M, Rahmani S (2015) Case report: combination therapy with mesenchymal stem cells and granulocyte-colony stimulating factor in a case of spinal cord injury. Basic Clin Neurosci 6(4):299

Dong HW (2008) The Allen reference atlas: a digital color brain atlas of the C57Bl/6J male mouse. Wiley

Droege G, Barker K, Astrin JJ, Bartels P, Butler C, Cantrill D, Coddington J, Forest F, Gemeinholzer B, Hobern D (2014) The global genome biodiversity network (GGBN) data portal. Nucleic Acids Res 42(D1): D607–D612

Duque-Correa MA, Maizels RM, Grencis RK, Berriman M (2020) Organoids—new models for host–helminth interactions. Trends Parasitol 36(2):170–181

Dwivedi S, Purohit P, Misra R, Pareek P, Goel A, Khattri S, Pant KK, Misra S, Sharma P (2017) Diseases and molecular diagnostics: a step closer to precision medicine. Indian J Clin Biochem 32(4):374–398

Ebrahimi-Barough S, Ai J, Payab M, Alavi-Moghadam S, Shokati A, Aghayan HR, Larijani B, Arjmand B (2020) Standard operating procedure for the good manufacturing practice-compliant production of human endometrial stem cells for multiple sclerosis. Methods Mol Biol 2286:199–212

Edwards JH (1963) The genetic basis of common disease. Am J Med 34(5):627–638

Efstratiadis G, Platsas I, Koukoudis P, Vergoulas G (2007) Interventional nephrology: a new subspecialty of nephrology. Hippokratia 11(1):22

Feder ME, Bennett AF, Huey RB (2000) Evolutionary physiology. Annu Rev Ecol Syst 31(1):315–341

Ferrua F, Aiuti A (2017) Twenty-five years of gene therapy for ADA-SCID: from bubble babies to an approved drug. Hum Gene Ther 28(11):972–981

Fordtran JS, Armstrong WM, Emmett M, Kitchens Jr LW, Merrick BA (2004) The history of internal medicine at Baylor University Medical Center, part 1. Baylor University Medical Center Proceedings, Taylor & Francis, Abingdon

Forsblad d'Elia H, Carlsten H (2006) Hormone replacement therapy in rheumatoid arthritis. Curr Rheumatol Rev 2(3):251–260

Fuller J (2017) The new medical model: a renewed challenge for biomedicine. CMAJ 189(17):E640–E641

Ganong WF (1995) Review of medical physiology. Mcgraw-Hill, New York

Gilany K, Masroor MJ, Minai-Tehrani A, Mani-Varnosfaderani A, Arjmand B (2019a) Metabolic profiling of the mesenchymal stem cells' secretome. In: B. Arjmand (ed) Genomics, proteomics, and metabolomics. Stem Cell Biology and Regenerative Medicine. Humana, Cham. https://doi.org/10.1007/978-3-030-27727-7_3

Gilany K, Mohamadkhani A, Chashmniam S, Shahnazari P, Amini M, Arjmand B, Malekzadeh R, Nobakht Motlagh Ghoochani BF (2019b) Metabolomics analysis of the saliva in patients with chronic hepatitis B using nuclear magnetic resonance: a pilot study. Iran J Basic Med Sci 22(9):1044–1049. https://doi.org/10.22038/ijbms.2019.36669.8733. PMID: 31807248; PMCID: PMC6880533

Gilany K et al (2019c) Lipidomics of adipogenic differentiation of mesenchymal stem cells. In: B Arjmand (ed) Genomics, proteomics, and metabolomics. Stem Cell Biology and Regenerative Medicine. Humana, Cham. https://doi.org/10.1007/978-3-030-27727-7_7

Ginn SL, Amaya AK, Alexander IE, Edelstein M, Abedi MR (2018) Gene therapy clinical trials worldwide to 2017: an update. J Gene Med 20(5):e3015

Glazer AN, Nikaido H (2007) Microbial biotechnology: fundamentals of applied microbiology. Cambridge University Press, Cambridge

Gluvic Z, Sudar E, Tica J, Jovanovic A, Zafirovic S, Tomasevic R, Isenovic ER (2015) Effects of levothyroxine replacement therapy on parameters of metabolic syndrome and atherosclerosis in hypothyroid patients: a prospective pilot study. Int J Endocrinol. 2015:147070

Goodarzi P, Aghayan HR, Soleimani M, Norouzi-Javidan-A, Mohamadi-Jahani F, Jahangiri S, Emami-Razavi SH, Larijani B, Arjmand B (2014) Stem cell therapy for treatment of epilepsy. Acta Med Iran:651–655

Goodarzi P, Aghayan HR, Larijani B, Soleimani M, Dehpour A-R, Sahebjam M, Ghaderi F, Arjmand B (2015) Stem cell-based approach for the treatment of Parkinson's disease. Med J Islam Repub Iran 29:168

Goodarzi P, Alavi-Moghadam S, Sarvari M, Beik AT, Falahzadeh K, Aghayan H, Payab M, Larijani B, Gilany K, Rahim F (2018a) Adipose tissue-derived stromal cells for wound healing. In: Cell biology and translational medicine, vol 4. Springer, Berlin, pp 133–149

Goodarzi P, Falahzadeh K, Aghayan H, Jahani FM, Payab M, Gilany K, Rahim F, Larijani B, Beik AT, Adibi H (2018b) GMP-compliant human fetal skin fibroblasts for wound healing. Arch Neurosci 5(3):468–497

Goodarzi P, Larijani B, Alavi-Moghadam S, Tayanloo-Beik A, Mohamadi-Jahani F, Ranjbaran N, Payab M, Falahzadeh K, Mousavi M, Arjmand B (2018c) Mesenchymal stem cells-derived exosomes for wound regeneration. In: Cell biology and translational medicine, vol 4. Springer, pp 119–131

Goodarzi P, Alavi-Moghadam S, Payab M, Larijani B, Rahim F, Gilany K, Bana N, Tayanloo-Beik A, Heravani NF, Hadavandkhani M (2019a) Metabolomics analysis of mesenchymal stem cells. Int J Mol Cell Med 8(Suppl 1):30

Goodarzi P, Falahzadeh K, Aghayan H, Payab M, Larijani B, Alavi-Moghadam S, Tayanloo-Beik A, Adibi H, Gilany K, Arjmand B (2019b) Therapeutic abortion and ectopic pregnancy: alternative sources for fetal stem cell research and therapy in Iran as an Islamic country. Cell Tissue Bank 20(1):11–24

Goodarzi P, Aghayan HR, Payab M, Larijani B, Alavi-Moghadam S, Sarvari M, Adibi H, Khatami F, Heravani NF, Hadavandkhani M (2019c) Human fetal skin fibroblast isolation and expansion for clinical application. Springer, Epidermal Cells, pp 261–273

Goodarzi P, Payab M, Alavi-Moghadam S, Larijani B, Rahim F, Bana N, Sarvari M, Adibi H, Heravani NF, Hadavandkhani M (2019d) Development and validation of Alzheimer's disease animal model for the purpose of regenerative medicine. Cell Tissue Bank 20(2):141–151

Griffiths AJF, Miller JH, Suzuki DT, et al (2000) Genetics and the organism: introduction. In: An introduction to genetic analysis, 7th edn. W. H. Freeman, New York

Guo C, Guo G, Zhou X, Chen Y, Han Z, Yang C, Zhao S, Su H, Lian Z, Leung PS (2019) Long-term outcomes of autologous peripheral blood stem cell transplantation in patients with cirrhosis. Clin Gastroenterol Hepatol 17(6):1175–1182.e1172

Gwee MC, Samarasekera D, Chay-Hoon T (2010) Role of basic sciences in 21st century medical education: An Asian perspective. Med Sci Educator 20(3):300–308

Hajar R (2012) The air of history: early medicine to Galen (part I). Heart Views 13(3):120

Hansell DM, Lynch DA, McAdams HP, Bankier AA (2009) Imaging of diseases of the chest E-book. Elsevier Health Sciences, Amsterdam

Haris M, Yadav SK, Rizwan A, Singh A, Wang E, Hariharan H, Reddy R, Marincola FM (2015) Molecular magnetic resonance imaging in cancer. J Transl Med 13(1):1–16

Hasanzad M, Larijani B (2019) The pathway from gene therapy to genome editing: a nightmare or dream. Int J Mol Cell Med 8(2):69–70

Hashemi M, Fallah A, Aghayan HR, Arjmand B, Yazdani N, Verdi J, Ghodsi SM, Miri SM, Hadjighassem M (2016) A new approach in gene therapy of glioblastoma multiforme: human olfactory ensheathing cells as a novel carrier for suicide gene delivery. Mol Neurobiol 53(8):5118–5128

Hayes D Jr, Kopp BT, Hill CL, Lallier SW, Schwartz CM, Tadesse M, Alsudayri A, Reynolds SD (2019) Cell therapy for cystic fibrosis lung disease: regenerative basal cell amplification. Stem Cells Transl Med 8(3):225–235

Horgan RP, Kenny LC (2011) 'Omic'technologies: genomics, transcriptomics, proteomics and metabolomics. Obstetri Gynaecol 13(3):189–195

Hosseini MS, Roudsari PP, Gilany K, Goodarzi P, Payab M, Tayanloo-Beik A, Larijani B, Arjmand B (2020) Cellular dust as a novel hope for regenerative cancer medicine. Adv Exp Med Biol. 1288:139–160

Hu Y, An Q, Sheu K, Trejo B, Fan S, Guo Y (2018) Single cell multi-omics technology: methodology and application. Front Cell Develop Biol 6:28

Ibrahim R, Pasic M, Yousef GM (2016) Omics for personalized medicine: defining the current we swim in. Expert Rev Mol Diagn 16(7):719–722

Jackson M, Marks L, May GH, Wilson JB (2020) Correction: the genetic basis of disease. Essays Biochem 64(4):681–681

Jamshidi-Kia F, Lorigooini Z, Amini-Khoei H (2018) Medicinal plants: past history and future perspective. J Herbmed Pharmacol 7(1):1–7

Jimenez-Palomares M, Cristobal A, Ruiz MCD (2020) Organoids models for the study of cell-cell interactions. In: Cell interaction-regulation of immune responses, disease development, and management strategies. Intech Open, London

Kang TW, Kim ST, Byun HS, Jeon P, Kim K, Kim H, Lee JI (2009) Morphological and functional MRI, MRS, perfusion and diffusion changes after radiosurgery of brain metastasis. Eur J Radiol 72(3):370–380

Kanz C, Aldebert P, Althorpe N, Baker W, Baldwin A, Bates K, Browne P, van den Broek A, Castro M, Cochrane G (2005) The EMBL nucleotide sequence database. Nucleic Acids Res 33(suppl_1):D29–D33

Karp G (2009) Cell and molecular biology: concepts and experiments. Wiley, Chichester

Kaushik G, Ponnusamy MP, Batra SK (2018) Concise review: current status of three-dimensional organoids as preclinical models. Stem Cells 36(9):1329–1340

Khan MG (2005) Encyclopedia of heart diseases. Elsevier, Amsterdam

Khatami F, Payab M, Sarvari M, Gilany K, Larijani B, Arjmand B, Tavangar SM (2019) Oncometabolites as biomarkers in thyroid cancer: a systematic review. Cancer Manag Res 11:1829

Kim J, Koo B-K, Knoblich JA (2020) Human organoids: model systems for human biology and medicine. Nat Rev Mol Cell Biol 21(10):571–584

Kirschner MW, Marincola E, Teisberg EO (1994) The role of biomedical research in health care reform. Science 266(5182):49–51

Kogut M (1977) The uses of biochemistry in clinical medicine. Biochem Educ 5(1):12–14

Kumar D, Saini N, Jain N, Sareen R, Pandit V (2013) Gold nanoparticles: an era in bionanotechnology. Expert Opin Drug Deliv 10(3):397–409

Larijani B, Aghayan H-R, Goodarzi P, Arjmand B (2014) GMP-grade human fetal liver-derived mesenchymal stem cells for clinical transplantation. In: Stem cells and good manufacturing practices. Springer, Berlin, pp 123–136

Larijani B, Aghayan H, Goodarzi P, Mohamadi-Jahani F, Norouzi-Javidan A, Dehpour AR, Fallahzadeh K, Azam Sayahpour F, Bidaki K, Arjmand B (2015) Clinical grade human adipose tissue-derived mesenchymal stem cell banking. Acta Med Iran 53(9):540–546

Larijani B, Goodarzi P, Payab M, Alavi-Moghadam S, Rahim F, Bana N, Abedi M, Arabi M, Adibi H, Gilany K (2019a) Metabolomics and cell therapy in diabetes mellitus. Int J Mol Cell Med 8(Suppl1):41

Larijani B, Goodarzi P, Payab M, Tayanloo-Beik A, Sarvari M, Gholami M, Gilany K, Nasli-Esfahani E, Yarahmadi M, Ghaderi F (2019b) The design and application of an appropriate parkinson's disease animal model in regenerative medicine. Adv Exp Med Biol 1341:89–105

Larijani B, Heravani NF, Alavi-Moghadam S, Goodarzi P, Rezaei-Tavirani M, Payab M, Gholami M, Razi F, Arjmand B (2020) Cell therapy targets for autism spectrum disorders: hopes, challenges and future directions. Adv Exp Med Biol. 1341:107–124

Lee J, Hwang D (2020) Single-cell multiomics: technologies and data analysis methods. Exp Mol Med 52(9):1428–1442

Lee-Kong SA, Feingold DL (2017) Basic colonoscopic interventions: cold, hot biopsy techniques, submucosal injection, clip application, snare biopsy. In: Advanced colonoscopy and endoluminal surgery. Springer, Berlin, pp 91–95

Lesk A (2019) Introduction to bioinformatics. Oxford University Press

Li M, Izpisua Belmonte JC (2019) Organoids—preclinical models of human disease. N Engl J Med 380(6):569–579

Major RH (1954) A history of medicine. In: A history of medicine. Charles C. Thomas

Maximino C, van der Staay FJ (2019) Behavioral models in psychopathology: epistemic and semantic considerations. Behav Brain Funct 15(1):1–11

Mayer B (2011) Bioinformatics for omics data: methods and protocols. Springer, Berlin

Mehrparavar B, Minai-Tehrani A, Arjmand B, Gilany K (2019) Metabolomics of male infertility: a new tool for diagnostic tests. J Reprod Infertil 20(2):64

Mehrparvar B, Chashmniam S, Nobakht F, Amini M, Javidi A, Minai-Tehrani A, Arjmand B, Gilany K (2020) Metabolic profiling of seminal plasma from teratozoospermia patients. J Pharm Biomed Anal 178:112903

Mukherjee A, Bhattacharya J, Moulick RG (2020) Nanodevices: the future of medical diagnostics. Nano BioMedicine. Springer, Berlin, pp 371–388

Murray PR, Rosenthal KS, Pfaller MA (2020) Medical microbiology e-book. Elsevier Health Sciences, Amsterdam

Nass SJ, Levit LA, Gostin LO (2009) The value, importance, and oversight of health research. In: Beyond the HIPAA privacy rule: enhancing privacy, improving health through research. National Academies Press (US), Washington, D.C.

Nguyen PK, Rhee J-W, Wu JC (2016) Adult stem cell therapy and heart failure, 2000 to 2016: a systematic review. JAMA Cardiol 1(7):831–841

Organization WH (2002) Traditional medicine in Asia. WHO regional office for South-East Asia

Organization, WH (2018) Human African trypanosomiasis, symptoms, diagnosis and treatment. WHO Regional Office for South East Asia

Packer J, Trapnell C (2018) Single-cell multi-omics: an engine for new quantitative models of gene regulation. Trends Genet 34(9):653–665

Pal GK (2007) Importance of biomedicine and biomedical research. 27:1–2

Pan E, Bogumil D, Cortessis V, Yu S, Nieva J (2020) A systematic review of the efficacy of preclinical models of lung cancer drugs. Front Oncol 10:591

Patten BM (1954) Human embryology. J Nerv Ment Dis 119(5):463

Payab M, Goodarzi P, Heravani NF, Hadavandkhani M, Zarei Z, Falahzadeh K, Larijani B, Rahim F, Arjmand B (2018) Stem cell and obesity: current state and future perspective. In: Cell biology and translational medicine, vol 2. Springer, Berlin, pp 1–22

Payab M, Abedi M, Heravani NF, Hadavandkhani M, Arabi M, Tayanloo-Beik A, Hosseini MS, Gerami H, Khatami F, Larijani B (2020) Brown adipose tissue transplantation as a novel alternative to obesity treatment: a systematic review. Int J Obes 45:109–121

Pence HE, Williams A (2010) ChemSpider: an online chemical information resource. ACS Publications, Washington, D.C.

Quezada H, Guzmán-Ortiz AL, Díaz-Sánchez H, Valle-Rios R, Aguirre-Hernández J (2017) Omics-based biomarkers: current status and potential use in the clinic. Boletín Médico Del Hospital Infantil de México (English Edition) 74(3):219–226

Quirke V, Gaudillière J-P (2008) The era of biomedicine: science, medicine, and public health in Britain and France after the second world war. Med Hist 52(4): 441–452

Raghavendra P, Pullaiah T (2018) Advances in cell and molecular diagnostics. Academic Press

Rahim F, Arjmand B (2017) Stem cell clinical trials for multiple sclerosis: the past, present, and future. In: Neurological regeneration. Springer, Berlin, pp 159–172

Rahim F, Arjmand B, Shirbandi K, Payab M, Larijani B (2018a) Stem cell therapy for patients with diabetes: a systematic review and meta-analysis of metabolomics-based risks and benefits. Stem Cell Investig 5:40

Rahim S, Rahim F, Shirbandi K, Haghighi BB, Arjmand B (2018b) Sports injuries: diagnosis, prevention, stem cell therapy, and medical sport strategy. Springer, Tissue engineering and regenerative medicine, pp 129–144

Rahim F, Arjmand B, Larijani B, Goodarzi P (2018) Stem cells treatment to combat Cancer and genetic disease: from stem cell therapy to gene-editing correction. In: Stem cells for cancer and genetic disease treatment. Springer, Berlin, pp 29–59

Rahim F, Arjmand B, Tirad R, Malehi AS (2018d) Stem cell therapy for multiple sclerosis. Cochrane Database Syst Rev 2018(6):0–3

Ramírez HM, Plasencia EÁ (2018) Interventional rheumatology, an unsettled issue. Reumatol Clin 14(1):2

Ramsden JJ (2016) Chapter 11 – bionanotechnology. In: Ramsden JJ (ed) Nanotechnology, 2nd edn. William Andrew Publishing, Oxford, pp 263–278

Rauta PR, Mohanta YK, Nayak D (2019) Nanotechnology in biology and medicine: research advancements & future perspectives. CRC Press, Boca Raton, FL

Roudsari PP, Alavi-Moghadam S, Rezaei-Tavirani M, Goodarzi P, Tayanloo-Beik A, Sayahpour FA, Larijani B, Arjmand B (2020a) The outcome of stem cell-based therapies on the immune responses in rheumatoid arthritis. Adv Exp Med Biol 1326:159–186

Roudsari PP, Alavi-Moghadam S, Payab M, Sayahpour FA, Aghayan HR, Goodarzi P, Mohamadi-Jahani F, Larijani B, Arjmand B (2020b) Auxiliary role of mesenchymal stem cells as regenerative medicine soldiers to attenuate inflammatory processes of severe acute respiratory infections caused by COVID-19. Cell Tissue Bank 21(3):405–425

Saadatpour Z, Bjorklund G, Chirumbolo S, Alimohammadi M, Ehsani H, Ebrahiminejad H, Pourghadamyari H, Baghaei B, Mirzaei H, Sahebkar A (2016) Molecular imaging and cancer gene therapy. Cancer Gene Ther:1–5

Saadatpour Z, Rezaei A, Ebrahimnejad H, Baghaei B, Bjorklund G, Chartrand M, Sahebkar A, Morovati H,

Mirzaei H, Mirzaei H (2017) Imaging techniques: new avenues in cancer gene and cell therapy. Cancer Gene Ther 24(1):1–5

Saberi H, Moshayedi P, Aghayan H-R, Arjmand B, Hosseini S-K, Emami-Razavi S-H, Rahimi-Movaghar V, Raza M, Firouzi M (2008) Treatment of chronic thoracic spinal cord injury patients with autologous Schwann cell transplantation: an interim report on safety considerations and possible outcomes. Neurosci Lett 443(1):46–50

Samir J, Rizzetto S, Gupta M, Luciani F (2020) Exploring and analysing single cell multi-omics data with VDJView. BMC Med Genet 13(1):1–9

Scearce-Levie K, Sanchez PE, Lewcock JW (2020) Leveraging preclinical models for the development of Alzheimer disease therapeutics. Nat Rev Drug Discov 19(7):447–462

Schneider MV, Orchard S (2011) Omics technologies, data and bioinformatics principles. In: Bioinformatics for omics data. Springer, Berlin, pp 3–30

Shah R, Latham SB, Khan SA, Shahreyar M, Hwang I, Jovin IS (2018) A comprehensive meta-analysis of stem cell therapy for chronic angina. Clin Cardiol 41(4):525–531

Sharma M, Khurana SP (2018) Biomedical engineering: the recent trends. In: Omics technologies and bio-engineering. Elsevier, pp 323–336

Shirian S, Ebrahimi-Barough S, Saberi H, Norouzi-Javidan A, Mousavi SMM, Derakhshan MA, Arjmand B, Ai J (2016) Comparison of capability of human bone marrow mesenchymal stem cells and endometrial stem cells to differentiate into motor neurons on electrospun poly (ε-caprolactone) scaffold. Mol Neurobiol 53(8):5278–5287

Soleimani M, Aghayan HR, Goodarzi P, Hagh MF, Lajimi AA, Saki N, Jahani FM, Javidan AN, Arjmand B (2016) Stem cell therapy–approach for multiple sclerosis treatment. Arch Neurosci 3(1):e21564

Stryer L, Berg JM Tymoczko JL (2002) Biochemistry, 5th edn. WH Freeman & Co Ltd., New York

Swanson TA (2018) Biochemistry, molecular biology, and genetics. Oxford University Press, Hong Kong

Swanson K, Dodd MR, VanNess R, Crossey M (2018) Improving the delivery of healthcare through clinical diagnostic insights: a valuation of laboratory medicine through "clinical lab 2.0". J Appl Lab Med 3(3): 487–497

Tan M (2016) Cell and molecular biology for diagnostic and therapeutic technology. J Phys Conf Ser 694: 012001

Tayanloo-Beik A, Sarvari M, Payab M, Gilany K, Alavi-Moghadam S, Gholami M, Goodarzi P, Larijani B, Arjmand B (2020) OMICS insights into cancer histology; metabolomics and proteomics approach. Clin Biochem 84:13–20

Tseng J, Choi EA, Matthews JB (2019) Chronic pancreatitis. Shackelford's surgery of the alimentary Tract, 2 Volume Set, Elsevier, Amsterdam, pp 1085–1096

Valentinuzzi ME, Ertek S, Zanutto BS (2017) The future of bioengineering: possible new areas. Elsevier, Amsterdam

Verma IM, Weitzman MD (2005) Gene therapy: twenty-first century medicine. Annu Rev Biochem 74:711–738

Wade DT, Halligan PW (2004) Do biomedical models of illness make for good healthcare systems? BMJ 329(7479):1398–1401

Wang X, Peer D, Petersen B (2013) Molecular and cellular therapies: new challenges and opportunities. Mol Cell Ther. 1:1. BioMed Central

Weatherall D (1998) The future role of molecular and cell biology in medical practice in the tropical countries. Br Med Bull 54(2):489–501

Weatherall D, Greenwood B, Chee HL, Wasi P (2006) Science and technology for disease control: past, present, and future. In: Disease control priorities in developing countries, vol 2. Oxford University Press, Oxford, pp 119–138

Wei Q, Huang H (2013) Chapter five – Insights into the role of cell-cell junctions in physiology and disease. In: Jeon KW (ed) International review of cell and molecular biology. Academic Press, Amsterdam, pp 187–221, 306

Weiland A, Blankenstein AH, Van Saase JL, Van der Molen HT, Jacobs ME, Abels DC, Köse N, Van Dulmen S, Vernhout RM, Arends LR (2015) Training medical specialists to communicate better with patients with medically unexplained physical symptoms (MUPS). A randomized, controlled trial. PLoS One 10(9):e0138342

West CP, Dupras DM (2012) General medicine vs subspecialty career plans among internal medicine residents. JAMA 308(21):2241–2247

Williams ES, Silverman LM (2018) Molecular diagnosis of human disease. In: Molecular pathology. Elsevier, Amsterdam, pp 691–707

Withers PC (1992) Comparative animal physiology. Saunders College Pub, Philadelphia

Yadav D (2015) Relevance of bioinformatics in the era of omics driven research. J Next Generat Sequen Applicat 2(1):e102

Yan Z, Stewart ZA, Sinn PL, Olsen JC, Hu J, McCray PB Jr, Engelhardt JF (2015) Ferret and pig models of cystic fibrosis: prospects and promise for gene therapy. Hum Gene Ther Clin Dev 26(1):38–49

Yan J, Risacher SL, Shen L, Saykin AJ (2018) Network approaches to systems biology analysis of complex disease: integrative methods for multi-omics data. Brief Bioinform 19(6):1370–1381

Yong KW, Choi JR, Mohammadi M, Mitha AP, Sanati-Nezhad A, Sen A (2018) Mesenchymal stem cell therapy for ischemic tissues. Stem Cells Int 2018:8179075

Zhang Y, Liu X, Wang S, Li L, Dou S (2017) Bio-nanotechnology in high-performance supercapacitors. Adv Energy Mater 7(21):1700592

Adv Exp Med Biol - Cell Biology and Translational Medicine (2022) 17: 191–211
https://doi.org/10.1007/5584_2022_728
© Springer Nature Switzerland AG 2022
Published online: 12 July 2022

Mesenchymal Stem Cell-Derived Extracellular Vesicles: Progress and Remaining Hurdles in Developing Regulatory Compliant Quality Control Assays

Jessie Kit Ern Chua, Jiaxi Lim, Le Hui Foong, Chui Yang Mok, Hsiang Yang Tan, Xin Yee Tung, Thamil Selvee Ramasamy, Vijayendran Govindasamy ⓘ, Kong-Yong Then, Anjan Kumar Das, and Soon-Keng Cheong

Abstract

Regenerative medicine is shaping into a new paradigm and could be the future medicine driven by the therapeutic capabilities shown by mesenchymal stem cell-derived extracellular vesicles (MSC-EVs). Despite the advantages and promises, the therapeutic effectiveness of MSC-EVs in some clinical applications is restricted due to inconsistent manufacturing process and the lack of stringent quality control (QC) measurement. In particular, QC assays which are crucial to confirm the safety, efficacy, and quality of MSC-EVs available for end use are poorly designed. Hence, in this review, characterization of MSC-EVs and quality control guidelines for biologics are presented, with special attention given to the description of technical know-how in developing QC assays for MSC-EVs adhering to regulatory guidelines. The remaining challenges surrounding the development of potency and stability of QC assays are also addressed.

J. K. E. Chua, J. Lim, L. H. Foong, C. Y. Mok, H. Y. Tan, X. Y. Tung, V. Govindasamy (✉), and K.-Y. Then
Cryocord, 1, Bio X Centre, Persiaran Cyber Point Selatan, Cyberjaya, Cyberjaya, Selangor, Malaysia
e-mail: vijayendran@cryocord.com.my;
kongthen@gmail.com

T. S. Ramasamy
Stem Cell Biology Laboratory, Department of Molecular Medicine, Faculty of Medicine, Universiti Malaya, Kuala Lumpur, Malaysia

A. K. Das
Maharajah Agrasen Hospital, Siliguri, West Bengal, India

S.-K. Cheong
Faculty of Medicine & Health Sciences, Universiti Tunku Abdul Rahman (UTAR), Kajang, Selangor, Malaysia

Keywords

Bioprocessing · Conditioned media · Exosome · Microvesicles · Regenerative medicine · Secretomes

Abbreviations

μm	Micrometers
AD	Adipose tissue
AFM	Atomic force microscopy
BM	Bone marrow
CB	Cord blood

CDSCO	Central Drugs Standard Control Organisation
DLS	Dynamic light scattering
ELISA	Enzyme-linked immunosorbent assay
EM	Electron microscopy
EMA	European Medicines Agency
EVs	Extracellular vesicles
FC	Flow cytometry
FDA	Food and Drug Administration
HBV	Hepatitis B virus
HCV	Hepatitis C virus
HIV	Human immunodeficiency virus
ICH	International Council for Harmonisation of Technical Requirements for Pharmaceuticals for Human Use
ISEV	International Society for Extracellular Vesicles
MFDS	Ministry of Food and Drug Safety
MISEV	Minimal information for studies of extracellular vesicles
MSC-EVs	Mesenchymal stem cell-derived extracellular vesicles
MSC-Exo	Mesenchymal stem cell-derived exosome
MSC-MVs	Mesenchymal stem cell-derived microvesicles
MSCs	Mesenchymal stem cells
MVBs	Multivesicular bodies
MVs	Microvesicles
MWCO	Molecular weight cutoff
nm	Nanometers
NTA	Nanoparticle tracking analysis
PCR	Polymerase chain reaction
PMDA	Pharmaceuticals and Medical Devices Agency
QC	Quality control
qPCR	Quantitative polymerase chain reaction
RPS	Resistive pulse sensing
RT-PCR	Reverse transcription polymerase chain reaction
SEM	Scanning electron microscopy
sEVs	Small extracellular vesicles
TEM	Transmission electron microscopy
TRPS	Tunable resistive pulse sensing
TUNEL	Terminal deoxynucleotidyl transferase dUTP Nick-End Labeling
UC	Umbilical cord
WB	Western blot

1 Introductions

Mesenchymal stem cell secretome, the extracellular vehicles (MSC-EVs), has sparked attention in the last few years because of their increasing biological relevance in normal physiology and disease states (Xunian and Kalluri 2020; Hartjes et al. 2019). They have broader therapeutic effects due to its ability to carry a considerable number of functional therapeutic molecules in the form of mRNA, micro-RNAs, long-coding RNAs, DNA, and metabolites (Maumus et al. 2020). This unique characteristic allows MSC-EVs to be utilized in many forms such as the replacement of live cells in stem cell transplantation or as a vehicle in delivering specific targeted therapeutic agents (Maumus et al. 2020).

Though up to date more than 40 MSC-EV studies have been conducted broadly in preclinical animal models, however, only a few have been approved for clinical studies in which most of them are in the not yet recruiting or available phases and are confined to wound healing process and autoimmune, neurological, and cardiovascular diseases (Witwer et al. 2019; Lee et al. 2021). Inconsistent bioprocessing of EV production especially in large scale-up and lack of standardized quality control (QC) assay are the key factors downgrading the therapeutic effects of MSC-EVs products (Witwer et al. 2019; Nguyen et al. 2020). While attention has been given in addressing the issues related to the former, researchers still have difficulty in designing a proper QC assay to align with regulatory requirements.

MSC-EVs are classified as medicinal biological products by leading regulatory agencies, which means that MSC-EV-based products need to be approved by competent authorities before administration for clinical use. For the approval of MSC-EV products to be considered, compliance to Good Manufacturing Practices (GMP) and strict regulatory practices during the in-process QC of product, finished product QC, and stability studies are crucial to ensure the safety, efficacy, and quality of the MSC-EVs available to end use. Current MSC-EV QC assays are centralized to address the characterization as there is a certain degree of overlap in terms of sizes and markers between the various subtypes (exosomes, microvesicles, apoptotic bodies). While a certain degree of attention is given to address the QC-related assays on physical properties, safety, and purity, potency assays that overall predict the biological function of the EVs are often neglected (Ludwig et al. 2019). Further, most of these QC assays were developed without proper analytical validation resulting in inconsistency in release criteria.

Hence, this review will address the progress and challenges in designing regulatory compliant QC assays that cover major key elements such as identity, purity, safety, and potency. We anticipate that a validated QC assay will lead to the better therapeutic efficacy of MSC-EVs.

2 Biological Characterization of MSC-EVs

MSC-EVs are non-replicated lipid bilayer vesicles and generally secrete into the extracellular environment (Witwer et al. 2019). Typically, EVs are divided into three subtypes – apoptotic bodies, microvesicles (MVs), and exosomes – which are generally distinguished by their sizes (Fig. 1). Exosome being the smallest of all of them with a diameter size between 40 and 120 nanometers (nm), has the main role in regulating intercellular communication carrying specific biomolecular information such as RNAs, proteins, and lipids in their intracellular compartments and transporting them to the target cells (Andaloussi et al. 2019; Colombo et al. 2014). Exosome can be easily distinguished from other subtypes by characterising marker proteins such as tetraspanins (CD9, CD63, CD81, CD82), membrane transport proteins (annexins, Rab), heat shock proteins (Hsp60, Hsp70, Hsp90), and multivesicular bodies (MVBs) formation proteins (Alix, TSG101) (Zhang et al. 2019) that are found on its membrane surface. This is followed by MVs which are in the range of size of 50 to 1000 nm and have similar function and content as exosomes (Andaloussi et al. 2019). MVs also mediate intercellular communication via the delivery of contents to recipient cells and contain cytosolic and plasma membrane-associated proteins, mRNAs, miRNAs, nucleic acids and lipids in their intracellular compartments (Andaloussi et al. 2019; Doyle and Wang 2019). The protein markers present on the membrane of MVs are

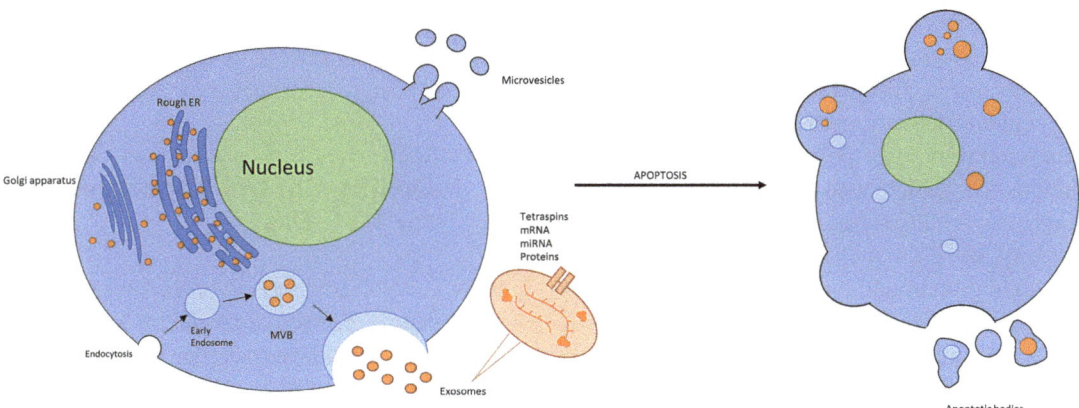

Fig. 1 The biogenesis and content of extracellular vesicles which include the subtypes of extracellular vesicles, microvesicles, and apoptotic bodies

integrins, selectins, and CD40 ligands that function as adhesion molecules (Andaloussi et al. 2019). Apoptotic bodies are larger EVs with a size between 500 and 2000 nm and formed during apoptosis. It contains a portion of a dying cell known as nuclear fractions and cell organelles and can be recognized by having large amounts of phosphatidylserine (Andaloussi et al. 2019).

Due to the overlapping of sizes, characterization markers, and functions between the subtypes of EVs, the International Society for Extracellular Vesicles (ISEV) has published minimal criteria for studies of extracellular vesicles (MISEV) in defining EVs, especially those from MSCs (Thery et al. 2018). According to guideline which is known as MISEV 2014 (later being adopted as MISEV 2018), MSC-EVs that are sized greater than 200 nanometers (nm) in diameter are termed large EVs while EVs that are smaller than 200 nm in diameter are termed small EVs (sEVs) and are named based on its contained biochemical compounds or their originating cell type (Thery et al. 2018).

ISEV further suggested that each research of EVs should be (1) outlined by quantitative measurement of EV source (e.g., total amount of conditioned medium, initial cell seeding, and final cell count); (2) characterized to the utmost practicable to determine the quantity of EVs (e.g., quantification of particles, protein, lipids, nucleic acid); (3) verified for the presence of components specific for particular EVs subtypes (e.g., characterization by protein composition or surface marker); and (4) examined for the presence of non-vesicular, co-isolated components. In other words, EVs should be characterised by the content such as protein or particle concentration, minimum three positives and one negative protein marker and two analytical measures of single extracellular vesicle. The positive protein markers of EVs include transmembrane and cytosolic protein while negative protein markers are composed of apolipoproteins A1/2 and albumin. In addition, EVs subtype can be characterized by subcellular compartments such as the nucleus, mitochondria, or extracellular protein with biological function like growth factors, cytokines or the extracellular matrix (Thery et al. 2018).

2.1 Current Scenario on MSC-EV Characterization Assays

Most of the characterization assays related to downstream activities of MSC-EVs are largely based on identifying and quantifying (Table 1). Apart from directly looking at the number of particles with the same range of sizes, the assays in this category also measure various components that are present in the EVs such as the specific protein markers and the contents of nucleic acids and proteins. Currently, nanoparticle tracking analysis (NTA) is the most commonly used method to determine the size and the number of particles followed by dynamic light scattering (DLS) and resistive pulse sensing (RPS). On the other hand, single vesicle analysis by electron microscope (EM) is more popularly used than atomic force microscopy (AFM) in most research to characterize the morphology and structure of EVs. In terms of evaluating protein marker quantification and expression, a majority of researchers used Bradford, Western blot (WB), or flow cytometry (FC) analysis. While most of the researchers are using compendial testing which is also known as pharmacopoeia standards to check on safety, only a handful of research had focused on reporting on potency testing. Further, despite many studies providing quantitative values, none of the research reported on the ratio (e.g., protein/particle, lipid/particle, or lipid/protein) to estimate EV purity which has been recommended by the ISEV.

3 Quality Control Regulation for MSC-EVs

Quality control consists of procedures done to warrant that the service or product fulfills certain requirements or achieves a certain degree of quality (Sachan et al. 2014). The Food Drug and Administration (FDA) defines biologics as products that can be used in the diagnosis,

Table 1 QC-related assays being used in MSC-EV research

Author	Cell origin	Phase	Disease	Isolation/ purification	QC assay Identity/quantity	Purity	Safety	Potency
Nassar et al. (2016)	Human CB-MSC-EVs	Phase II/III	Chronic kidney disease	Differential centrifugation Density gradient ultracentrifugation	Total protein content TEM SEM FC			
Zhang et al. (2018)	Human UC-MSC-Exo	Phase I	Refractory macular holes	Differential centrifugation	Total protein content SEM WB			
Shi et al. (2021)	Human AD-MSC-EVs	Phase I	Lung injury	Differential centrifugation PEG concentration	Total protein content TEM NTA WB		Gram stain – Microscopy Sterility test – BacT/ALERT Mycoplasma – qPCR Endotoxin – Limulus assay	
Gatti et al. (2011)	Human MSC-MVs	Preclinical	Acute and chronic kidney injury	Differential centrifugation	Total protein content TEM SEM DLS		Endotoxin – Limulus assay	
Bruno et al. (2012)	MSC-MVs	Preclinical	Acute kidney injury	Differential centrifugation	Total protein content TEM SEM DLS FC		Endotoxin – Limulus assay	
Li et al. (2012)	Human UC-MSC-Exo	Preclinical	Liver fibrosis	Differential centrifugation Ultrafiltration	Total protein content TEM WB			
Arslan et al. (2013)	MSC-Exo	Preclinical	Myocardial ischemia/ reperfusion injury	HPLC Ultrafiltration Tangential flow filtration				

(continued)

Table 1 (continued)

Author	Cell origin	Phase	Disease	Isolation/ purification	QC assay Identity/quantity	Purity	Safety	Potency
Bian et al. (2013)	Human BM-MSC-Evs	Preclinical	Myocardial infarction	Differential centrifugation	Total protein content TEM FC WB			Cell proliferation assay Cell migration assay Tube formation assay
Zhou et al. (2013)	Human UC-MSC-Exo	Preclinical	Cisplatin-induced nephrotoxicity	Density gradient centrifugation Ultrafiltration	TEM WB			
Zhu et al. (2013)	Human BM-MSC-MVs	Preclinical	Acute lung injury	Differential centrifugation	Total protein content SEM			
Chen et al. (2014)	Murine BM-MSC-MVs	Preclinical	Pulmonary arterial hypertension	Differential centrifugation Ultracentrifugation	Total protein content TEM NTA FC			
Zhang et al. (2014)	Human UC-MSC-MVs	Preclinical	Renal ischemia/ reperfusion injury	Differential centrifugation	Total protein content TEM FC	TUNEL assay		
Cruz et al. (2015)	Human BM-MSC-EVs	Preclinical	Allergic airway inflammation	Differential centrifugation Ultracentrifugation	Total protein content TEM NTA			
Doeppner et al. (2015)	Human BM-MSC-Exo	Preclinical	Poststroke neuroregeneration	Ultrafiltration PEG precipitation Ultracentrifugation	Total protein content NTA WB		Bacterial contamination – PCR and infectious serology	
Monsel et al. (2015)	Human BM-MSC-MVs	Preclinical	Severe pneumonia	Ultracentrifugation	Total protein content SEM WB			

Reference	Source	Stage	Disease model	Isolation method	Characterization	Safety	Functional assay
Teng et al. (2015)	Murine BM-MSC-Exo	Preclinical	Myocardial infarction	Precipitation (ExoQuick-TC)	Total protein content, TEM, FC		Cell proliferation assay, Tube formation assay
Zhang et al. (2015a)	Human UC-MSC-Exo	Preclinical	Cutaneous wound healing	Differential centrifugation, Ultrafiltration	Total protein content, TEM, NTA, WB		
Zhang et al. (2015b)	Murine BM-MSC-Exo	Preclinical	Traumatic brain injury	Precipitation (ExoQuick)	Total protein content, TEM, NTA, WB		
Zhao et al. (2015)	Human UC-MSC-Exo	Preclinical	Acute myocardial ischemic injury	Differential centrifugation, MWCO concentration, Density gradient centrifugation, Ultracentrifugation	Total protein content, TEM, NTA, WB		
Lin et al. (2016)	AD-MSC-Exo	Preclinical	Renal acute ischemia/reperfusion injury	Ultracentrifugation	EM, WB		
Ophelders et al. (2016)	Human BM-MSC-EVs	Preclinical	Preterm hypoxic-ischemic brain injury	Ultrafiltration, PEG precipitation, Ultracentrifugation	Total protein content, NTA, WB	Tested for the presence of bacteria, viruses, and endotoxins	
Tamura et al. (2016)	Murine BM-MSC-Exo	Preclinical	Liver injury	Differential centrifugation, Ultrafiltration, Ultracentrifugation	Total protein content, TEM, TRPS, FC		
Zhang et al. (2016)	BM-MSC-Exo	Preclinical	Myocardial repair	Differential centrifugation, Precipitation (ExoQuick-TC)	Total protein content, TEM, FC, WB		Cell proliferation assay, Cell migration assay, Tube formation assay

(continued)

Table 1 (continued)

Author	Cell origin	Phase	Disease	Isolation/ purification	QC assay Identity/quantity	Purity	Safety	Potency
Zou et al. (2016)	Human MSC-EVs	Preclinical	Renal ischemic reperfusion injury	Ultracentrifugation	Total protein content TEM FC NTA			
Bai et al. (2017)	Human UC-MSC-Exo	Preclinical	Autoimmune uveitis	Differential centrifugation Ultrafiltration	Total protein content EM WB			
de Castro et al. (2017)	Human AD-MSC-EVs	Preclinical	Allergic asthma inflammation	Differential centrifugation Ultracentrifugation	Total protein content SEM DLS			
Drommelschmidt et al. (2017)	Human BM-MSC-EVs	Preclinical	Inflammation-induced preterm brain injury	Ultrafiltration PEG precipitation Ultracentrifugation	NTA WB		HIV, HCV, HBV – Multiplex PCR Microbiological contamination – BacTAlert bottles	
Gangadaran et al. (2017)	Murine BM-MSC-EVs	Preclinical	Hindlimb ischemia	Differential centrifugation Density gradient ultracentrifugation	NTA TEM WB			Cell migration assay Cell proliferation assay Tube formation assay
Haga et al. (2017)	BM-MSC-EVs	Preclinical	Lethal hepatic failure	Differential centrifugation Ultracentrifugation	Total protein content NTA TEM			
Mao et al. (2017)	Human UC-MSC-Exo	Preclinical	Inflammatory bowel disease	Differential centrifugation Density gradient centrifugation Ultrafiltration	Total protein content TEM NTA WB			

Song et al. (2017)	Human UC-MSC-Exo	Preclinical	Sepsis	Differential centrifugation Ultrafiltration	Total protein content TEM WB	Cell migration assay
Stone et al. (2017)	Human UC-MSC-EVs	Preclinical	Lung ischemic/ reperfusion injury	Differential centrifugation	Total protein content NTA Imaging FC	
Wang et al. (2017)	MSC-EVs		Myocardial infarction	Ultracentrifugation Differential centrifugation	Total protein content TEM WB	Cell proliferation assay Cell migration assay Tube formation assay
Ahn et al. (2018)	Human UC-MSC-EVs	Preclinical	Neonatal hyperoxic lung injury	Differential centrifugation Ultracentrifugation	TEM NTA	
Bandeira et al. (2018)	AD-MSC-EVs	Preclinical	Silicosis	Differential centrifugation Ultracentrifugation	Total protein content TEM/SEM NTA FC	
Cho et al. (2018)	Human AD-MSC-Exo	Preclinical	Atopic dermatitis	Differential centrifugation	TEM NTA WB FC	Cell-based assay
Jiang et al. (2018)	Human UC-MSC-Exo	Preclinical	Liver injury	Differential centrifugation Ultracentrifugation MWCO concentration Density gradient centrifugation Ultrafiltration	TEM NTA FC	

(continued)

Table 1 (continued)

Author	Cell origin	Phase	Disease	Isolation/purification	QC assay Identity/quantity	Purity	Safety	Potency
Sun et al. (2018a)	Human UC-MSC-Exo	Preclinical	Spinal cord injury	Differential centrifugation MWCO concentration Ultracentrifugation Ultrafiltration	Total protein content TEM DLS WB			
Sun et al. (2018b)	Human MSC-Exo	Preclinical	Type 2 diabetes mellitus	Differential centrifugation Ultrafiltration	Total protein content WB TEM NTA			
Wu et al. (2018)	Human UC-MSC-Exo	Preclinical	Inflammatory bowel disease	Differential centrifugation Density gradient centrifugation Ultrafiltration	Total protein content TEM NTA WB			
Hao et al. (2019)	Human BM-MSC-EVs	Preclinical	Lung injury	Differential centrifugation Ultracentrifugation	Total protein content SEM NTA FC WB			
Shi et al. (2019)	Human UC-MSC-Exo	Preclinical	Acute myocardial infarction	Differential centrifugation MWCO concentration ExoQuick-TC	Total protein content TEM NTA WB			
Shiue et al. (2019)	Human UC-MSC-Exo	Preclinical	Nerve injury-induced pain	Differential centrifugation Ultrafiltration	Total protein content TEM WB FC			
Varkouhi et al. (2019)	Human UC-MSC-EVs	Preclinical	Acute lung injury	Differential centrifugation	Total protein content TEM FC			

EM electron microscope, *TEM* transmission electron microscope, *SEM* scanning electron microscope, *NTA* nanoparticle tracking analysis, *DLS* dynamic light scattering, *TRPS* tunable resistive pulse sensing, *FC* flow cytometry, *WB* Western blot, *HPLC* high-performance liquid chromatography prevention, and treatment of medical disorders (Code of Federal Regulations 2021). Hence, biological products, such as MSC-EVs, are subjected to several QC regulations, which cover aspects such as identity, quantity, purity, sterility, potency, and stability. Each country has its own regulatory body, governing the quality of biological products with its own set of QC guidelines (Table 2).

3.1 Establishing QC Assays for EV-Based Products

Many characterization assays are established and being used routinely in MSC-EV research work; however, the main stumbling block is that only a handful of these assays are being developed into a proper QC assay. In this regard, the International Council for Harmonisation of Technical Requirements for Pharmaceuticals for Human Use (ICH) has proposed the feasibility of developing any assays into QC assays with specific acceptance criteria to ensure the assays' reproducibility, reliability, and therapeutic value. To develop a good QC assay, ICH Q2 (R1) guideline highlights the need to conduct validation experiments based on the eight important parameters which include linearity, specificity, range, accuracy, robustness, precision, quantitation limit, and detection limit (Table 3). Here, we briefly explain

Table 2 Guideline subsections of each regulatory aspect for biological products by several regulatory bodies

Regulatory aspects	Guideline subsection by each regulatory body				
	FDA (USA) – Code of Federal Regulations Title 21, Subchapter F, Chap. 1, Part 610 (Code of Federal Regulations 2021)	EMA (Europe) – ICH Q5C – Quality of Biotechnological Products: Stability Testing of Biotechnological/ Biological Products and ICH Q6B – Specifications: Test Procedures and Acceptance Criteria for Biotechnological/ Biological Products (European Medicines Agency 1994a, 1994b)	MFDS (Korea) – Regulation on Approval and Review of Biological Products (Ministry of Food and Drug Safety 2003)	PMDA (Japan) – Guideline for the Quality, Safety, and Efficacy Assurance of Follow-On Biologics (Pharmaceuticals and Medical Devices Agency 2009)	CDSCO (India) – Biosimilar Guideline 2016 (Central Drugs Standard Control Organisation 2016)
Identity	610.14 identity	–	Article 28 (review criteria for biologics)	–	6.3.2 product characterization
Quantity	–	Q6B 2.1.5 quantity	–	6. Specifications and test procedures	6.3.1 analytical methods
Purity	610.13 purity	Q6B 2.1.4 purity, impurities, and contaminants	Article 28 (review criteria for biologics)	6. Specifications and test procedures	6.3.2 product characterization
Sterility	610.12 sterility	Q6B 2.1.4 purity, impurities, and contaminants	Article 28 (review criteria for biologics)	6. Specifications and test procedures	–
Potency	610.10 potency	Q6B 2.1.2 biological activity	Article 28 (review criteria for biologics)	6. Specifications and test procedures	6.3.2 product characterization
Stability	–	Q5C	–	4.4 stability testing	6.3.4 stability

Table 3 List of QC assays corresponding to ICHQ2 (R1) guidelines

Author	Assays' category	Assays	Specificity	Linearity	Range	Accuracy	Precision	Detection limit	Quanitation limit	Robustness	Acceptance range	Cost	Technicality
			ICH validation criteria based on ICH Q2 (R1)								Other parameters		
Adan et al. (2016); Maas et al. (2015)	Identity/ quantity	Flow cytometry (FC)	/	/	/	×	/	×	/	/	Percentage of positive or negative expression markers	↕	→
Choudhary and Ka (2017)		Scanning electron microscopy (SEM)	/	/	/	/	/	×	/	×	The morphology and structure, either intact or not intact	←	←
Dragovic et al. (2011); Maas et al. (2015)		Nanoparticle tracking analysis (NTA)	/	/	/	×	×	/	/	/	The amount of particle size	→	→
Ghosh et al. (2014); Hartjes et al. (2019)		Western blot (WB)	/	/	/	/	×	×	/	/	Either presence or absence of the bands	→	→
Koritzinsky et al. (2017); Zhao et al. (2015)		Bradford assay	×	/	/	/	/	×	/	×	Detect the presence and concentration of protein in the sample of 5–50 µg/ml	→	→
Lu et al. (2017); Williams and Carter (1996)		Transmission electron microscopy (TEM)	/	/	/	/	/	×	×	/	The morphology and structure, either intact or not intact	←	←
Sharma et al. (2018)		Atomic force microscopy (AFM)	/	/	/	/	×	×	/	/	The amount of particle size	↕	→

Reference	Safety	Assay										Description		
Doeppner et al. (2015)	Safety	Infectious serology	x	/	/	x	/	/	/	x	x	Positive result from this test is able to detect the presence of certain antibodies	↑	↑
Enderle et al. (2015); Garibyan and Avashia (2013); Montero-Calle et al. (2021)		Quantitative polymerase chain reaction (qPCR) for mycoplasma	/	/	/	/	/	x	/	x	x	Able to determine the exact CT value and also either the presence or absence of the bands for mycoplasma	↔	↑
Iwanaga (2007); Mehmood (2019)		Limulus assay	x	/	/	/	/	/	/	x	x	Gelation reaction once exposed to endotoxin from gram negative bacteria	↑	↑
U.S. Food and Drug Administration (2018)		BacT/ALERT	/	/	/	x	/	/	/	x	x	The culture will change to yellow with the presence of microorganism contamination	↑	↑
Tripathi and Sapra (2021)		Microscopy gram stain	/	/	/	/	x	x	/	x	x	Detect the presence of contamination of either gram positive or gram negative microorganisms	↑	↑

Legend: ↑ = high, ↓ = low, ↔ = moderate

the role of each of these parameters and how important they are in shaping up a good QC assay.

Specificity is defined as the capacity of an assay to assess the existence of components that may be anticipated to be present within the analyte. For instance, in terms of MSC-EVs for identity or quantity purposes, specificity denotes the size of MSC-EVs and typically carries markers such as CD63, CD9, and CD81. In this regard, assays such as NTA, Bradford, WB, or FC can be used to validate the specificity of MSC-EVs. Other assays such as scanning electron microscopy (SEM), transmission electron microscopy (TEM), or AFM may be used; however, we reckon that the assays are less specific as compared to the former assays due to the limitation of the instrument itself which only visualizes the morphology and structures of MSC-EVs. In terms of safety-related assays, specificity refers to the ability to differentiate MSC-EVs from contaminants such as body fluids given that we derived the exosomes from the blood and qPCR-based assays tend to have higher specificity as compared to other assays (Ludwig et al. 2019).

This is followed by the validation of the linearity of an assay which is defined as the ability to acquire experimental results that have a directly proportional relationship to the concentration of analyte within a sample. Generally, the acceptance coefficient of determination (R-squared) is 0.99, and one who performs assays such as NTA, Bradford, total protein content, WB, and qPCR has to validate the linearity of these assays. Next, range is defined as the range between the lower and upper concentrations of analyte within a sample. With this in mind, AFM has the highest range as it can detect particle size as close as 1 nm up until 120 nm. On the other hand, accuracy is the degree of closeness between the experimental value and the value acknowledged as the conventional true value or approved reference value. For example, FC does not meet the requirement of accuracy criterion for MSC-EVs as it cannot determine the sample concentration accurately due to the swarming effect and its insensitivity toward lower size range solutes.

The precision of an assay demonstrates the proximity of agreement among a set of results obtained from several samplings of a sample under prearranged conditions. Moreover, as stated in the ICH guidelines, precision is to be tested using homogenous and authentic samples; if not possible, artificially prepared samples can be used for the investigation. To give an example for the precision criterion, Hartjes et al. and Kurian et al. stated that although AFM has a better range, it has a very low reproducibility as the technique is highly dependent on the sample size, such that it can only image a maximum height of 10–20 micrometers (μm) within a total scanning area of 150 × 150 μm (Hartjes et al. 2019; Kurian et al. 2021). The validation for precision is further divided into three subcategories, namely, repeatability, intermediate precision, and reproducibility. Precision under the identical operational conditions over a short period is expressed by repeatability and requires at least nine determinations that cover a defined range for the assay or a minimum of six determinations at 100% test concentration. Finally, intermediate precision is expressed within-laboratories variations which include different days, analysts, and equipment. Reproducibility is expressed by the prevision between laboratories which is usually as a mean for methodology standardization. In short, techniques that qualify for the categories of range, accuracy, and precision should be able to detect a range of different size EVs while maintaining optimum accuracy and precision.

Penultimately, the detection limit of a test is referred to as the least amount of analyte within a sample which could be detected. For example, techniques that qualify for this category should have the ability to detect the presence of EVs although in a low concentration. Other than that, the quantitation limit of a test is the least amount of analyte within a sample that can be precisely and accurately quantified. This parameter is mainly used for low concentrations of compounds in a sample and to determine impurities or degradation products within an analyte. For example, the low penetration of EM beam and vacuum conditions required the sample to be ultrathin and completely dry affecting the morphology of EVs. Finally, assays that qualify for the robustness category should show reliability with a deliberate variation. For example, the low scanning speed of AFM requires a longer time to obtain an accurate image. This leads to thermal drift causing variation in image quality that can affect the analytical condition which in turn produces an invalid result. In a nutshell, to ensure the repeatability and precise result, a combination of several QC assay techniques should be implemented so that the overall result obtained will be able to fulfill the ICH validation guidelines. For example, the Bradford test, TEM, SEM, FC, and qPCR can be used in the quality control process along with NTA as they complement each other, hence producing reliable results.

4   Remaining Challenges

Hurdles persist among currently available QC assays such as challenges in developing specific potency assays and ascertaining the life span of EVs. Potency assay consists of biological (in vitro or in vivo) or nonbiological assays which test the specific biological capabilities of the product (USFDA 2011). Establishing QC-related potency assays for MSC-EVs is more challenging compared to other pharmaceutical or biological products due to several factors as listed below:

1. Differences in EV preparations may result in enrichment of different components within MSC-EVs resulting in a change of therapeutic outcomes (Gimona et al. 2021).
2. Different donors of MSC-EVs have different biological properties and components which may change the therapeutic outcome (Gimona et al. 2021).
3. The mechanism of therapeutic potential of MSC-EVs is still vague as it can be involved in more than one different pathological process making it difficult to predict the potency of MSC-EVs (Gimona et al. 2021).
4. MSC-EVs from different MSC sources can vary in therapeutic potency such that

suppression of T-cell proliferation is higher in AD-MSC-EVs than BM-MSC-EVs (Adlerz et al. 2020).
5. The spatiotemporal site of action remains unknown making it challenging to determine the precise biodistribution of MSC-EVs within a cell/tissue (Gimona et al. 2021).

Once the dynamic biological activities of EVs have been identified, currently available QC assays can be customized to comply with potency assay requirements. For example, Bruno and colleagues found that combining RT-PCR, enzyme-linked immunosorbent assay (ELISA), immunosorbent assays, and antibody assays was able to measure the full potency of the MSC-EVs but only for acute kidney injury (2009). Another study on myocardial ischemia injury showed that using a combination of enzymatic assays was also able to determine the potency of the MSC-EVs (Lai et al. 2010). Within these two examples, the combination of currently available assays such as RT-PCR, ELISA, and enzymatic assays can comply with the potency assay requirements such that it is specific, accurate, precise, and quantitative. However, due to the diverse attributes of MSC-EVs, combinations of different types of potency assays can only be specific to their respective pathological processes.

Another important aspect that needs to be considered is the life span which is orchestrated by stability data collectively contributed by factors such as temperature, light, and handling procedure. Nevertheless, a poorly designed QC assay may result in contradicting outcomes. For example, a study showed that 4 days' preserved exosomes at -80 °C are not stable as compared to freshly prepared ones (Maroto et al. 2017). Surprisingly, Jeyaram and Jay reported that the optimum storage condition for exosomes was -80 °C, contradicting the former study (Jeyaram and Jay 2017). This shows the important validation of the analytical procedure for each selected QC assay to ensure consistent results.

Further, majority of the purity tests that detect only endotoxins and mycoplasma contamination often neglect potential viruses contamination. Viruses are capable of enclosing themself into

EVs due to convergence of pathways (van der Grein et al. 2018). It is also known that the size ranges of EVs and viruses are similar which could lead to EV preparation being susceptible to viral contamination (Gyorgy et al. 2011). However, implementing QC to detect the presence of viral contamination is an uphill task because most methods, such as the PCR multiplex assay, are not product-specific. To test for viral contamination, short fragments of DNA are needed to complement specific parts of transcribed viral DNA. In short, current methods are only capable of detecting specific target viruses, and development for nonspecific target assays for viral detection should be carried out.

5   Conclusion

Despite these challenges, there is still room to improve QC assays for MSC-EVs. For QC tests that involve categories such as identity, purity, and quantity, apart from complying with ICH guidelines, efforts also should be taken to establish a similar QC procedure between various laboratories around the world which leads to a standard reference. Such an initiative can also be arranged by the ISEV as part of a compliance program. The issue remains that the enrichment method for isolation of EVs still remains with challenges that are difficult to keep up with as there is a lack of consensus concerning isolation steps for EVs (Stam et al. 2021). Additionally, the preparation of MSC-EVs remains expensive and has limited scalability, and if these preparations were scaled up, QC assays for MSC-EVs would be difficult to maintain. Moreover, the key to developing a potency assay is determining and mapping the pathological processes in preclinical animal models as it will lead to a better understanding of MSC-EVs' potency. In terms of extending the life span of EVs, lyophilization of EVs can be adopted. For example, a study by Charoenviriyakul and colleagues demonstrated that EVs could be lyophilized with the use of trehalose without affecting their stability and structure which allows exosomes to be preserved at room temperature which is useful for many

applications (Charoenviriyakul et al. 2018). However, immortalized MSC-derived EVs would have special considerations for QC assays as they divide infinitely and express unique gene patterns that sometimes cannot be found in regular EVs (Rohde et al. 2019). With this, it is possible to develop separate assay criteria for MSC-derived EVs and immortalized MSC-derived EVs. Furthermore, special surface markers such as tetraspanins on MSC-derived EVs should be taken into consideration when developing an identification assay as it provides specificity only to EVs. With the combination of these suggestions and improvements, precision of a standardized QC assay can be developed to not only serve as a gold standard for MSC-EVs research but to also enhance its therapeutic value.

Conflict of Interest Kong-Yong Then and Soon-Keng Cheong are directors of CryoCord Sdn Bhd and declare direct share interest in the company, whereas all other authors declare no conflict of interest.

Author Contribution Statement J.K.E.C., L.H.F., C.Y. M., H.Y.T., and X.Y.T. contributed to the writing (original and final drafting) of the manuscript. V.G. contributed to the conceptualization and writing (review and editing) of the manuscript. T.S.R., K.Y.T., A.K.D., and S.K.C. contributed to the writing (review and editing) of the manuscript.

Data Availability Statement Data sharing is not applicable to this article as no datasets were generated or analyzed during the current study.

References

Adan A, Alizada G, Kiraz Y, Baran Y, Nalbant A (2016) Flow cytometry: basic principles and applications. Crit Rev Biotechnol 37(2):163–176. https://doi.org/10. 3109/07388551.2015.1128876

Adlerz K, Patel D, Rowley J, Ng K, Ahsan T (2020) Strategies for scalable manufacturing and translation of MSC-derived extracellular vesicles. Stem Cell Res 48:1–9. https://doi.org/10.1016/j.scr.2020.101978

Ahn SY, Park WS, Kim YE, Sung DK, Sung SI, Ahn JY, Chang YS (2018) Vascular endothelial growth factor mediates the therapeutic efficacy of mesenchymal stem cell-derived extracellular vesicles against neonatal hyperoxic lung injury. Exp Mol Med 50(4):1–12. https://doi.org/10.1038/s12276-018-0055-8

Andaloussi SE, Imre M, Breakefield OX, Wood MJ (2019) Extracellular vesicles: biology and emerging therapeutic opportunities. Nat Rev 12(1):347–357. https://doi.org/10.1038/nrd3978

Arslan F, Lai RC, Smeets MB, Akeroyd L, Choo A, Aguor ENE, Timmers L, van Rijen HV, Doevendans PA, Pasterkamp G, Lim SK, de Kleijn DP (2013) Mesenchymal stem cell-derived exosomes increase ATP levels, decrease oxidative stress and activate PI3K/Akt pathway to enhance myocardial viability and prevent adverse remodeling after myocardial ischemia/reperfusion injury. Stem Cell Res 10(3):301–312. https://doi.org/10.1016/j.scr.2013.01.002

Bai L, Shao H, Wang H, Zhang Z, Su C, Dong L, Yu B, Chen X, Li X, Zhang X (2017) Effects of mesenchymal stem cell-derived exosomes on experimental autoimmune uveitis. Sci Rep 7(1):1–11. https://doi.org/10.1038/s41598-017-04559-y

Bandeira E, Oliveira H, Silva JD, Menna-Barreto RFS, Takyia CM, Suk JS, Witwer KW, Paulaitis ME, Hanes J, Rocco PRM, Morales MM (2018) Therapeutic effects of adipose-tissue-derived mesenchymal stromal cells and their extracellular vesicles in experimental silicosis. Respir Res 19(104):1–10. https://doi.org/10.1186/s12931-018-0802-3

Bian S, Zhang L, Duan L, Wang X, Min Y, Yu H (2013) Extracellular vesicles derived from human bone marrow mesenchymal stem cells promote angiogenesis in a rat myocardial infarction model. J Mol Med 92(4):387–397. https://doi.org/10.1007/s00109-013-1110-5

Bruno S, Grange C, Deregibus MC, Calogero RA, Saviozzi S, Collino F, Morando L, Busca A, Falda M, Bussolati B, Tetta C, Camussi G (2009) Mesenchymal stem cell-derived microvesicles protect against acute tubular injury. J Am Soc Nephrol 20(5):1053–1067. https://doi.org/10.1681/asn.2008070798

Bruno S, Grange C, Collino F, Deregibus MC, Cantaluppi V, Biancone L, Tetta C, Camussi G (2012) Microvesicles derived from mesenchymal stem cells enhance survival in a lethal model of acute kidney injury. PLoS One 7(3):e33115. https://doi.org/10.1371/journal.pone.0033115

Central Drugs Standard Control Organisation (2016) Guidelines on similar biologics: regulatory requirements for marketing authorization in India, New Delhi, India

Charoenviriyakul C, Takahashi Y, Nishikawa M, Takakura Y (2018) Preservation of exosomes at room temperature using lyophilization. Int J Pharm 553(1–2):1–7. https://doi.org/10.1016/j.ijpharm.2018.10.032

Chen J, An R, Liu Z, Wang J, Chen S, Hong M, Liu J, Xiao M, Chen Y (2014) Therapeutic effects of mesenchymal stem cell-derived microvesicles on pulmonary arterial hypertension in rats. Acta Pharmacol Sin 35(9):1121–1128. https://doi.org/10.1038/aps.2014.61

Cho BS, Kim JO, Ha DH, Yi YW (2018) Exosomes derived from human adipose tissue-derived mesenchymal stem cells alleviate atopic dermatitis. Stem Cell Res Ther 9:187. https://doi.org/10.1186/s13287-018-0939-5

Choudhary OP, Ka P (2017) Scanning electron microscope: advantages and disadvantages in imaging components. Int J Curr Microbiol App Sci 6(5):1877–1882. https://doi.org/10.20546/ijcmas.2017.605.207

Code of Federal Regulations (2021) General Biological Products Standards, 21 C.F.R. § 610.10. Office of the Federal Register (United States)

Colombo M, Raposo G, Théry C (2014) Biogenesis, secretion, and intercellular interactions of exosomes and other extracellular vesicles. Annu Rev Cell Dev Biol 30(1):255–289. https://doi.org/10.1146/annurev-cellbio-101512-122326

Cruz FF, Borg ZD, Goodwin M, Sokocevic D, Wagner DE, Coffey A, Antunes M, Robinson KL, Mitsialis SA, Kourembanas S, Thane K, Hoffman AM, McKenna DH, Rocco PRM, Weiss DJ (2015) Systemic administration of human bone marrow-derived mesenchymal stromal cell extracellular vesicles ameliorates aspergillus hyphal extract-induced allergic airway inflammation in immunocompetent mice. Stem Cells Transl Med 4(11):1302–1316. https://doi.org/10.5966/sctm.2014-0280

de Castro LL, Xisto DG, Kitoko JZ, Cruz FF, Olsen PC, Redondo PAG, Ferreira TPT, Weiss DJ, Martins MA, Morales MM, Rocco PRM (2017) Human adipose tissue mesenchymal stromal cells and their extracellular vesicles act differentially on lung mechanics and inflammation in experimental allergic asthma. Stem Cell Res Ther 8(1):151. https://doi.org/10.1186/s13287-017-0600-8

Doeppner TR, Herz J, Gorgens A, Schlechter J, Ludwig A-K, Radtke S, de Miroschedji K, Horn PA, Giebel B, Hermann DM (2015) Extracellular vesicles improve post-stroke neuroregeneration and prevent postischemic immunosuppression. Stem Cells Transl Med 4(10):1131–1143. https://doi.org/10.5966/sctm.2015-0078

Doyle L, Wang M (2019) Overview of extracellular vesicles, their origin, composition, purpose, and methods for exosome isolation and analysis. Cell 8(7):727. https://doi.org/10.3390/cells8070727

Dragovic RA, Gardiner C, Brooks AS, Tannetta DS, Ferguson DJP, Hole P, Carr B, Redman CWG, Harris AL, Dobson PJ, Harrison P, Sargent IL (2011) Sizing and phenotyping of cellular vesicles using nanoparticle tracking analysis. Nanomedicine 7(6):780–788. https://doi.org/10.1016/j.nano.2011.04.003

Drommelschmidt K, Serdar M, Bendix I, Herz J, Bertling F, Prager S, Keller M, Ludwig A-K, Duhan V, Radtke S, de Miroschedji K, Horn PA, van de Looij Y, Giebel B, Felderhoff-Müser U (2017) Mesenchymal stem cell-derived extracellular vesicles ameliorate inflammation-induced preterm brain injury. Brain Behav Immun 60:220–232. https://doi.org/10.1016/j.bbi.2016.11.011

Enderle D, Spiel A, Coticchia CM, Berghoff E, Mueller R, Schlumpberger M, Sprenger-Haussels M, Shaffer JM, Lader E, Skog J, Noerholm M (2015) Characterization of RNA from exosomes and other extracellular vesicles isolated by a novel spin column-based method. PLoS

One 10(8):e0136133. https://doi.org/10.1371/journal. pone.0136133

European Medicines Agency (1994a) ICH topic Q5C quality of biotechnological products: stability testing of biotechnological/biological products, Amsterdam, Netherlands

European Medicines Agency (1994b) Topic Q 6 B specifications: test procedures and acceptance criteria for biotechnological/biological products. Netherlands, Amsterdam

Gangadaran P, Rajendran RL, Lee HW, Kalimuthu S, Hong CM, Jeong SY, Lee S-W, Lee J, Ahn B-C (2017) Extracellular vesicles from mesenchymal stem cells activates VEGF receptors and accelerates recovery of hindlimb ischemia. J Control Release 264:112–126. https://doi.org/10.1016/j.jconrel.2017.08.022

Garibyan L, Avashia N (2013) Polymerase chain reaction. J Investig Dermatol 133(3):1–4. https://doi.org/10.1038/jid.2013.1

Gatti S, Bruno S, Deregibus MC, Sordi A, Cantaluppi V, Tetta C, Camussi G (2011) Microvesicles derived from human adult mesenchymal stem cells protect against ischaemia-reperfusion-induced acute and chronic kidney injury. Nephrol Dialysis Transplant 26(5):1474–1483. https://doi.org/10.1093/ndt/gfr01

Ghosh R, Gilda J, Gomes A (2014) The necessity of and strategies for improving confidence in the accuracy of western blots. Expert Rev Proteomics 11(5):549–560. https://doi.org/10.1586/14789450.2014.939635

Gimona M, Brizzi MF, Choo ABH, Dominici M, Davidson SM, Grillari J, Hermann DM, Hill AF, de Kleijn D, Lai RC, Lai CP, Lim R, Monguió-Tortajada M, Muraca M, Ochiya T, Ortiz LA, Toh WS, Yi YW, Witwer KW, Giebel B (2021) Critical considerations for the development of potency tests for therapeutic applications of mesenchymal stromal cell-derived small extracellular vesicles. Cytotherapy 23(5):373–380. https://doi.org/10.1016/j.jcyt.2021.01.001

Gyorgy B, Szabó TG, Pásztói M, Pál Z, Misják P, Aradi B, László V, Pállinger É, Pap E, Kittel Á, Nagy G, Falus A, Buzás EI (2011) Membrane vesicles, current state-of-the-art: emerging role of extracellular vesicles. Cell Mol Life Sci 68(16):2667–2688. https://doi.org/10.1007/s00018-011-0689-3

Haga H, Yan IK, Takahashi K, Matsuda A, Patel T (2017) Extracellular vesicles from bone marrow-derived mesenchymal stem cells improve survival from lethal hepatic failure in mice. Stem Cells Transl Med 6(4):1262–1272. https://doi.org/10.1002/sctm.16-0226

Hao Q, Gudapati V, Monsel A, Park JH, Hu S, Kato H, Lee JH, Zhou L, He H, Lee JW (2019) Mesenchymal stem cell–derived extracellular vesicles decrease lung injury in mice. J Immunol 203(7):1961–1972. https://doi.org/10.4049/jimmunol.1801534

Hartjes TA, Mytnyk S, Jenster GW, van Steijn V, van Royen ME (2019) Extracellular vesicle quantification and characterization: common methods and emerging approaches. Bioengineering (Basel, Switzerland) 6(1):7. https://doi.org/10.3390/bioengineering6010007

Iwanaga S (2007) Biochemical principle of limulus test for detecting bacterial endotoxins. Proc Japan Acad Ser B 83(4):110–119. https://doi.org/10.2183/pjab.83.110

Jeyaram A, Jay SM (2017) Preservation and storage stability of extracellular vesicles for therapeutic applications. AAPS J 20(1). https://doi.org/10.1208/s12248-017-0160-y

Jiang W, Tan Y, Cai M, Zhao T, Mao F, Zhang X, Xu W, Yan Z, Qian H, Yan Y (2018) Human umbilical cord MSC-derived exosomes suppress the development of CCl4-induced liver injury through antioxidant effect. Stem Cells Int 2018:1–11. https://doi.org/10.1155/2018/6079642

Koritzinsky EH, Street JM, Star RA, Yuen PST (2017) Quantification of exosomes. J Cell Physiol 232(7):1587–1590. https://doi.org/10.1002/jcp.25387

Kurian TK, Banik S, Gopal D, Chakrabarti S, Mazumder N (2021) Elucidating methods for isolation and quantification of exosomes: a review. Mol Biotechnol 63(4):249–266. https://doi.org/10.1007/s12033-021-00300-3

Lai RC, Arslan F, Lee MM, Sze NSK, Choo A, Chen TS, Salto-Tellez M, Timmers L, Lee CN, El Oakley RM, Pasterkamp G, de Kleijn DPV, Lim SK (2010) Exosome secreted by MSC reduces myocardial ischemia/reperfusion injury. Stem Cell Res 4(3):214–222. https://doi.org/10.1016/j.scr.2009.12.003

Lee B-C, Kang I, Yu K-R (2021) Therapeutic features and updated clinical trials of mesenchymal stem cell (MSC)-derived exosomes. J Clin Med 10(4):711. https://doi.org/10.3390/jcm10040711

Li T, Yan Y, Wang B, Qian H, Zhang X, Shen L, Wang M, Zhou Y, Zhu W, Li W, Xu W (2012) Exosomesderived from human umbilical cord mesenchymal stem cells alleviate liver fibrosis. Stem Cells Dev 22(6):845–854. https://doi.org/10.1089/scd.2012.0395

Lin KC, Yip HK, Shao PL, Wu SC, Chen KH, Chen YT, Yang CC, Sun CK, Kao GS, Chen SY, Chai HT, Chang CL, Chen CH, Lee MS (2016) Combination of adipose-derived mesenchymal stem cells (ADMSC) and ADMSC-derived exosomes for protecting kidney from acute ischemia–reperfusion injury. Int J Cardiol 216:173–185. https://doi.org/10.1016/j.ijcard.2016.04.061

Lu K, Li H, Yang K, Wu J, Cai X, Zhou Y, Li C (2017) Exosomes as potential alternatives to stem cell therapy for intervertebral disc degeneration: in-vitro study on exosomes in interaction of nucleus pulposus cells and bone marrow mesenchymal stem cells. Stem Cell Res Ther 8(1). https://doi.org/10.1186/s13287-017-0563-9

Ludwig N, Whiteside TL, Reichert TE (2019) Challenges in exosome isolation and analysis in health and disease. Int J Mol Sci 20(19):4684. https://doi.org/10.3390/ijms20194684

Maas SLN, de Vrij J, van der Vlist EJ, Geragousian B, van Bloois L, Mastrobattista E, Schiffelers RM, Wauben MHM, Broekman MLD, Nolte-'t Hoen ENM (2015) Possibilities and limitations of current technologies for quantification of biological extracellular vesicles and synthetic mimics. J Control Release 200:87–96. https://doi.org/10.1016/j.jconrel.2014.12.041

Mao F, Wu Y, Tang X, Kang J, Zhang B, Yan Y, Qian H, Zhang X, Xu W (2017) Exosomes derived from human umbilical cord mesenchymal stem cells relieve inflammatory bowel disease in mice. Biomed Res Int 2017:1–12. https://doi.org/10.1155/2017/5356760

Maroto R, Zhao Y, Jamaluddin M, Popov VL, Wang H, Kalubowilage M, Zhang Y, Luisi J, Sun H, Culbertson CT, Bossmann SH, Motamedi M, Brasier AR (2017) Effects of storage temperature on airway exosome integrity for diagnostic and functional analyses. J Extracell Vesicles 6(1):1359478. https://doi.org/10.1080/20013078.2017.1359478

Maumus M, Rozier P, Boulestreau J, Jorgensen C, Noël D (2020) Mesenchymal stem cell-derived extracellular vesicles: opportunities and challenges for clinical translation. Front Bioeng Biotechnol 8:e997. https://doi.org/10.3389/fbioe.2020.00997

Mehmood Y (2019) What is limulus Amebocyte lysate (LAL) and its applicability in endotoxin quantification of pharma products. In: Growing and handling of bacterial cultures. https://doi.org/10.5772/intechopen.81331

Ministry of Food and Drug Safety (2003) Regulation on approval and review of biological products (no. 2015–104), Cheongju, Korea

Monsel A, Zhu Y, Gennai S, Hao Q, Hu S, Rouby J-J, Rosenzwajg M, Matthay MA, Lee JW (2015) Therapeutic effects of human mesenchymal stem cell–derived microvesicles in severe pneumonia in mice. Am J Respir Crit Care Med 192(3):324–336. https://doi.org/10.1164/rccm.201410-1765oc

Montero-Calle A, Aranguren-Abeigon I, Garranzo-Asensio M, Poves C, Fernández-Aceñero MJ, Martínez-Useros J, Sanz R, Dziaková J, Rodriguez-Cobos J, Solís-Fernández G, Povedano E, Gamella M, Torrente-Rodríguez RM, Alonso-Navarro M, de los Ríos V, Casal JI, Domínguez G, Guzman-Aranguez A, Peláez-García A, Pingarrón JM (2021) Multiplexed biosensing diagnostic platforms detecting autoantibodies to tumor-associated antigens from exosomes released by CRC cells and tissue samples showed high diagnostic ability for colorectal cancer. Engineering 7(10):1393–1412. https://doi.org/10.1016/j.eng.2021.04.026

Nassar W, El-Ansary M, Sabry D, Mostafa MA, Fayad T, Kotb E, Temraz M, Saad A-N, Essa W, Adel H (2016) Umbilical cord mesenchymal stem cells derived extracellular vesicles can safely ameliorate the progression of chronic kidney diseases. Biomater Res 20(1). https://doi.org/10.1186/s40824-016-0068-0

Nguyen VVT, Witwer KW, Verhaar MC, Strunk D, van Balkom BWM (2020) Functional assays to assess the therapeutic potential of extracellular vesicles. J Extracell Vesicles 10(1):e12033. https://doi.org/10.1002/jev2.12033

Ophelders DRMG, Wolfs TGAM, Jellema RK, Zwanenburg A, Andriessen P, Delhaas T, Ludwig A-K, Radtke S, Peters V, Janssen L, Giebel B, Kramer BW (2016) Mesenchymal stromal cell-derived extracellular vesicles protect the fetal brain after hypoxia-ischemia. Stem Cells Transl Med 5(6):754–763. https://doi.org/10.5966/sctm.2015-0197

Pharmaceuticals and Medical Devices Agency (2009) Guideline for the quality, safety, and efficacy assurance of follow-on biologics (PFSB/ELD Notification No. 0304007). Japan

Rohde E, Pachler K, Gimona M (2019) Manufacturing and characterization of extracellular vesicles from umbilical cord–derived mesenchymal stromal cells for clinical testing. Cytotherapy 21(6):581–592. https://doi.org/10.1016/j.jcyt.2018.12.006

Sachan S, Nigam U, Gangwar P, Sood R (2014) An overview of pharmaceutical and biological product quality control. J Drug Deliv Ther 4(2):167–168. https://doi.org/10.22270/jddt.v4i2.780

Sharma S, LeClaire M, Gimzewski JK (2018) Ascent of atomic force microscopy as a nanoanalytical tool for exosomes and other extracellular vesicles. Nanotechnology 29(13):132001. https://doi.org/10.1088/1361-6528/aaab06

Shi Y, Yang Y, Guo Q, Gao Q, Ding Y, Wang H, Xu W, Yu B, Wang M, Zhao Y, Zhu W (2019) Exosomes derived from human umbilical cord mesenchymal stem cells promote fibroblast-to-myofibroblast differentiation in inflammatory environments and benefit cardioprotective effects. Stem Cells Dev 28(12):799–811. https://doi.org/10.1089/scd.2018.0242

Shi M, Yang Q, Monsel A, Yan J, Dai C, Zhao J, Shi G, Zhou M, Zhu X, Li S, Li P, Wang J, Li M, Lei J, Xu D, Zhu Y, Qu J (2021) Preclinical efficacy and clinical safety of clinical-grade nebulized allogenic adipose mesenchymal stromal cells-derived extracellular vesicles. J Extracell Vesicles 10(10):e12134. https://doi.org/10.1002/jev2.12134

Shiue S-J, Rau R-H, Shiue H-S, Hung Y-W, Li Z-X, Yang KD, Cheng J-K (2019) Mesenchymal stem cell exosomes as a cell-free therapy for nerve injury–induced pain in rats. Pain 160(1):210–223. https://doi.org/10.1097/j.pain.0000000000001395

Song Y, Dou H, Li X, Zhao X, Li Y, Liu D, Ji J, Liu F, Ding L, Ni Y, Hou Y (2017) Exosomal miR-146a contributes to the enhanced therapeutic efficacy of interleukin-1β-primed mesenchymal stem cells against sepsis. Stem Cells 35(5):1208–1221. https://doi.org/10.1002/stem.2564

Stam J, Bartel S, Bischoff R, Wolters JC (2021) Isolation of extracellular vesicles with combined enrichment methods. J Chromatogr B 1169:122604. https://doi.org/10.1016/j.jchromb.2021.122604

Stone ML, Zhao Y, Robert Smith J, Weiss ML, Kron IL, Laubach VE, Sharma AK (2017) Mesenchymal stromal cell-derived extracellular vesicles attenuate lung ischemia-reperfusion injury and enhance reconditioning of donor lungs after circulatory death. Respir Res 18(1). https://doi.org/10.1186/s12931-017-0704-9

Sun G, Li G, Li D, Huang W, Zhang R, Zhang H, Duan Y, Wang B (2018a) HucMSC derived exosomes promote functional recovery in spinal cord injury mice via attenuating inflammation. Mater Sci Eng C 89:194–204. https://doi.org/10.1016/j.msec.2018.04.006

Sun Y, Shi H, Yin S, Ji C, Zhang X, Zhang B, Wu P, Shi Y, Mao F, Yan Y, Xu W, Qian H (2018b) Human mesenchymal stem cell derived exosomes alleviate type 2 diabetes mellitus by reversing peripheral insulin resistance and relieving β-cell destruction. ACS Nano 12(8):7613–7628. https://doi.org/10.1021/acsnano.7b07643

Tamura R, Uemoto S, Tabata Y (2016) Immunosuppressive effect of mesenchymal stem cell-derived exosomes on a concanavalin A-induced liver injury model. Inflamm Regen 36(1). https://doi.org/10.1186/s41232-016-0030-5

Teng X, Chen L, Chen W, Yang J, Yang Z, Shen Z (2015) Mesenchymal stem cell-derived exosomes improve the microenvironment of infarcted myocardium contributing to angiogenesis and anti-inflammation. Cell Physiol Biochem 37(6):2415–2424. https://doi.org/10.1159/000438594

Thery C, Witwer KW, Aikawa E, Alcaraz MJ, Anderson JD, Andriantsitohaina R, Antoniou A, Arab T, Archer F, Atkin-Smith GK, Ayre DC, Bach J-M, Bachurski D, Baharvand H, Balaj L, Baldacchino S, Bauer NN, Baxter AA, Bebawy M, Beckham C et al (2018) Minimal information for studies of extracellular vesicles 2018 (MISEV2018): a position statement of the international society for extracellular vesicles and update of the MISEV2014 guidelines. J Extracell Vesicles 7(1):1535750. https://doi.org/10.1080/20013078.2018.1535750

Tripathi N, Sapra A (2021) Gram staining. In: PubMed https://www.ncbi.nlm.nih.gov/books/NBK562156/#:~:text=The%20Gram%20staining%20is%20one

U.S. Food & Drug Administration (2011) Potency tests for cellular and gene therapy products, Maryland, United States of America

U.S. Food & Drug Administration (2018) BK170142: BacT/ALERT BPA culture bottle: BacT/ALERT BPN culture bottle, Maryland, United States of America

van der Grein SG, Defourny KAY, Slot EFJ, Nolte-'t Hoen ENM (2018) Intricate relationships between naked viruses and extracellular vesicles in the crosstalk between pathogen and host. Semin Immunopathol 40(5):491–504. https://doi.org/10.1007/s00281-018-0678-9

Varkouhi AK, Jerkic M, Ormesher L, Gagnon S, Goyal S, Rabani R, Masterson C, Spring C, Chen PZ, Gu FX, dos Santos CC, Curley GF, Laffey JG (2019) Extracellular vesicles from interferon-γ–primed human umbilical cord mesenchymal stromal cells reduce escherichia coli–induced acute lung injury in rats. Anesthesiology 130(5):778–790. https://doi.org/10.1016/j.bbadis.2017.02.023

Wang N, Chen C, Yang D, Liao Q, Luo H, Wang X, Zhou F, Yang X, Yang J, Zeng C, Wang WE (2017) Mesenchymal stem cells-derived extracellular vesicles, via miR-210, improve infarcted cardiac function by promotion of angiogenesis. Biochim Biophys Acta (BBA) - Mol Basis Dis 1863(8):2085–2092. https://doi.org/10.1016/j.bbadis.2017.02.023

Williams DB, Carter CB (1996) The transmission electron microscope. Transmission Electron Microscopy:3–17. https://doi.org/10.1007/978-1-4757-2519-3_1

Witwer KW, Van Balkom BWM, Bruno S, Choo A, Dominici M, Gimona M, Hill AF, De Kleijn D, Koh M, Lai RC, Mitsialis SA, Ortiz LA, Rohde E, Asada T, Toh WS, Weiss DJ, Zheng L, Giebel B, Lim SK (2019) Defining mesenchymal stromal cell (MSC)-derived small extracellular vesicles for therapeutic applications. J Extracell Vesicles 8(1):1609206. https://doi.org/10.1080/20013078.2019.1609206

Wu Y, Qiu W, Xu X, Kang J, Wang J, Wen Y, Tang X, Yan Y, Qian H, Zhang X, Xu W, Mao F (2018) Exosomes derived from human umbilical cord mesenchymal stem cells alleviate inflammatory bowel disease in mice through ubiquitination. Am J Transl Res 10(7):2026–2036

Xunian Z, Kalluri R (2020) Biology and therapeutic potential of mesenchymal stem cell-derived exosomes. Cancer Sci 111(9). https://doi.org/10.1111/cas.14563

Zhang G, Zou X, Miao S, Chen J, Du T, Zhong L, Ju G, Liu G, Zhu Y (2014) The anti-oxidative role of micro vesicles derived from human Wharton-jelly mesenchymal stromal cells through NOX2/gp91(phox) suppression in alleviating renal ischemia-reperfusion injury in rats. PLoS One 9(3):e92129. https://doi.org/10.1371/journal.pone.0092129

Zhang B, Wang M, Gong A, Zhang X, Wu X, Zhu Y, Shi H, Wu L, Zhu W, Qian H, Xu W (2015a) HucMSC-exosome mediated-Wnt4 signaling is required for cutaneous wound healing. Stem Cells (Dayton, Ohio) 33(7):2158–2168. https://doi.org/10.1002/stem.1771

Zhang Y, Chopp M, Meng Y, Katakowski M, Xin H, Mahmood A, Xiong Y (2015b) Effect of exosomes derived from multipluripotent mesenchymal stromal cells on functional recovery and neurovascular plasticity in rats after traumatic brain injury. J Neurosurg 122(4):856–867. https://doi.org/10.3171/2014.11.jns14770

Zhang Z, Yang J, Yan W, Li Y, Shen Z, Asahara T (2016) Pretreatment of cardiac stem cells with exosomes derived from mesenchymal stem cells enhances myocardial repair. J Am Heart Assoc 5(1). https://doi.org/10.1161/jaha.115.002856

Zhang X, Liu J, Yu B, Ma F, Ren X, Li X (2018) Effects of mesenchymal stem cells and their exosomes on the healing of large and refractory macular holes. Graefes Arch Clin Exp Ophthalmol 256(11):2041–2052. https://doi.org/10.1007/s00417-018-4097-3

Zhang Y, Liu Y, Liu H, Tang WH (2019) Exosomes: biogenesis, biologic function and clinical potential. Cell Biosci 9(1). https://doi.org/10.1186/s13578-019-0282-2

Zhao Y, Sun X, Cao W, Ma J, Sun L, Qian H, Zhu W, Xu W (2015) Exosomes derived from human umbilical cord mesenchymal stem cells relieve acute myocardial ischemic injury. Stem Cells Int 2015:1–12. https://doi.org/10.1155/2015/761643

Zhou Y, Xu H, Xu W, Wang B, Wu H, Tao Y, Zhang B, Wang M, Mao F, Yan Y, Gao S, Gu H, Zhu W, Qian H (2013) Exosomes released by human umbilical cord mesenchymal stem cells protect against cisplatin-induced renal oxidative stress and apoptosis in vivo and in vitro. Stem Cell Res Ther 4(2):34. https://doi.org/10.1186/scrt194

Zhu Y, Feng X, Abbott J, Fang X, Hao Q, Monsel A, Qu J, Matthay MA, Lee JW (2013) Human mesenchymal stem cell microvesicles for treatment of Escherichia coli endotoxin-induced acute lung injury in mice. Stem Cells 32(1):116–125. https://doi.org/10.1002/stem.1504

Zou X, Gu D, Xing X, Cheng Z, Gong D, Zhang G, Zhu Y (2016) Human mesenchymal stromal cell-derived extracellular vesicles alleviate renal ischemic reperfusion injury and enhance angiogenesis in rats. Am J Transl Res 8(10):4289–4299

Adv Exp Med Biol - Cell Biology and Translational Medicine (2022) 17: 213–225
https://doi.org/10.1007/5584_2022_718
© Springer Nature Switzerland AG 2022
Published online: 14 July 2022

Pharmacological Effects of Caffeic Acid and Its Derivatives in Cancer: New Targeted Compounds for the Mitochondria

Haydeé Bastidas, Gabriel Araya-Valdés, Gonzalo Cortés, José A. Jara, and Mabel Catalán

Abstract

Cancer is a complex pathology of great heterogeneity and difficulty that makes the constant search for new therapies necessary. A major advance on the subject has been made by focusing on the development of new drugs aimed to alter the metabolism of cancer cells, by generating a disruption of mitochondrial function. For this purpose, several new compounds with specific mitochondrial action have been tested, leading successfully to cell death. Recently, attention has centered on a group of natural compounds present in plants named polyphenols, among which is caffeic acid, a polyphenol that has proven to be a powerful antitumoral agent and a prominent compound for studies focused on the development of new therapies against cancer.

In this review, we revised the antitumoral capacity and mechanisms of action of caffeic acid and its derivatives, with special emphasis in a new class of caffeic acid derivatives that target mitochondria by chemical binding to the lipophilic cation triphenylphosphonium.

H. Bastidas, G. Araya-Valdés, G. Cortés, and
M. Catalán (✉)
Clinical and Molecular Pharmacology Program, Institute of Biomedical Sciences (ICBM), Faculty of Medicine, Universidad de Chile, Santiago, Chile
e-mail: mabelcatalan@u.uchile.cl

J. A. Jara
Institute for Research in Dental Sciences (ICOD), Faculty of Dentistry, Universidad de Chile, Santiago, Chile

Graphical Abstract

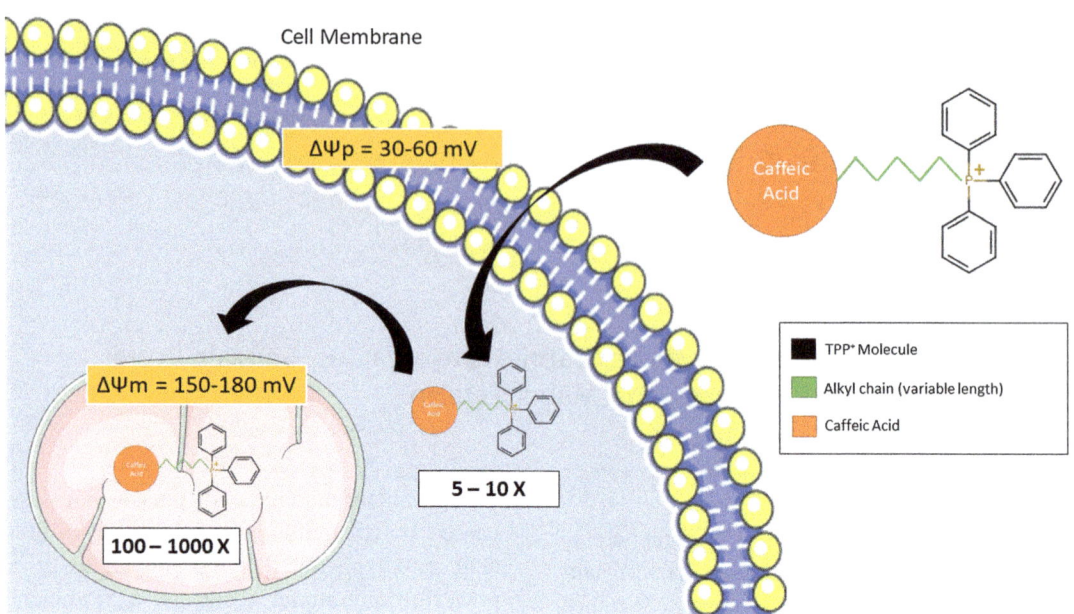

Keywords

Caffeic acid · Cancer · Cancer therapy · Mitochondria · Triphenylphosphonium cation

Abbreviations

α-KG	Alpha-ketoglutarate
$\Delta\Psi p$	Plasma membrane potential
$\Delta\Psi m$	Membrane potential of the MII
2-HG	2-Hydroxyglutarate
AMPK	AMP-activated protein kinase
CA	Caffeic acid
CAPE	Caffeic acid phenethyl ester
CAPPE	Caffeic acid phenylpropyl ester
Cyt c	Cytochrome c
DR5	Death receptor 5
DRP-1	Dynamin-related protein 1
ETC	Electron transport chain
FDA	Food and Drug Administration
GSH-Px	Glutathione peroxidase
HIF	Hypoxia-induced factor
IMM	Inner mitochondrial membrane
IC_{50}	Inhibitory concentration 50
MDIVI-1	Mitochondrial division inhibitor 1
MitoCaA	Mitochondriotropic caffeic acid
MitoCA	Mitochondriotropic cinnamic acid
MitoFA	Mitochondriotropic ferulic acid
Mitop-CoA	Mitochondriotropic p-coumaric acid
mIDH	Mutant isocitrate dehydrogenase
MMP	Mitochondrial membrane potential
MMP2	Matrix metalloproteinase 2
MMP9	Matrix metalloproteinase 9
mtDNA	Mitochondrial DNA
mtROS	Mitochondrial ROS
mTOR	Mechanistic target of rapamycin
NFκB	Nuclear factor kappa B
O_2^-	Superoxide anion
OH-	Hydroxyl radical
PI3-K/Akt	Phosphoinositide 3-kinase/protein kinase B
p38	Mitogen-activated protein kinase
ROS	Reactive oxygen species
SI	Selectivity indexes
SODs	Superoxide dismutases
TCA	Tricarboxylic acid
TRAIL	Tumor necrosis factor-related apoptosis-inducing ligand
TPP$^+$	Triphenylphosphonium
VEGF	Vascular endothelial growth factor

1   Background

Noncommunicable diseases represent nowadays the main cause of death in almost every country in the world. Between them, cancer emerges as one of the most significant by reporting an incidence of 24.5 million new cases and 9.6 million deaths in 2017 (Fitzmaurice et al. 2019). In Chile, studies indicate from a total of 106,388 deaths registered in 2017, 27,504 were due to cancer, therefore representing 25.9% of all deaths (Comité Nacional de Estadísticas Vitales 2017) and placing this disease among the pathologies responsible for the largest number of deaths country and worldwide.

Cancer has a great heterogeneity in its origin, course, and response to treatment, which is one of the greatest challenges to achieve, due to different patient response, time-tumor progression, type of tissue, genetic and epigenetic mutations, and metabolic alterations (Prasetyanti and Medema 2017). Due to the high degree of variability, various types of treatments have been developed, chemotherapy being one of the most used of those with clinical efficacy. For instance, main drugs used in chemotherapy are classified according to the type of molecular target and mechanism of action in alkylating agents that cause direct damage to DNA – antimetabolites which correspond to analogs of endogenous molecules that are necessary for DNA and RNA synthesis; mitosis inhibitors which act by altering the formation of the mitotic spindle; and topoisomerase inhibitors that are enzymes that regulate the DNA supercoiling. However, there have been reports of several side effects and drug resistance; therefore, the greatest issue is to improve effectiveness and selectivity against cancerous cells by providing other chemical alternatives. As a result, over the last few years, the search for new pharmacological targets has become relevant in order to develop novel drugs with specific action. Within them, mitochondria have turned out to exhibit special characteristics, allowing the design of selectively targeted drugs (Frattaruolo et al. 2020). As is widely known, this organelle plays a fundamental role in the cell, by participating in vital functions such as ATP production, cycle control, production of metabolic proteins, and cell signaling and death, in addition to being closely related with cellular metabolic stress since it is responsible for reactive oxygen species (ROS) production (Grasso et al. 2020).

2   Mitochondria in Cancer Cells

Cancer cells present various genetic, epigenetic, and metabolic alterations, which lead to the loss of normal cell functions and the acquisition of abnormal survival and proliferation capacities. In this reprogramming phenomenon, the mitochondrion plays a fundamental role by transforming its normal bioenergetic metabolism into an elevated glycolytic metabolism that triggers mitochondrial dysfunction, which is known as the Warburg effect (Grasso et al. 2020; Anderson et al. 2018a).

In cancer, the mitochondria manifest a series of alterations as mutations in mitochondrial DNA (mtDNA), affecting the synthesis of enzymes that participate in the tricarboxylic acid (TCA) cycle and the synthesis of complexes that participates in the electron transport chain (ETC). As a consequence, the production of reductor electron equivalents NADH and FADH2 is altered, resulting in the accumulation of TCA cycle intermediates depending on the affected enzyme. These intermediates can act as oncometabolites, as is the case of fumarate and succinate (Grasso et al. 2020). In addition, tumoral cells present a high rate of ROS production, which is a key feature in carcinogenesis as it correlates with the progression of malignancy, as ROS toxicity may induce the alteration of intracellular pathway signaling (Idelchik et al. 2017). The overproduction of ROS induces the accelerated metabolism and the redox imbalance that cancer cells possess, which prevents the neutralization of ROS. When cells are immersed in hypoxia conditions or a high metabolic rate, electrons can leak from complexes I and III of the ETC and conjugate with O_2 forming a superoxide anion (O_{2-}), which is the main cellular ROS. In this sense, if O_{2-}

cannot be neutralized, it will lead to the oxidation of cellular macromolecules such as DNA, lipids, and proteins (Grasso et al. 2020). As it is well known, ROS neutralization is given by cellular antioxidant mechanisms such as superoxide dismutases (SODs) – which catalyze the reaction of production of H_2O_2 from O_{2-} – and the catalase (CAT) and glutathione peroxidase (GSH-Px) enzymes that catalyze reactions where H_2O is produced from H_2O_2– (Idelchik et al. 2017).

Additionally, mitochondria in tumoral cells are characterized by presenting an abnormal inner mitochondrial membrane (IMM) potential that is higher when compared to normal cells, therefore presenting a more electronegative charge (Kalyanaraman et al. 2018; Modica-Napolitano and Weissig 2015).

3 Antineoplastic Agents Targeting Mitochondrial Function

Given the fact that mitochondria participate in multiple metabolic functions, researchers may employ different strategies to develop more effective drugs, such as targeting the TCA cycle, ETC, mitochondrial biogenesis, or the mitochondrial apoptotic pathway (Frattaruolo et al. 2020; Grasso et al. 2020; Anderson et al. 2018a). Among drugs under study or between those already approved by the US Food and Drug Administration (FDA) are those that operate on the TCA cycle such as CB-839, CPI-613, ivosidenib (AG-120), enasidenib (AG-221), and vorasidenib (AG-881) (Frattaruolo et al. 2020; Grasso et al. 2020; Anderson et al. 2018a, b; Konteatis et al. 2020) (Fig. 1).

In this sense, it has been described that CB-839 is a specific inhibitor of glutaminase. This enzyme converts glutamine into glutamate, in what is supposed to be a physiological process in cells, upregulated in cancer. Excess of glutamate has been associated with an increase in cell proliferation and malignancy (Grasso et al. 2020; Chen and Cui 2015). Additionally, CPI-613 corresponds to a lipoate analog that inhibits the pyruvate dehydrogenase and α-ketoglutarate dehydrogenase enzymes, therefore preventing the incorporation to the TCA cycle of carbons derived from glucose or glutamine (Anderson et al. 2018a; Stuart et al. 2014). Further,

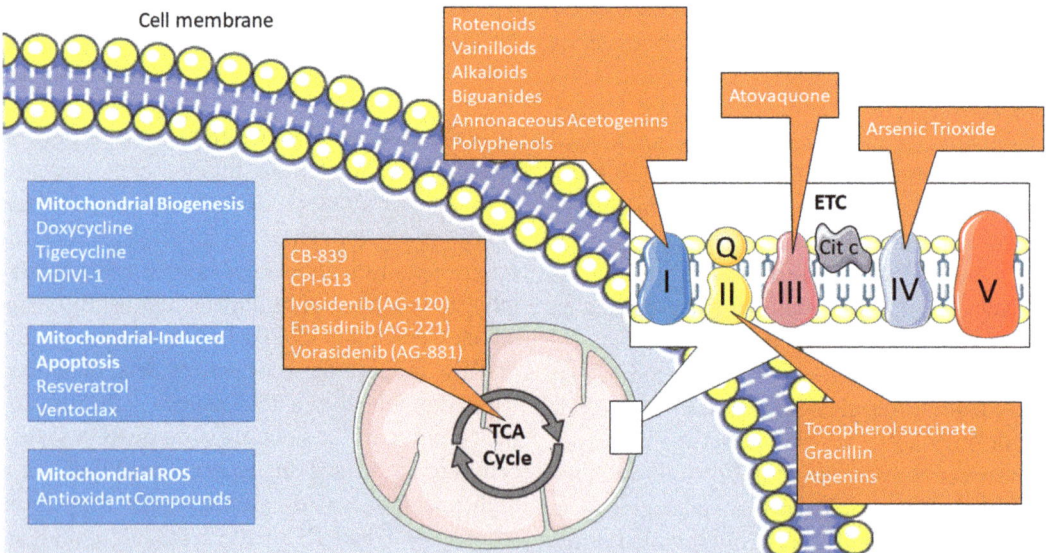

Fig. 1 Agents with antineoplastic effect that act on the mitochondria with action on biogenesis, apoptosis and mitochondrial ROS, TCA cycle, and CTE in their different complexes are described

ivosidenib, enasidenib, and vorasidenib work by inhibiting mutant isocitrate dehydrogenase (mIDH) enzymes. Under normal conditions, isocitrate dehydrogenase should catalyze the conversion of isocitrate to α-ketoglutarate (α-KG); however, when mutated it can catalyze the conversion of α-KG to the oncometabolite 2-hydroxyglutarate (2-HG) (Frattaruolo et al. 2020; Anderson et al. 2018a, b; Abou Dalle and DiNardo 2018). Particularly, ivosidenib inhibits the mutant mIDH1 enzyme, enasidenib inhibits mIDH2, and vorasidenib inhibits both mIDH1 and mIDH2, as described above (Konteatis et al. 2020; Abou Dalle and DiNardo 2018; Popovici-Muller et al. 2018). The mutual mechanism of action consists of blocking the active site that catalyzes the conversion reaction of α-KG to 2-HG (Anderson et al. 2018a, b).

Another strategy that has been developed is to target the different complexes of the ETC (Frattaruolo et al. 2020; Urra et al. 2017; Ashton et al. 2018). As a result, there are several series of drugs described as inhibitors of complex I that can be classified into rotenoids, vanilloids, alkaloids, biguanides, annonaceous acetogenins, and polyphenols (Urra et al. 2017). The latter have been recently attracted attention for their multiple beneficial effects, including antitumoral action (Zhou et al. 2016). Complex I-targeted drugs can inhibit complex I in a competitive or noncompetitive way, being common for competitive compounds to have a hydroquinone/quinone structure, while noncompetitive compounds, such as metformin and other biguanides, can bind non-competitively to different domains of complex I (Frattaruolo et al. 2020). Drugs targeting complex II include α-tocopherol succinate, gracillin, and atpenins. These agents have been described to increase ROS production, leading to apoptosis in cancer cells (Frattaruolo et al. 2020). Among the complex III inhibitor drugs is atovaquone, a ubiquinone analog that can competitively inhibit complex III of ETC as a result of its structural similarity to CoQ10 (Fiorillo et al. 2016). Finally, between drugs whose action is focused on complex IV, the more relevant is arsenic trioxide, a drug used to treat acute promyelocytic leukemia (Ashton et al. 2018).

In addition, researchers have also developed pharmacological compounds interfering with mitochondrial biogenesis, either by preventing the processes of transcription and translation of mtDNA or by altering the fission and fusion dynamics of this organelle (Frattaruolo et al. 2020; Anderson et al. 2018a). Between those that interfere with the DNA-translation process are doxycycline and tigecycline, compounds commonly used as antibiotics that have shown antitumoral effects in many cancer cell lines. As bacteriostatics, they can exert their effect by binding to the 30S ribosomal subunit of bacteria due to the structural similarities between this subunit and mitochondrial 28S ribosomal subunit. Consequently they can block the entry of aminoacyl-tRNA to the A-site of the ribosome, inhibiting the process of elongation and translation of mitochondrial proteins (Dong et al. 2019; Protasoni et al. 2018). Likewise, mitochondrial division inhibitor 1 (MDIVI-1) and indomethacin inhibit mitochondrial fission – an increased process in cancer cells – by blocking dynamin-related protein 1 (DRP-1), which affects mitochondrial dynamics (Frattaruolo et al. 2020; Anderson et al. 2018a; Mazumder et al. 2019).

Another pharmacological strategy is to restore mitochondrial-induced apoptosis, as occurs with resveratrol and venetoclax by inducing mitochondrial release of cytochrome c (Cyt c), therefore initiating the apoptotic cascade by activating the caspase pathway (Frattaruolo et al. 2020; Anderson et al. 2018a).

In recent years, drugs that can inhibit mitochondrial ROS (mtROS) production have been proposed as new strategy, especially those based on antioxidant compounds capable of selectively targeting the mitochondria (Grasso et al. 2020). The accumulation of mtROS produces several alterations in various signaling pathways, activating survival and proliferation factors such as hypoxia-induced factor 1 (HIF-1) and hypoxia-induced factor 2 (HIF-2). As a consequence, it has been described that cells present an elevated angiogenesis and glycolytic enzyme activity that allow the maintenance of ATP production for the tumor cell, despite of increased mtROS (Dickerson et al. 2017). Although excessive mtROS production is a key process in the development of cancer since it leads to the oxidation of cellular macromolecules (Grasso et al. 2020) and malignant cell transformation, it has been largely observed that very high increases in its production cause the death of cancer cells (Idelchik et al. 2017; Dickerson et al. 2017). However, mechanisms of action of many drugs that causes mtROS accumulation have not yet been thoroughly elucidated.

4 Polyphenols as Antitumoral Agents

Nowadays, the approach of antitumoral therapy focuses on the beneficial effects of polyphenols, a group of compounds that contain two or more phenolic groups. Given their chemical structure, polyphenols have been extensively described as antioxidant, anti-inflammatory, antimicrobial, and antitumoral agents (Zhou et al. 2016). They can be classified into different categories based on their structure and number of phenolic rings, such as stilbenes, lignans, phenolic alcohols, flavonoids, and phenolic acids. This last group can be subdivided into derivatives of hydroxybenzoic acid or hydroxycinnamic acid (Quiñones et al. 2012). Within hydroxycinnamic acid derivatives

is caffeic acid (CA), which exhibits powerful anti-inflammatory, antitumoral, and antioxidant effects, thus controlling oxidative stress by free radicals that constitutes a key process in cancer progression (Caffeic acid 2020). The biological effects exhibited by CA are closely related to its chemical structure. Therefore, its powerful antioxidant effects are associated with the presence of a catechol group in its structure, which is known for its great reducing capacity resulting from two hydroxyl groups (Damasceno et al. 2017).

Antioxidants can be classified as primary or secondary according to their mechanism of action. If they react directly with the radical, they are classified as primary antioxidants, while if they have an indirect effect on radicals, they are classified as secondary antioxidants (Damasceno et al. 2017). CA exhibits both mechanisms. As a primary antioxidant, CA has demonstrated direct neutralization of free radicals by donating protons from its hydroxyl groups to form stable compounds that are not able to produce oxidative damage to cell structures. In the process, CA acquires a semiquinone structure when it has one oxidized group and then an o-quinone structure when both groups are (Damasceno et al. 2017) (Fig. 2). As a secondary antioxidant, CA acts through the chelation of transition metals, which catalyze the decomposition of H_2O_2 in

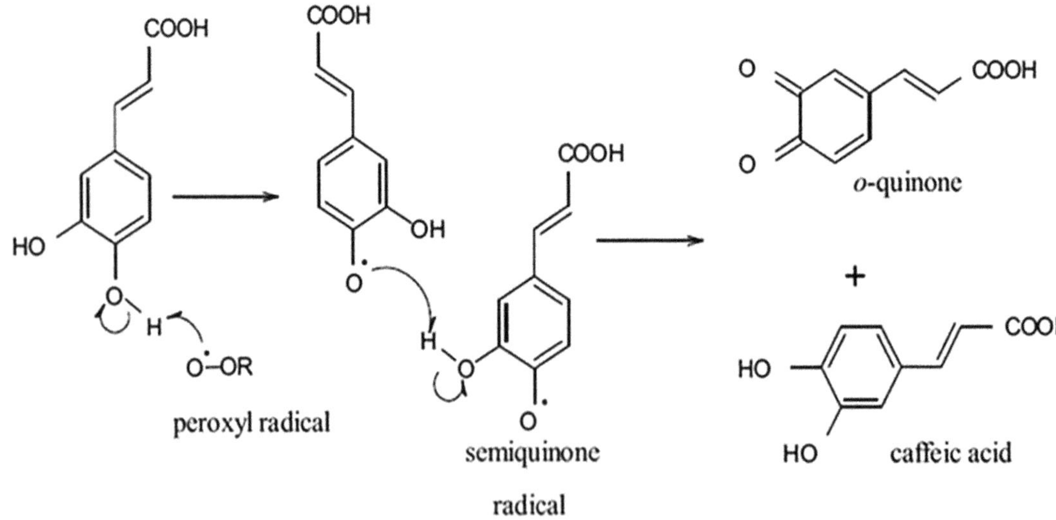

Fig. 2 Possible mechanism involved in the antioxidant activity of caffeic acid. (Damasceno et al. 2017)

hydroxyl radical (OH-). This radical has a great redox potential, which is why it produces oxidative damage in the cell by reacting with lipids, proteins, and nucleic acids (Damasceno et al. 2017). In this process, CA also undergoes structural transformation to semiquinone and o-quinone.

As mentioned above, high concentration CA can behave as prooxidants, and it is mainly to this effect that their effective antitumoral and proapoptotic capacities are associated (Damasceno et al. 2017). Therefore, the administration of high concentrations of CA to cancer cells results in the production of very high quantities of free radicals in presence of O_2 or Cu^{+2}, thus causing extensive oxidative damage and consequently triggering death in tumor cells, without significant side effects on non-cancerous cells, since they have a normal antioxidant balance (Damasceno et al. 2017).

5 Caffeic Acid and Its Derivatives with Antitumoral Action

Numerous studies have shown that CA and its derivatives have proven to be effective on different types of cancer, exhibiting antiproliferative, proapoptotic, antiangiogenic, and antimetastatic effects (Chiang et al. 2014; Monteiro Espíndola et al. 2019; Kabała-Dzik et al. 2018). As in colorectal cancer, CA derivatives such as caffeic acid phenethyl ester (CAPE) and caffeic acid phenylpropyl ester (CAPPE) (Fig. 3) have been described with antiproliferative effects by inducing cell cycle arrest as resulting from the suppression of the mechanistic target of rapamycin (mTOR) and the phosphoinositide 3-kinase/protein kinase B (PI3-K/Akt) signaling pathways (Chiang et al. 2014). Both targets induce cell proliferation and are overexpressed in cancer. The activation of the PI3-K/Akt pathway improves cell proliferation by increasing levels of cyclin D1, which is involved in the progression of the cell cycle from G1 to S phase, hence the antiproliferative effect resulting from the inhibition of this pathway induced by these derivatives.

In addition, CAPE and CAPPE produce an increased activity of the AMP-activated protein kinase (AMPK), which is defined as an energy sensor involved in the maintenance of cellular energy homeostasis. Increased AMPK activity is inversely associated with the risk of cancer by

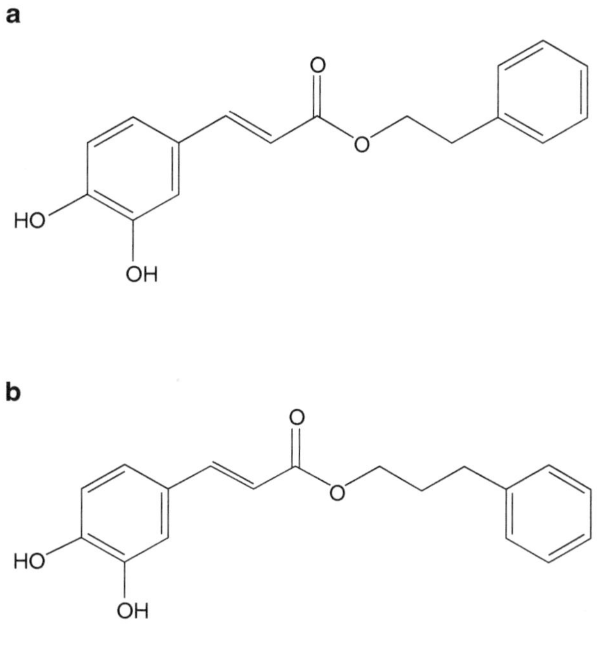

Fig. 3 Chemical structure of CAPE (**a**) and CAPPE (**b**). (Chiang et al. 2014)

suppressing mTOR activity and increasing apoptosis in cancer cells (Chiang et al. 2014).

As indicated by Monteiro et al., studies in hepatocarcinoma have demonstrated the antiangiogenic and antimetastatic capacity of CA and CAPE. The latter has shown the activation of intrinsic and extrinsic apoptotic pathways (Monteiro Espíndola et al. 2019). In addition, the antiangiogenic capacity of CAPE is given by the inhibition of HIF-1α, which has an increased expression in tumor cells due to the hypoxic environment in which they develop, thus increasing the expression of vascular endothelial growth factor (VEGF) (Monteiro Espíndola et al. 2019). Furthermore, CAPE has an antimetastatic capacity exerted through inhibiting the expression of matrix metalloproteinase 2 (MMP2) and matrix metalloproteinase 9 (MMP9), by suppressing the expression of nuclear factor kappa B (NFκB) (Monteiro Espíndola et al. 2019). These molecules are known for their role in the degradation of the extracellular matrix, which is why they can promote metastasis as a result of their overexpression in cancer.

Additionally, CAPE successfully altered the mitochondrial membrane potential (MMP) in vitro, causing the release of Cyt c from the MMI. This event increases the activation of caspase 9, promoting apoptosis through the intrinsic pathway. Moreover, CAPE also activated the extrinsic pathway of apoptosis that is mediated by the apoptosis-inducing ligand related to tumor necrosis factor-related apoptosis-inducing ligand (TRAIL), through upregulation of the death receptor 5 (DR5) resulting from the activation of mitogen-activated protein kinase (p 38) (Monteiro Espíndola et al. 2019).

In studies with breast adenocarcinoma conducted by Kabała-Dzik et al., CA and CAPE have demonstrated a dose- and time-dependent cytotoxic effect, showing greater effects at 48 h rather than 24 h, with a concentration of 100 μM for CA compared to 10 μM for CAPE (Kabała-Dzik et al. 2018). CAPE showed a more powerful cytotoxic capacity than CA, presenting a lower IC_{50} (inhibitory concentration 50) value than the one exhibited by CA (Kabała-Dzik et al. 2018). CA and CAPE also induced the inhibition of migratory capacity in a dose-dependent manner, showing again that CAPE was more potent than CA (Kabała-Dzik et al. 2018).

6 Mitochondriotropic Derivatives of Caffeic Acid

Given the extensive evidence promoting caffeic acid and its derivatives as powerful antitumoral agents, the idea of creating mitochondria-targeted compounds based on CA emerged to maximize the arrival at their site of action (Teixeira et al. 2018). To achieve this goal, several methods have been developed including those based on compounds linked to triphenylphosphonium (TPP^+) (Zielonka et al. 2017).

TPP^+ is a lipophilic cation widely used to direct various molecules to the mitochondria since its delocalized positive charge allows selective accumulation within this organelle. This process occurs in two phases. First, the compound enters the cell guided by the electrical attraction generated by a negative plasma membrane potential ($\Delta\Psi p$) of 30–60 mV. Second, it enters the mitochondria as a result of the even more negative membrane potential of the MMI ($\Delta\Psi m$), which is between 150 and 180 mV. Therefore, TPP^+ acts as a driving force for its accumulation in the mitochondria against the concentration gradient, reaching intracellular concentrations 100–1000 times higher. This feature leads to micromolar range concentrations of these compounds, thus achieving millimolar concentrations within the mitochondria (Zielonka et al. 2017). Furthermore, the incorporation of an alkyl chain serves as a link between TPP^+ and the compound of interest, giving different degrees of lipophilicity depending on its length, being more lipophilic with a longer chain length and vice versa (Zielonka et al. 2017) (Fig. 4).

There are only a few studies of CA derivatives linked to a mitochondrial target delivery system such as TPP^+. Consequently, Teixeira et al. developed a series of mitochondriotropic antioxidants derived from CA named "AntiOxCINs." These antioxidants are constituted from a primary compound of CA linked to TPP^+ which they have

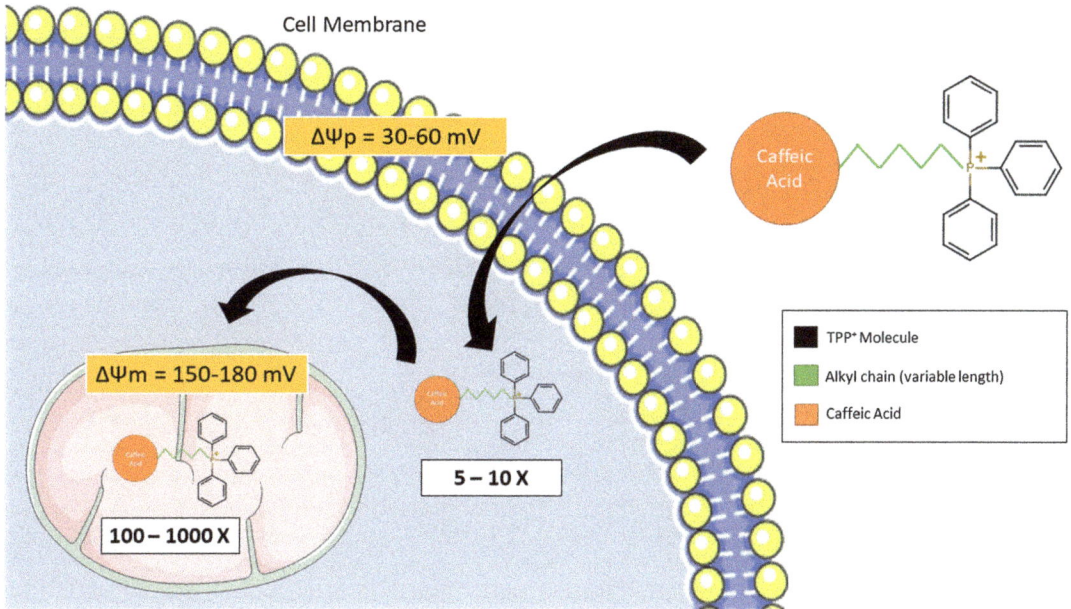

Fig. 4 General structure of TPP$^+$-based mitochondriotropic compounds and their intracellular and intramitochondrial accumulation driven by differences in plasma membrane and inner mitochondrial membrane potential, respectively

called "compound N° 1." Then, they synthetized novel compounds that present the catechol characteristic of CA, named as "compounds N° 22," "N° 23," and "N° 24." In addition, they synthetized another group of compounds derived from pyrogallol group named as "compounds N° 25," "N° 26," and "N° 27." Both groups of new compounds are identical, except for the length of the alkyl chain which can be of six, eight, or ten carbons, as shown in Fig. 5 (Teixeira et al. 2017). The results obtained show that AntiOxCINs present a high chelating capacity on transition metals (even similar to the activity exhibited by EDTA, a known chelating agent), and an increased mitochondrial uptake and ability to prevent mitochondrial membrane lipid peroxidation, compared to compound N° 1. In addition, the compounds were shown to induce the opening of the mitochondrial permeability transition pore (mPTP) (Teixeira et al. 2017), which has a known role in cell death. Its opening causes the mitochondrial release of Cyt c, thus initiating apoptotic signaling pathways (Grasso et al. 2020). Additional tests were performed on compounds N° 24 and N° 25, as they proved to be the most promising

ones. The results illustrated their ability to prevent ROS and Fe^{+2}-related cytotoxicity in HepG2 cells at concentrations of 2.5 μM (compound N° 24) and 100 μM (compound N° 25), within incubation periods of 48 h. In addition, these derivatives did not induce proapoptotic changes related to nuclear morphology or mitochondrial depolarization in normal cells, suggesting their safe use in every type of cells (Teixeira et al. 2017).

Furthermore, in a study developed by Li et al. (2017), mitochondriotropic compounds based on hydroxycinnamic acid derivatives were synthetized, named "MitoHCAs" (Li et al. 2017). The mitochondriotropic character of these compounds was also achieved by binding to TPP$^+$ (Li et al. 2017). In relation to these novel compounds, four new mitochondriotropic compounds based on p-coumaric (Mitop-CoA), caffeic (MitoCaA), ferulic (MitoFA), and cinnamic (MitoCA) acids were synthesized (Fig. 6) and were subjected to several tests evaluating their antioxidant and antiproliferative capacities (Li et al. 2017). The results obtained showed that MitoCaA was the most powerful antioxidant since it inhibits lipid peroxidation

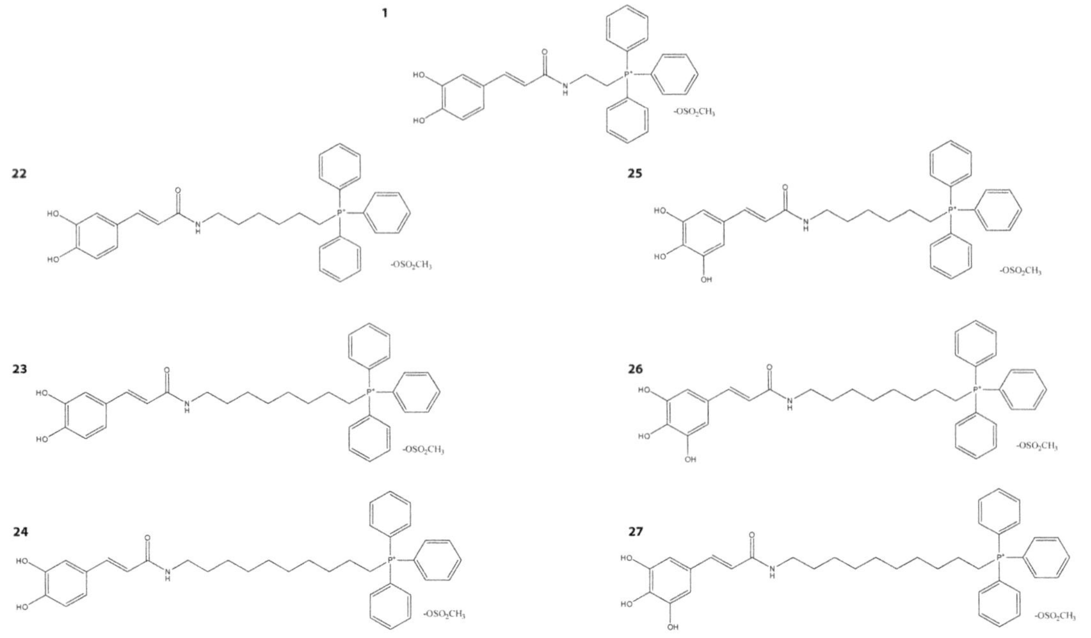

Fig. 5 Chemical structure of the caffeic acid mitochondriotropic derivatives "AntiOxCINs" developed by Teixeira et al. (2017). Compounds N° 1, 22, 23, 24, 25, 26 and 27

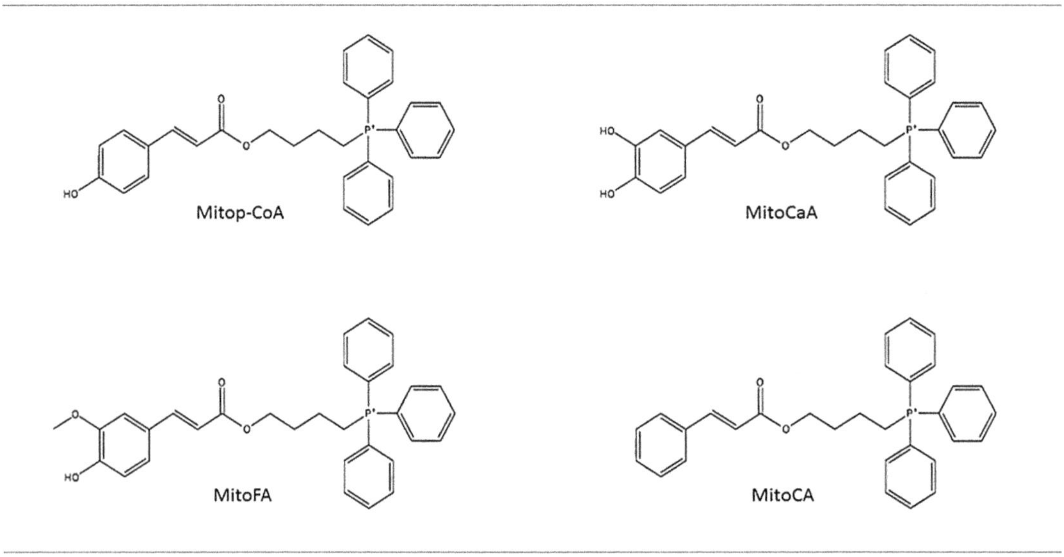

Fig. 6 Chemical structure of the hydroxycinnamic acid derivatives "MitoHCAs" developed by Li et al. (2017)

and exhibits a concentration-dependent behavior (Li et al. 2017). Subsequently, similar results were obtained when comparing the antioxidant capacity of MitoHCAs against endogenous ROS (H_2O_2) (Li et al. 2017). In addition – in order to comprehend the antioxidant mechanism by which MitoHCAs reduce H_2O_2 – results showed that MitoHCAs do not have direct radical elimination capacity and neither exert changes in the production of mitochondrial O_2^-. However, they do affect the expression and activity of antioxidant enzymes: MitoCaA, MitoFA, and Mitop-CoA increased significantly the activity of the GSH-Px and CAT enzymes, suggesting this is the mechanism by which MitoHCAs decrease H_2O_2 (Li et al. 2017). Furthermore, both MitoCaA and Mitop-CoA were found to have an inhibitory effect on mitochondrial SOD, while MitoFA did not (Li et al. 2017). Additionally, antiproliferative capacity of MitoHCAs against human hepatoma HepG2 cells and normal cell lines (human liver L02 and WI38 diploid human fibroblasts) was also evaluated, resulting in a selective inhibition of cell viability over HepG2 cancer cells compared to normal cells (Li et al. 2017). MitoCA, MitoFA, and MitoCaA compounds demonstrated the most potent antiproliferative capacity against HepG2 cells at 48 h of cell treatment (Li et al. 2017). In addition, selectivity indexes (SI) for MitoHCAs were calculated, which correspond to the quotient between the antiproliferative activity of the compounds in normal cells and the same activity but in HepG2 cancer cells. The compounds with the higher SI were MitoCA, MitoFA, and MitoCaA. The latter was determined as the most selective (Li et al. 2017). Since MitoCaA was the compound with the highest antioxidant and antiproliferative capacity, additional studies were performed. The results showed that MitoCaA caused mitochondrial fragmentation in HepG2 cells in a dose-dependent manner, causing donut-shaped morphology and a discontinuous mitochondrial network at a concentration of 20 μM. Under the same experimental conditions, the compound failed to produce mitochondrial damage in normal L02 cells (Li et al. 2017), although apoptotic assays did demonstrate that MitoCaA possesses a dose-dependent apoptotic effect (Li et al. 2017). MitoCaA was able to induce the apoptosis mechanism through the opening of mPTP and its consequent release of mitochondrial Cyt c (Li et al. 2017). These results differ from the tests performed by Teixeira et al., where it was reported that CA-based mitochondriotropic derivatives would not have an effect on mPTP opening. The difference could be attributed to structural variability in molecular targets and mechanisms of action, despite they originated

from the same compound, which requires further studies.

7   Conclusion

Polyphenols and especially CA have been described before as antitumoral, antimicrobial, and antioxidant agents. The novel mitochondriotropic derivatives of caffeic acid have shown to maximize their effects on tumor cells by allowing its selective accumulation within the cancer mitochondria, an organelle on which they exert their effects by altering the redox state and ultimately leading to mitochondrial-mediated apoptosis process. The effects of increased lipophilicity and the mitochondrial targeting product of the link to TPP$^+$ moiety are reflected in more potent compounds than those without TPP$^+$. The selectivity of these compounds for cancer cells is also highlighted, which could be given by the difference in MMP between normal and cancerous cells, the latter exhibiting a higher MMP that leads to the selective accumulation of mitochondrial derivatives. This would indicate a safe use of these compounds, also reducing side effects, a common problem in chemotherapy due to the low selectivity of these drugs. Therefore, in initial studies on tumor cell lines, mitochondriotropic derivatives of CA appear as promising and powerful antitumoral agents for future development of new molecules with targeted approach for cancer therapy, given their increased potency and selectivity.

Acknowledgments This review was supported by Fondo Nacional de Ciencia e Investigación (FONDECYT) grant 11160281 (M.C.).

Supplementary Materials Not applied.

Author Contributions H.B., G.A.V., and M.C. wrote the manuscript. H.B. and G.A.V. did all the figures and table. G.A.V., G.C., J.A.J., and MC edited the manuscript.

Conflicts of Interest The authors declare no conflict of interest.

Funding This research received no external funding.

References

Abou Dalle I, DiNardo CD (2018, Jul) The role of enasidenib in the treatment of mutant IDH2 acute myeloid leukemia. Ther Adv Hematol 9(7):163–173. https://doi.org/10.1177/2040620718777467

Anderson RG, Ghiraldeli LP, Pardee TS (2018a) Mitochondria in cancer metabolism, an organelle whose time has come? Biochim Biophys Acta Rev Cancer 1870(1):96–102. https://doi.org/10.1016/j.bbcan.2018.05.005. Elsevier B.V., Aug. 01, 2018

Anderson NM, Mucka P, Kern JG, Feng H (2018b) The emerging role and targetability of the TCA cycle in cancer metabolism. Protein Cell 9(2):216–237. https://doi.org/10.1007/s13238-017-0451-1. Higher Education Press, Feb. 01, 2018

Ashton TM, Gillies McKenna W, Kunz-Schughart LA, Higgins GS (2018) Oxidative phosphorylation as an emerging target in cancer therapy. Clin Cancer Res 24(11):2482–2490. https://doi.org/10.1158/1078-0432.CCR-17-3070. American Association for Cancer Research Inc., Jun. 01, 2018

***, "Caffeic acid | C9H8O4 – PubChem." https://pubchem.ncbi.nlm.nih.gov/compound/689043#section=Pharmacology-and-Biochemistry. Accessed 29 July 2020

Chen L, Cui H (2015) Targeting glutamine induces apoptosis: a cancer therapy approach. Int J Mol Sci 16(9):22830–22855. https://doi.org/10.3390/ijms160922830. MDPI AG, Sep. 22, 2015

Chiang E-PI et al (2014, Jun) Caffeic acid derivatives inhibit the growth of colon cancer: involvement of the PI3-K/Akt and AMPK signaling pathways. PLoS One 9(6):e99631. https://doi.org/10.1371/journal.pone.0099631

Comité Nacional de Estadísticas Vitales (2017) ANUARIO DE ESTADÍSTICAS VITALES, 2017 Período de información : 2017. Anu. Estad. Vitales

Damasceno SS, Dantas BB, Ribeiro-Filho J, Antônio D, Araújo M, da Costa JGM (2017) Chemical properties of caffeic and ferulic acids in biological system: implications in cancer therapy. A review. Curr Pharm Des 23(20):3015–3023. https://doi.org/10.2174/1381612822666161208145508

Dickerson T, Jauregui CE, Teng Y (2017) Friend or foe? Mitochondria as a pharmacological target in cancer treatment. Fut Med Chem 9(18):2197–2210. https://doi.org/10.4155/fmc-2017-0110. Future Medicine Ltd., Dec. 01, 2017

Dong Z et al (2019) Biological functions and molecular mechanisms of antibiotic tigecycline in the treatment of cancers. Int J Mol Sci 20(14). https://doi.org/10.3390/ijms20143577. MDPI AG, Jul. 02, 2019

Fiorillo M et al (2016, Jun) Repurposing atovaquone: targeting mitochondrial complex III and OXPHOS to eradicate cancer stem cells. Oncotarget 7(23):34084–34099. https://doi.org/10.18632/oncotarget.9122

Fitzmaurice C et al (2019, Dec) Global, regional, and national cancer incidence, mortality, years of life lost, years lived with disability, and disability-adjusted life-years for 29 cancer groups, 1990 to 2017: a systematic analysis for the global burden of disease study. JAMA Oncol 5(12):1749–1768. https://doi.org/10.1001/jamaoncol.2019.2996

Frattaruolo L, Brindisi M, Curcio R, Marra F, Dolce V, Cappello AR (2020) Targeting the mitochondrial metabolic network: a promising strategy in cancer treatment. Int J Mol Sci 21(17):1–21. https://doi.org/10.3390/ijms21176014. MDPI AG, Sep. 01, 2020

Grasso D, Zampieri LX, Capelôa T, Van De Velde JA, Sonveaux P (2020) Mitochondria in cancer. Cell Stress 4(6):114–146. https://doi.org/10.15698/cst2020.06.221. Isfahan University of Medical Sciences (IUMS), Jun. 01, 2020

Idelchik M d PS, Begley U, Begley TJ, Melendez JA (2017, Dec) Mitochondrial ROS control of cancer. Semin Cancer Biol 47:57–66. https://doi.org/10.1016/j.semcancer.2017.04.005

Kabała-Dzik A, Rzepecka-Stojko A, Kubina R, Wojtyczka RD, Buszman E, Stojko J (2018, Dec) Caffeic acid versus caffeic acid phenethyl ester in the treatment of breast cancer MCF-7 cells: migration rate inhibition. Integr Cancer Ther 17(4):1247–1259. https://doi.org/10.1177/1534735418801521

Kalyanaraman B et al (2018) A review of the basics of mitochondrial bioenergetics, metabolism, and related signaling pathways in cancer cells: therapeutic targeting of tumor mitochondria with lipophilic cationic compounds. Redox Biol 14:316–327. https://doi.org/10.1016/j.redox.2017.09.020. Elsevier B.V., Apr. 01, 2018

Konteatis Z et al (2020, Feb) Vorasidenib (AG-881): a first-in-class, brain-penetrant dual inhibitor of mutant IDH1 and 2 for treatment of glioma. ACS Med Chem Lett 11(2):101–107. https://doi.org/10.1021/acsmedchemlett.9b00509

Li J et al (2017, Jan) Synthesis of hydroxycinnamic acid derivatives as mitochondria-targeted antioxidants and cytotoxic agents. Acta Pharm Sin B 7(1):106–115. https://doi.org/10.1016/j.apsb.2016.05.002

Mazumder S et al (2019, May) Indomethacin impairs mitochondrial dynamics by activating the PKCζ-p38-DRP1 pathway and inducing apoptosis in gastric cancer and normal mucosal cells. J Biol Chem 294(20):8238–8258. https://doi.org/10.1074/jbc.RA118.004415

Modica-Napolitano JS, Weissig V (2015) Treatment strategies that enhance the efficacy and selectivity of mitochondria-targeted anticancer agents. Int J Mol Sci 16(8):17394–17421. https://doi.org/10.3390/ijms160817394. MDPI AG, Jul. 29, 2015

Monteiro Espíndola KM et al (2019) Chemical and pharmacological aspects of caffeic acid and its activity in hepatocarcinoma. Front Oncol 9(JUN):541. https://doi.org/10.3389/fonc.2019.00541. Frontiers Media S. A., 2019

Popovici-Muller J et al (2018, Apr) Discovery of AG-120 (Ivosidenib): a first-in-class mutant IDH1 inhibitor for the treatment of IDH1 mutant cancers. ACS Med Chem Lett 9(4):300–305. https://doi.org/10.1021/acsmedchemlett.7b00421

Prasetyanti PR, Medema JP (2017) Intra-tumor heterogeneity from a cancer stem cell perspective. Mol Canc 16(1):41. https://doi.org/10.1186/s12943-017-0600-4. BioMed Central Ltd., Feb. 16, 2017

Protasoni M, Kroon AM, Taanman JW (2018, Sep) Mitochondria as oncotarget: a comparison between the tetracycline analogs doxycycline and COL-3. Oncotarget 9(73):33818–33831. https://doi.org/10.18632/oncotarget.26107

Quiñones M, Miguel M, Aleixandre A (2012) Los polifenoles, compuestos de origen natural con efectos saludables sobre el sistema cardiovascular. Nutr Hosp organo Of la Soc Espa??ola Nutr Parenter y Enter 27(1):76–89. https://doi.org/10.3305/nh.2012.27.1.5418

Stuart SD et al (2014) A strategically designed small molecule attacks alpha-ketoglutarate dehydrogenase in tumor cells through a redox process. Cancer Metab 2(1):4. https://doi.org/10.1186/2049-3002-2-4

Teixeira J et al (2017) Development of a mitochondriotropic antioxidant based on caffeic acid: proof of concept on cellular and mitochondrial oxidative stress models. J Med Chem 60(16):7084–7098. https://doi.org/10.1021/acs.jmedchem.7b00741

Teixeira J, Deus CM, Borges F, Oliveira PJ (2018) Mitochondria: targeting mitochondrial reactive oxygen species with mitochondriotropic polyphenolic-based antioxidants. Int J Biochem Cell Biol 97:98–103. https://doi.org/10.1016/j.biocel.2018.02.007. Elsevier Ltd, Apr. 01, 2018

Urra FA, Muñoz F, Lovy A, Cárdenas C (2017) The mitochondrial complex(I)ty of cancer. Front Oncol 7 (JUN):118. https://doi.org/10.3389/fonc.2017.00118. Frontiers Media S.A., Jun. 08, 2017

Zhou Y et al (2016) Natural polyphenols for prevention and treatment of cancer. Nutrients 8(8). https://doi.org/10.3390/nu8080515. MDPI AG, Aug. 22, 2016

Zielonka J et al (2017) Mitochondria-targeted triphenylphosphonium-based compounds: syntheses, mechanisms of action, and therapeutic and diagnostic applications. Chem Rev 117(15):10043–10120. https://doi.org/10.1021/acs.chemrev.7b00042. American Chemical Society, Aug. 09, 2017

Adv Exp Med Biol - Cell Biology and Translational Medicine (2022) 17: 227–241
https://doi.org/10.1007/5584_2022_727
© Springer Nature Switzerland AG 2022
Published online: 27 July 2022

Systematically Assessing Natural Compounds' Wound Healing Potential with Spheroid and Scratch Assays

Gabriel Virador, Lisa Patel, Matthew Allen, Spencer Adkins, Miguel Virador, Derek Chen, Win Thant, Niloofar Tehrani, and Victoria Virador

Abstract

Understanding cellular processes involved in wound healing is very important given that there are diseases, such as diabetes, in which wounds do not heal. To model tissue regeneration, we focus on two cellular processes: cellular proliferation, to replace cells lost to the wound, and cell motility, activated at the wound edges. We address these two processes in separate, drug responsive, in vitro models. The first model is a scaffold-free three-dimensional (3D) spheroid model, in which spheroids grow larger – to a certain extent – with increased time in culture. The second model, the scratch wound assay, is focused on cell motility. In conjunction with collagen staining, it analyzes changes to the coverage of the wound edge and wound bed. Our workflow gives insights into candidate compounds for wound healing as we show using manuka honey (MH) as an example. Spheroids are responsive to oxidative damage by hydrogen peroxide (H_2O_2) which affects viability but mostly produces disaggregation. Conversely, MH supports spheroid health, shown by size measurements and viability. In two-dimensional scratch wound assays, MH helps close wounds with relative less collagen production and increases the loose cellular coverage adjacent to and within the wound. We use these methods in the undergraduate research laboratory as teaching and standardization tools, and we hope these will be useful in similar settings.

Keywords

Collagen stain · Manuka honey · Murine stem cells · Tissue regeneration · Viability

Abbreviations

DAPI   4′,6-Diamidino-2-phenylindole
DMSO   Dimethyl sulfoxide
EDTA   Ethylenediaminetetraacetic acid
MRI   Montpellier Ressources Imagerie
MTT   3-(4,5-Dimethylthiazol-2-yl)-2,5-diphenyl-2H-tetrazolium bromide
PBS   Phosphate buffered saline
SDS   Sodium dodecyl sulfate

G. Virador
Universidad de Navarra, Pamplona, Spain

L. Patel, M. Allen, S. Adkins, M. Virador, W. Thant,
N. Tehrani, and V. Virador (✉)
Montgomery College, Rockville, MD, USA
e-mail: vvirador@montgomerycollege.edu

D. Chen
Virginia-Maryland Regional College of Veterinary
Medicine, Blacksburg, VA, USA

1 Introduction

The science of wound healing has its roots in the human need for survival. Since ancient times, people from all over the world have found healing substances in their environment. In general, reports of natural compounds' healing abilities are limited to case studies, while more systematic assessment of their properties is reserved for purified fractions or individual active ingredients (Atanasov et al. 2021). Generally processes of regeneration after wound are described to include hemostasis (clot formation to limit blood loss), inflammation to remove damaged cells, proliferation with formation of immature tissue (granulation tissue) and contraction of the wound, and finally, maturation of the new tissue to restore tissue functionality (Rodrigues et al. 2019). To provide an in vitro system for assessment of such natural compounds in wound healing, we have focused on two aspects of tissue regeneration: cellular proliferation in a 3D in vitro model and cell motility measured in 2D scratch wound assays. We report here our efforts to standardize a workflow for in vitro testing of candidate natural products for wound healing. Prevalent animal models in the wound healing field can be thus displaced or substituted by mini tissue models that can be generated in vitro from relevant cells and contribute important scientific insights to tissue regeneration.

Spheroids have been chosen as the model 3D systems; as such, there are thousands of publications reporting their fabrication and attempts at standardization for application to high throughput assays (Brüningk et al. 2020; Virador et al. 2019). Our spheroids are made of NIH 3T3 cells (fibroblasts, according to ATCC information) which self-aggregate in the absence of extracellular matrix or scaffold. The term fibroblast describes a broad spectrum of cellular populations which are considered mesodermal cells, are not parenchymal, and have a prominent role in generating and maintaining the extracellular matrix; however, the distinction between fibroblasts and stromal cells is unclear and more

so when their murine origin is embryonic or neonate (Robey 2017). Regardless of these debates, fibroblasts are generally agreed to be precursors of various features of the mesenchyme; they can be activated in response to various signals and participate in inflammation (Buckley et al. 2001), wound healing, or cancer, by secreting growth factors and proteolytic enzymes (Kalluri 2016). The cells we used in this work are the original Swiss albino 3T3 isolated from mouse embryos by Todaro and Green (Todaro and Green 1963). While spheroids formed from the adipogenic NIH 3T3-L1 subset have been well characterized (Graham et al. 2019), there is limited information on how spheroids of the original Swiss albino cell line are formed and sustained in the absence of scaffold. Here we offer a summary of our observations on the formation of these spheroids which appears to be chiefly dependent on the plastic surface and well size.

We established a reproducible workflow to form spheroids in a 96-well format, as well as time and end points to screen for natural compounds for wound healing. By testing cell viability as an end point, we intended to find compounds which consistently increased spheroid viability in a set time. These compounds would then be taken to the well-established scratch wound healing assay in 2D to verify their ability to increase the cell motility needed to close the wound. We added collagen staining of the 48-h wounds to our procedure to demonstrate whether the compound increases or decreases collagen production. In our analysis of scratch wounds, we looked at the traditional ratio of wound closure, but we also included the analysis of the loose space around the wound, providing insight on potential effects of the compound on cellular subpopulations affected by the wound. As an example of our workflow applied to a natural compound, we present results obtained with MH, a kind of honey native to New Zealand, produced by bees which pollinate the manuka bush flower (*Leptospermum scoparium*) and for which there are many reports showing positive effects in wounds (Tashkandi 2021).

2　Materials and Methods

2.1　Cells and Culture Conditions

The cells we used in our experiments are NIH 3T3 fibroblasts (CCL-92, ATCC, Manassas, VA). Cells were expanded and multiple vials were kept frozen under liquid nitrogen. A set of experiments was conducted from one expanded vial. Cells were used for a low number of passages (typically less than 7). Cells were monitored for mycoplasma contamination by visually assessing DAPI (4′,6-diamidino-2-phenylindole) stained samples. For regular culture maintenance and to expand cells for 3D experiments, cells were grown in T75 flasks in DMEM with high glucose, supplemented with 10% v/v bovine calf serum and 1% penicillin-streptomycin, at 37° C in a humidified atmosphere with 5% CO_2. Flat bottom 96-well plates (Corning, NY) were used for monolayer cell viability tests and flat bottom 12-well plates (Corning, NY) for scratch wound assays. For 3D spheroids, polystyrene round bottom low adhesion plates (Greiner Bio1, Cat 650,970, Monroe, NC) or non-tissue culture treated polystyrene flat well multiwell plates (VWR, 10861–556 or 10,861–558) were used, and similar sized tissue culture treated polystyrene flat well plates were used for comparison (e.g., CELLTREAT Scientific Products, Pepperell, MA).

2.2　Spheroid Formation for Compound Screening

Subconfluent cells with more than 90% viability as assessed by Trypan Blue staining were passaged with a 0.48 mM EDTA in PBS rinse (VERSENE, ThermoFisher, Waltham, MA) and exposure to 0.25% Trypsin/EDTA. Cells were centrifuged at 700 RPM for 5 min and resuspended in fresh media before seeding onto various surfaces as detailed in figure legends. In round bottom low adherence 96-well plates, 50,000 cells per well in 50 µl media were seeded. Outer wells contained only 100 µl of PBS. When spheroids had formed (typically by day 3 after seeding), 50 µl media containing the test compounds were added to each well. Cultures were monitored daily, and viability was assessed by adding MTT at the end of 1 week.

2.3　Dissolving Natural Compounds: Spheroid Treatment

MH, a product of New Zealand, was purchased from Costco (Y.S Eco Bee Farms, Sheridan, IL LOT # 9178 – the same lot was used throughout the project), and a stock solution of 1 g/ml was prepared as follows: first, 4 g of MH were weighed and dissolved in 1 ml of deionized water. The stock vial was kept at 37 °C for a few hours to overnight. When the honey stock was a fine suspension/solution, the total volume was assessed, and then the final solution was diluted with deionized water to a final concentration of 1 g/ml (w:v). From this stock solution, half-log dilution series was prepared with deionized water.

Commercially available 3% hydrogen peroxide from a freshly opened bottle was used to produce a half-log dilution series using deionized water. Care was taken to use a very small volume of each compound stock to produce final concentrations in spheroid medium.

When spheroids had formed (typically by day 3 after seeding), 50 µl media containing the test compounds were added to each well. This brought the total volume in well to 100 µl. Cultures were monitored daily.

2.4　Cell Viability and Stains

The MTT reduction assay (Berridge and Tan 1993) was used as follows: each well containing one spheroid received 10 µl MTT (Sigma-Aldrich, St. Louis, MO) dissolved in water to a final concentration of 0.5 mg/ml MTT. The spheroids were incubated at 37 °C and 5% CO_2 for 45 min, and then solubilizing reagent (10% SDS in 0.01 M HCl) was added. The plate was maintained overnight at room temperature in the dark after which the amount of MTT formazan formed was measured at 450 nm in a Tristar2 Multimode Reader LB942 (Berthold Technologies, Oak Ridge, TN).

Mito tracker green (Molecular Probes) was dissolved in DMSO and added to the cultures at a 100 nM final concentration for 15 min in incubator prior to fixing.

Rhodamine-phalloidin (Molecular Probes) was dissolved in methanol and added to the cultures at nanomolar concentration according to manufacturer's recommendation. DAPI (Invitrogen) was dissolved in deionized water and added to the cultures according to manufacturer's recommendation.

Mason Trichrome stain to detect collagen in scratch wound assays was done as follows: Forty-eight hours post scratch, after documenting the live cell culture, cells were fixed with 4% paraformaldehyde, and then the protocol http://www.ihcworld.com/_protocols/special_stains/masson_trichrome.htm was followed.

2.5   Scratch Wounds

Subconfluent NIH 3T3 cells were seeded in 24-well plates at 10000 cells per well. Prior to seeding, a line was drawn on the underside of the wells to facilitate recording the same location of the wound over time. Two days later, medium was removed and a scratch was made by gliding a 100 µl pipette tip vertically through the center of the well, followed by a PBS rinse to eliminate floating debris and addition of media containing the test compounds. Pictures were taken immediately (time 0) at 24 and 48 h to document scratch reduction.

2.6   Imaging and Measurements

Fluorescence images were taken in an EVOS M5000 microscope with a highly sensitive 3.2 MP monochrome CMOS camera (2048 × 1536) with 3.45-µm pixel resolution (Thermo Fisher). For brightfield images, an EVOS XLCore (AMEX 1000, Invitrogen) was used. To measure spheroids, using brightfield or phase contrast, the focus was adjusted so that the images represented a circular cross section of the spheroid at its center, with the diameter of the circle matching that of the spheroid.

For scratch wounds, images were taken in the same area of each wound using the previously drawn marker and the wound as reference. The MRI Wound Healing plugin for ImageJ/Fiji® was used to analyze scratch wound images using the variance method (filter radius, 10; radius open, 4; min size, 10,000) https://github.com/MontpellierRessourcesImagerie/imagej_macros_and_scripts/wiki/Wound-Healing-Tool.

Original images, converted to 8-bit images, were used as input and processed with a fixed threshold value of 20 to find the area that contained no cells. Adjusting the selection threshold in the plugin settings up to a fixed threshold number of 100 allowed for distinction between empty space and loose cellular coverage. Ratio of closure was calculated by dividing the area given by ImageJ (in pixels) at the specific time by the area at time 0 in the same well.

To quantitate large cells, present in the wound area, ImageJ "analyze particles" plugin was used on 8-bit images with size (100-infinity) and circularity (0.50–1.00) settings. These settings were adopted after comparing the data obtained with results obtained by visual counting.

2.7   Histology

Spheroids were fixed in 4% paraformaldehyde (Electron Microscopy Sciences, Hatfield, PA) for 15 min, rinsed with PBS, and carefully placed between lens paper and then embedded and sectioned using routine protocols (summarized in https://www.corning.com/catalog/cls/documents/protocols/CLS-AN-431_DL.pdf). Hematoxylin- and eosin-stained sections were imaged in an EVOS XLCore microscope at 10 x magnification.

2.8   Statistics

Experiments had an N = 3 (three independent experiments) with six to eight biological replicates per condition unless indicated. Data were analyzed with Microsoft Excel and with

GraphPad Prism software (GraphPad, San Diego, USA). All the statistical analyses were performed with GraphPad Prism. Data are reported as mean $\pm$ SD *p $<$ 0.05, **p $<$ 0.01, ***p $<$ 0.0001.

3   Results

3.1   NIH 3T3 Spheroids Self-Assemble from Cell Sheets in the Absence of Extracellular Matrix

CCL-92 is a 3T3-Swiss albino embryonic cell line from *Mus musculus*, house mouse, which grows as adherent monolayers on flat tissue culture surfaces. In 24-well non-tissue culture treated polystyrene plates, cells formed spontaneous clusters of varied shapes ranging from small colonies (between 100 and 200 µm in diameter) to larger spheroids (300 and 800 µm in diameter depending on plate and time in culture). We compared morphological stages of these spontaneous clusters and timing in various flat bottom plates. For example, at day 10 after plating, the non-tissue culture treated plate (VWR) had formed small colonies in 6 out of 24 wells, spheroids in 7 out of 24 wells, and stretched ribbons in 5 wells, while the remaining 6 wells contained small pieces of cell sheet curling at the edges. Sarstedt flat bottom tissue culture plate showed similar adhesions and pulling sheets which took longer to detach, and only fragments of cell sheets or flat sheets with ruffled edges were observed at day 10. In tissue culture treated plates with special adhesion properties for difficult to grow cells, for example, CELLTREAT, the pulling from the edges was observed in 15/24 wells, but the cell sheets never completely detached and sheet or cell clusters did not form even after 3 weeks in culture (Fig. 1a, top panels), while various shaped tissue clusters appeared in the VWR non-tissue culture plate in the same time frame (Fig. 1a, bottom panels). Most of the larger spontaneous clusters had tubular morphologies, and some resembled incomplete toroids. Detailed observation of the stages of cluster formation indicated that 2D compact cell sheets formed in various regions of the plate with

one or more strong adhesions to the edges of the plate or to each other (Fig. 1b). In cases when the compact sheets had a stronger attachment to the plate, the edges curled (Fig. 1c), but in most cases complete rolling of the sheet edges occurred to form spheroids or pouches or wells filled with colonies and small spheroids (Fig. 1d). In some cases, adhesions on symmetrical ends of the cell sheet stretched and rolled over small colonies engulfing them (Fig. 1e). These observations agree with a recent report (Granato et al. 2017) of spontaneous clusters formed by human dermal myofibroblasts. Histology of typical tissue aggregates indicates that they possess a thick wall and some have a lumen (Fig. 1f–i).

Since it appeared that all different tissue aggregates formed in a similar fashion, we focused on documenting the formation of small spheroid pouches in 96-well round bottom plates. For reproducible spheroids, we found it was very important to put the plate after seeding on a flat surface at room temperature absent of any vibration, before putting it into the incubator, which is also proposed as a technique to reduce edge effect (Lundholt et al. 2003; White et al. 2019). First, we used Costar 7007 plates, made of ultra-low attachment polystyrene. In these plates, the spheroids did not appear to go through the process of folding but appeared spheroidal in shape from the day of seeding. These spheroids had an average size of 0.15 mm^2 at day 12 with an increase to 0.19 mm^2 at day 15 and a progressive decrease to 0.17 mm^2 by day 23 consistent with spheroid maturation (Graham et al. 2019). Based on these observations, we speculate that spheroids made in those ultralow attachment plates may be similar to those made from cancer cells described as a "closely packed, spherical geometry of cells" (Mueller-Klieser 2000), but they do not appear to have undergone sheet retraction and folding.

We screened other 96-well polystyrene plates and found that CellStar suspension plates produced tissue structures from cell sheets. By observing the same well at days 4, 5, and 7 post seeding, we could document that spheroid formation begins by folding, rolling, and finally detaching from symmetrical stress fiber adhesions

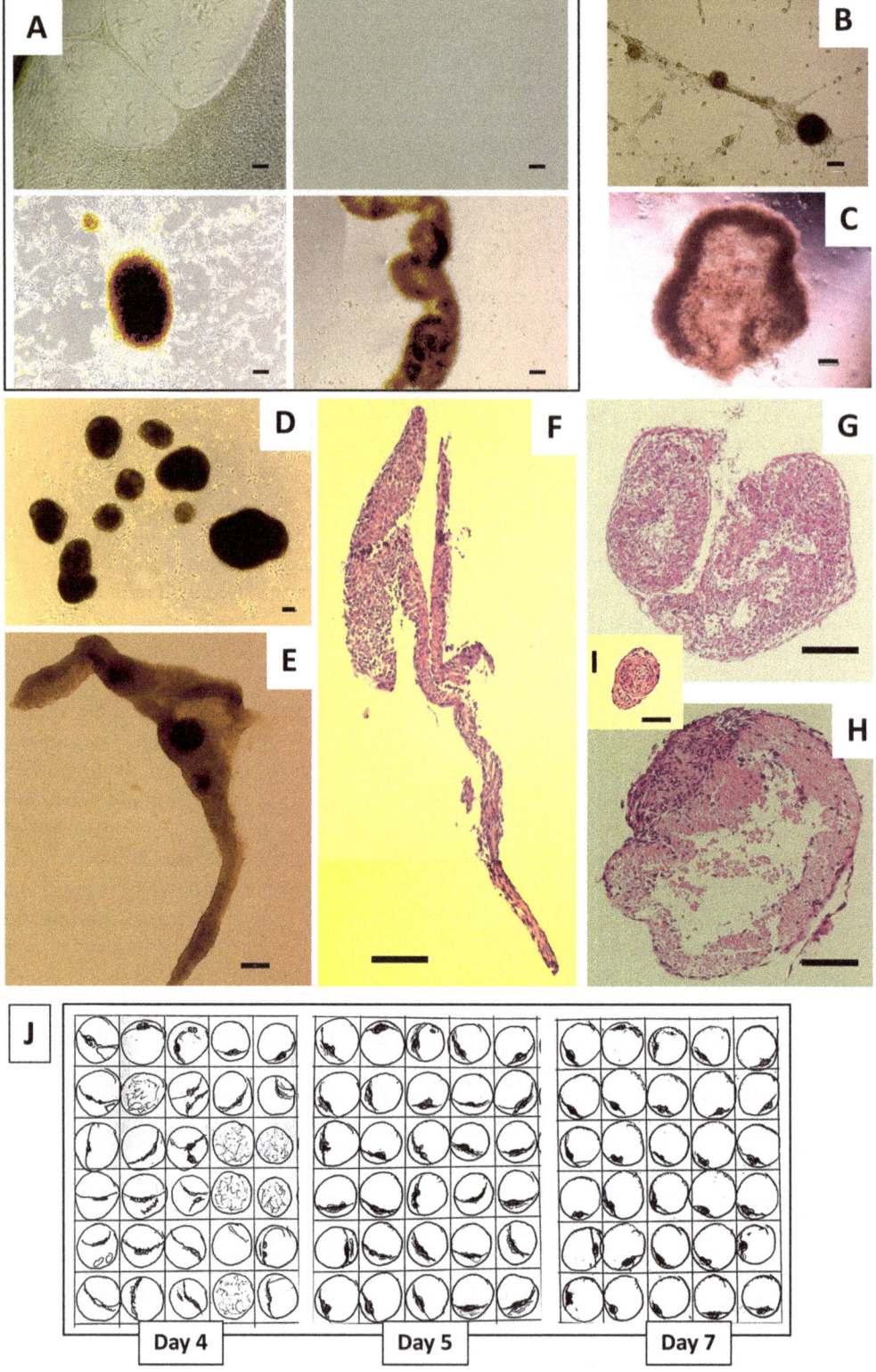

Fig. 1 Examples of tissue aggregates from cell sheets formed in low adhesion plates
(**a**) Three-week-old cultures of NIH3T3 cells (CCL92). Top panels show cells grown in CELLTREAT plates, left image shows the edge of the plate where sheets start pulling away, and right image shows tightly attached

to the concave well edges, thus undergoing similar stages to the non-tissue culture 24-well plates (Fig. 1j). Then, increased stretching was accompanied by rolling of the sheet like a candy wrapper, prior to formation of the spheroid, which resembled the closing of a pouch. In a small percentage of cases, flat sheets with ruffled edges were also observed in the 96-well plate due to imperfect pouch formation. Of note, these observations and drawings were part of a "Introduction to Scientific Research" class in a community college setting aimed at familiarizing students with cell biology and fostering scientific curiosity and systematic scientific skills.

3.2 Manuka Honey Is a Spheroid-Supporting Compound

To standardize 3D assays for multiwell plates, spheroids are preferred as they are more regular in size and reproducible. In preliminary experiments using MTT, we determined that 1-week treatment of spheroids without a medium change does not significantly alter their viability (data not shown) possibly because, once formed, spheroids have low metabolic activity compared with cells that are actively dividing (Granato et al. 2017; Fukushima et al. 2019).

Once we could reliably produce spheroids in 96-well plate and characterize their growth, we focused on standardizing methods to assess natural compounds reported as wound healing aids, many of which are part of botanical complex mixtures. To assess drug responsiveness of our model, we first chose H_2O_2, to induce oxidative stress (Sies 2017), and we compared its effect on confluent monolayers and on spheroids (Fig. 2a, b, c). As expected, the decrease of viability in response to H_2O_2 was more pronounced in monolayers. After testing a new candidate compound in monolayer, we proceeded to test spheroid drug responsiveness by assessing viability in a standardized manner in 96-well round bottom plates as shown in Fig. 2d.

Next MH with its antioxidant, antibacterial (Carter et al. 2016), and wound healing properties (Bulman et al. 2017; Frydman et al. 2020; Mokhtar et al. 2020; White 2016) was chosen to test our workflow. We found that when MH was present in the culture from the time of cell seeding (in a range of concentrations from 0.3 to 100 mg/ml), spheroids formed by day 3 in 60% of the wells compared to 20% in its absence. From this early data, we concluded that MH was a spheroid-supporting compound, whereas H_2O_2 was a spheroid-damaging compound, and this was supported by the viability data (Fig. 2e) where H_2O_2 produced a small but consistent decrease in viability while MH produced a dose-dependent small but consistent increase in viability suggesting a small increase in proliferation (Fig. 2f). Our observations indicated that spheroids responded to oxidative damage by initially disassembling, consistent with oxidative stress effects on tight junctions (Gangwar et al. 2017). Interestingly, spheroids were able to reassemble similarly to the findings of (Brüningk et al. 2020), and this reassembly was prevented by a second addition of H_2O_2 (data not shown). We speculate that this ability to reassemble may be one of the reasons for the more limited

---

Fig. 1 (continued) monolayer. Bottom panel, examples of cell clusters formed in wells of VWR non-tissue culture treated plates. (**b**) Colonies are formed as cell clusters pull away from each other. (**c**) An incomplete spheroid formed by rolling the edges of a flat sheet. (**d**) Small colonies and spheroids 200–500 μm in diameter. (**e**) Several colonies formed on a sheet are enveloped as the sheet rolls and gathers. (**f**) Hematoxylin-eosin stained section of a long cluster with tubular end. (**g,h**) Hematoxylin-eosin stained sections of typical 500 μm spheroids. (**i**) Hematoxylin-eosin stained section of a small 200 μm colony. (**j**) Detailed drawings of spheroid formation in 96-well low adhesion plate containing 50,000 cells per well; the same well is drawn at day 4, 5, and 7 post seeding. Scale bars, 100 μm

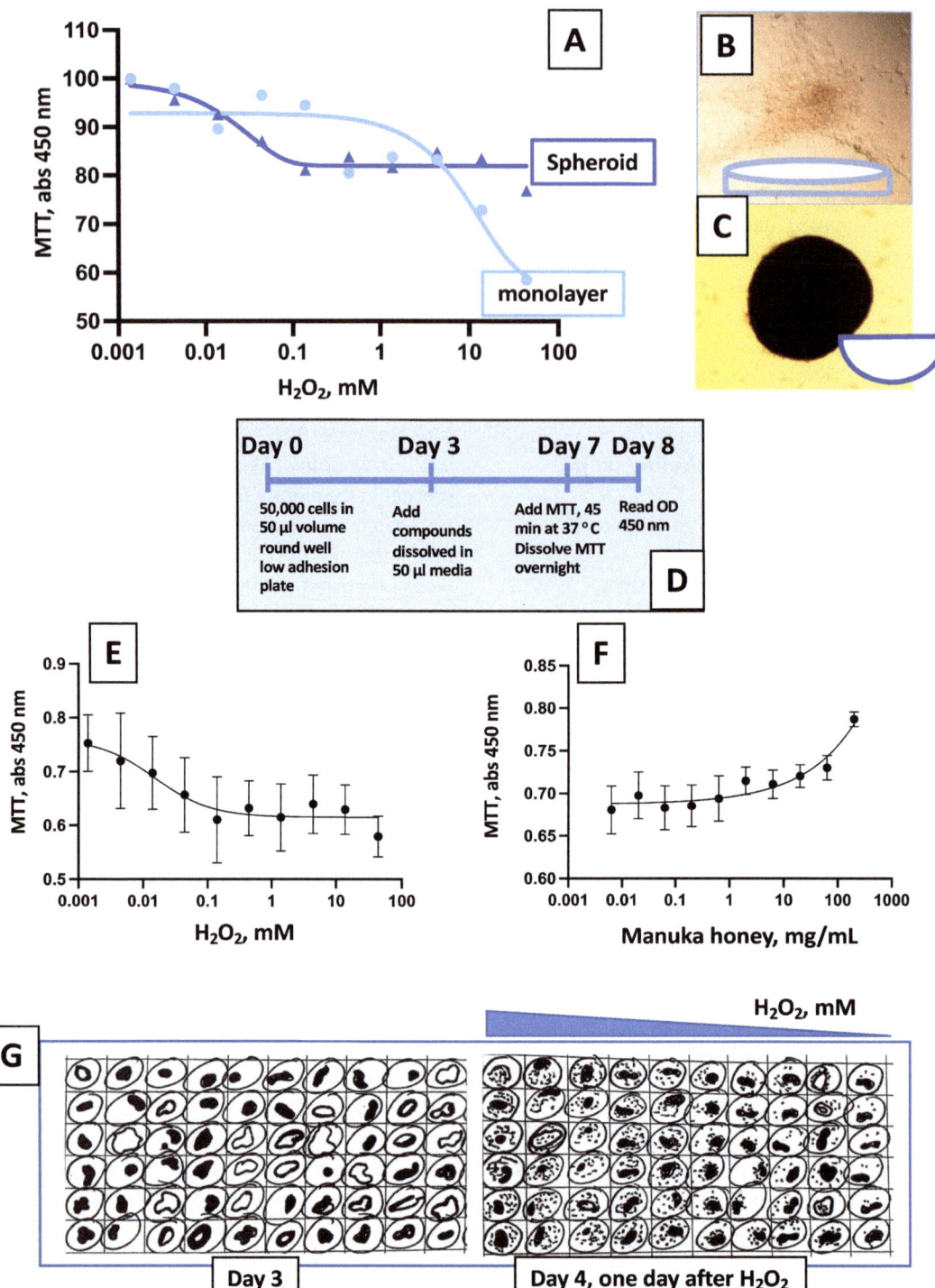

Fig. 2 Spheroids are responsive to drug treatments
Response of confluent monolayer of 10,000 cells per well (light blue line and panel B) vs. spheroids (dark blue line and panel C) to increasing concentrations of H_2O_2. D. Standardized treatment scheme applied to spheroids for all compounds tested. E. Cell viability in response to H_2O_2 treatment measured by MTT absorbance (450 nm). F. Cell viability in response to MH treatment measured by MTT absorbance (450 nm). Summary data from three independent experiments with six technical replicates for each concentration. G. H_2O_2 causes disaggregation of the spheroids. Left panel, spheroid plate at day 3 prior to adding H_2O_2; right panel, spheroid plate at day 4, after 24 h of addition of H_2O_2 at the same concentrations as shown in panel E

decrease in viability observed when comparing to H_2O_2 treated monolayers.

Next, we measured spheroid surface area at several concentrations of H_2O_2 and MH. We found that at low H_2O_2 concentrations when viability is decreasing, spheroids have smaller sizes than control (Fig. 3a, b), but at higher than 0.1 mM spheroids actually became bigger, perhaps due to swelling or to different patterns of shrinkage and regrowth as seen in other studies (Brüningk et al. 2020). For MH we found a small but consistent increase in spheroid size at the concentrations we tested (Fig. 3c, d).

3.3 Manuka Honey Increases Cell Motility in Scratch Wound Assays

After MH had been established as a spheroid-supporting compound, we used the scratch wound assay to investigate the compound's effect on cell motility. Our method to produce and document the scratch wounds was adapted from (Pinto et al. 2019). Scratch wound assays are typically analyzed by segmenting the images with imaging programs to document and quantitate the amount of closure of the wound in a given amount of time, generally 24 h. For example, the MRI wound healing plugin is extensively cited in these studies (Kauanova et al. 2021). Basically, the method highlights different areas in the image by replacing each pixel with the specified neighboring area through a variance filter with a specific radius to effectively differentiate the aspect of tissue from the empty areas (also see (Suarez-Arnedo et al. 2020)). These studies generally look for disappearance of the empty space in the wound bed. Using ImageJ plugin "MRI wound healing tool" to follow the decrease in empty space in the presence of MH, we found no significant difference (Fig. 4a).

We realized that analyzing the images of the empty space did not provide sufficient

Fig. 3 MH is a spheroid-supporting compound (**a**) Size change in spheroid in response to H_2O_2. (**b**) Representative spheroid pictures taken at day 7 of the treatment protocol left spheroid control, right spheroid with 0.44 mM H_2O_2. (**c**) Size change in spheroid in response to MH. (**d**) Left spheroid control, right spheroid with 6.25 mg/ml MH. Experiments were carried out twice with six technical replicates per concentration, and one representative experiment is shown

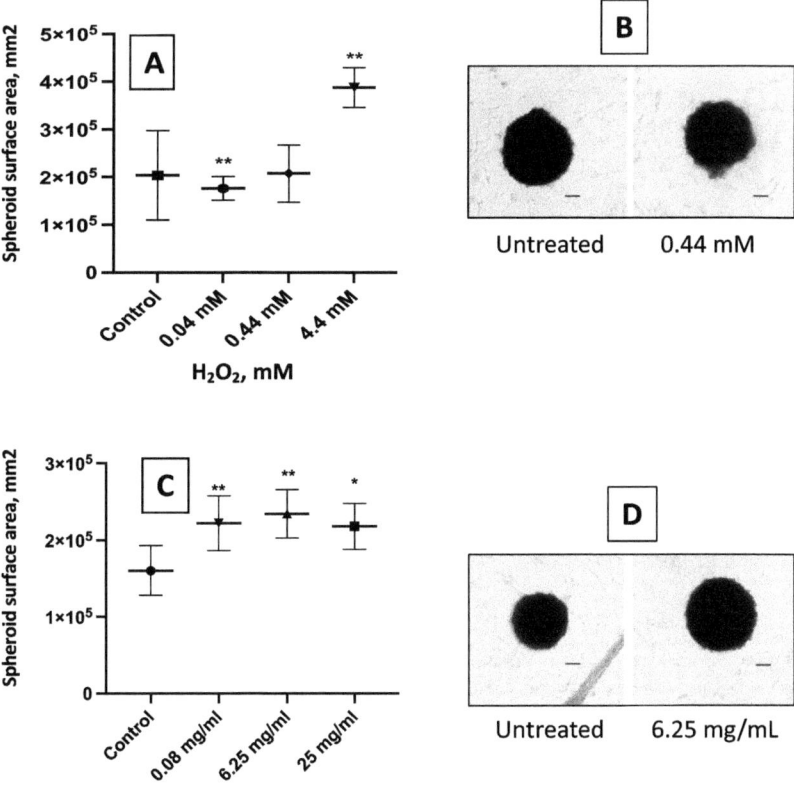

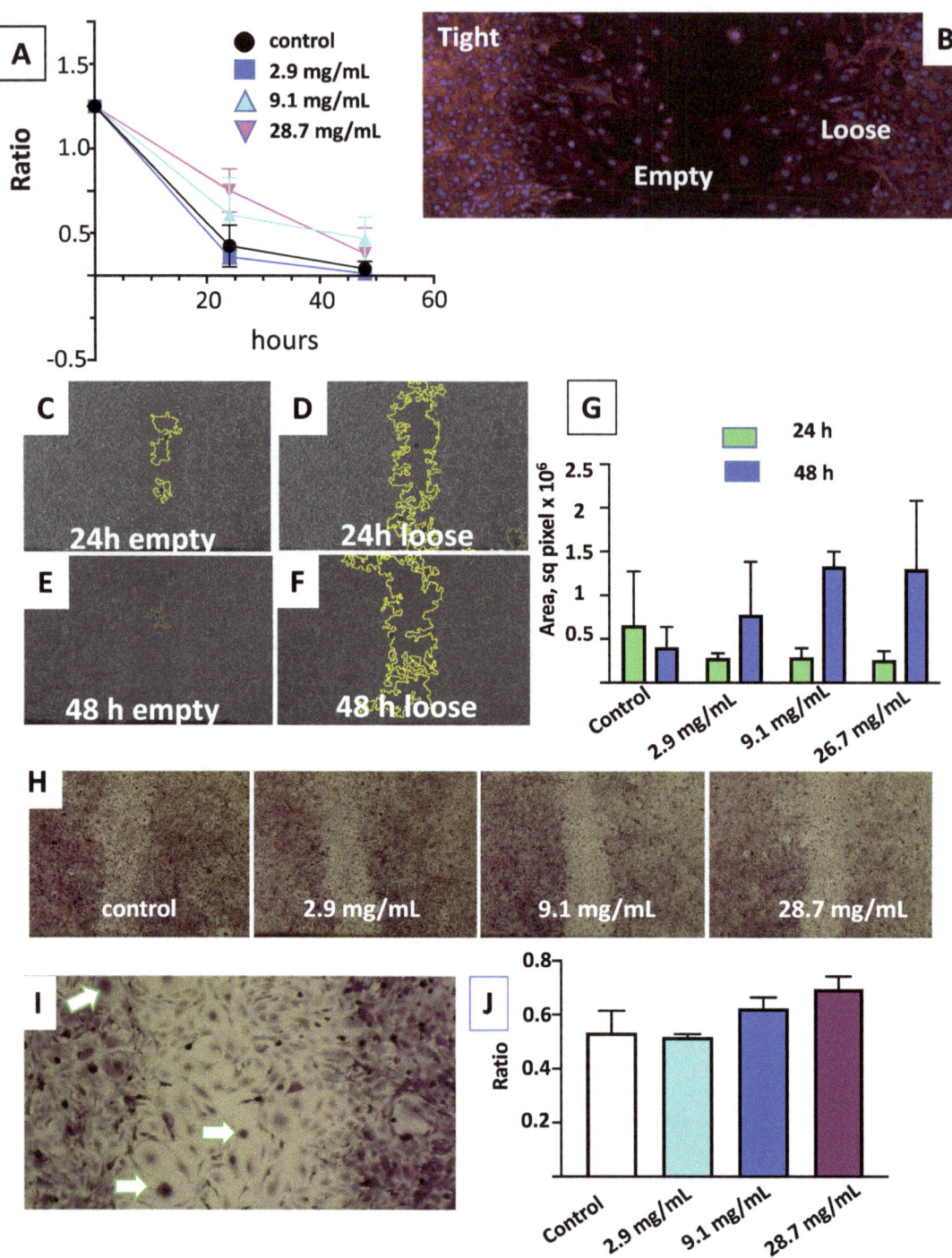

Fig. 4 Scratch wound assays treated with MH
(**a**) Closure of the wound, ratio of empty space (treated vs. control) at 24 and 48 h post wound. Unpaired t test, NS. (**b**) Rhodamine-phalloidin stained wound demonstrating three kinds of cellular coverage: empty, loose, and tight. (**c**) Cellular coverage quantified with ImageJ MRI wound healing tool by assigning fixed thresholds to each image; as example, 24-h scratch treated with 2.9 mg/ml honey, threshold 20 to quantify empty space. (**d**) Twenty-four-hour scratch treated with 2.9 mg/ml honey, threshold 100 to quantify loose space. (**e**) Forty-eight-hour scratch treated with 2.9 mg/ml honey, threshold

information regarding the closure of wounds because the wound margin did not completely disappear at 24 h or even at 48 h in our experiments, and this margin was clearly noticeable when we stained for collagen; therefore, we looked for alternative ways to analyze our images which could provide insights as to what effects the compound might have on the complex processes occurring around and within the wound.

With the MRI wound healing plugin, we first tried to divide the images into three regions based on visual assessment of their morphology which correlated with the pixel density obtained from ImageJ; we could thus distinguish between empty space; loose coverage, characterized by motile cells; and the remaining tight coverage, denser cellular coverage surrounding empty and loose coverage. In other words, loose-empty space represents visual culture heterogeneity away from the homogeneous tight space (Fig. 4b). As we saw no significant change in the tight coverage at 24 h, we focused on analyzing empty and loose coverage by the ImageJ variance method after choosing two thresholds that were representative of empty space and loose space (Fig. 4c–f). Next, we followed the empty and loose space at 24 and 48 h post wound with various concentrations of MH, and we found there was a trend toward increased loose space with MH treatment at 48 h (Fig. 3g), indicating an increase in culture heterogeneity. We wondered if that correlated with an increase in fibrotic cells and stained for collagen in the 48-h wounds. Interestingly, our results suggest that MH does not increase collagen but rather decreases it (Fig. 4h–i), and this supports recent reports of scarless healing with MH (Singh et al. 2018).

We wondered what cells could contribute to the measured increase in loose space and turned our attention to flat giant cells, which appear to be interpreted by the MRI tool as empty space within covered space. These may be myofibroblast precursor cells, or perhaps fibrocytes (Reilkoff et al. 2011; Tomasek et al. 2002), mesenchymal cells with features both of fibroblasts and macrophages involved in tissue remodeling. Visual assessment and quantitation of flat giant cells in scratch wound images within and outside the wound area using ImageJ suggested that the number of such cells increases in MH treated scratch wounds, particularly in the wound area.

Taken together, our results presented here give added quantitative evidence to qualitative and anecdotal studies of MH applications in human wound healing.

4 Discussion

4.1 Scaffold-Free Spheroids from NIH 3T3 Cells Form by Substrate-Specific Collective Cellular Processes from Cell Sheets

There are different forms of 3D cellular aggregates broadly grouped under the terms "spheroids" and "organoids" (a very good current review of the nomenclature is found in (Decarli et al. 2021)). Both kinds of aggregates may form spontaneously in culture from cellular populations which contain a large proportion of stem cells or early progenitors. Organoids, characterized by their ability to organize in a manner similar to their tissue of origin, usually contain multiple cell types (Clevers 2016); they arise in tissue culture either by disaggregation of tissues and limited culturing or from coaxing stem

◄

Fig. 4 (continued) 20 to quantify empty space. (**f**) Forty-eight-hour scratch treated with 2.9 mg/ml honey, threshold 100 to quantify loose space. (**g**) Loose minus empty space for each MH concentration. (**h**) Collagen staining 48 h post scratch in MH treated wounds. (**i**) Flat giant cells observed within and outside the margins of the wound. (**j**) Ratio of flat giant cells occupying the space within the wound to total flat giant cells in the image after 48 h. Data are representative of at least two independent experiments with three technical replicates per condition

cells into aggregation by modifications of their environment. The term spheroid is reserved for multicellular spherical shaped clusters of cells from homogeneous cell populations or cell lines of various tissue origins. Such cellular aggregates clump together and maintain their spherical shape with or without exogenous extracellular matrices. For difficult to culture cells, various scaffolds made of hydrogel materials are required (Caliari and Burdick 2016). The spheroids we describe here are very similar in their appearance to those shown by (Granato et al. 2017) formed using human dermal cells aggregated in hanging drops and collected in agarose coated plates, whereas our spheroids form and are maintained in the absence of scaffold. Similar to spheroids from cancer cell lines, commonly used in 3D screens for drug discovery, which are relatively easy to form and maintain scaffold-free, our spheroids are also good models for 3D screens because the absence of scaffold decreases complexity and variability for the screen. This study adds to the relatively scarce literature characterizing spheroid cultures with fibroblasts in the absence of hydrogel materials (Jorgenson et al. 2017; Graham et al. 2019).

It is clear that the stages of formation of spheroids differ greatly among various cell types (Smyrek et al. 2019; Livoti and Morgan 2010) and particularly if one compares cancer line derived spheroids to those originated from progenitor populations. Our work sheds light into a specific mode of formation of stem cell derived spheroids from collective rolling of cell sheets which is highly dependent on interaction with the surface. Using sarcoma cells on polyacrylamide substrates (Beaune et al. 2018) observed cell sheet collective behavior to be highly dependent on surface rigidity. We report here that, only using low adhesion plates, the stages of spheroid formation consisted of the production of a thin layer of cells spreading through the surface which pulled and broke in various places in the same time frame (roughly within 1 min) to produce a collective rolling edge of cells that quickly gathered to a pouch to close a spheroid if the surface area was small. With the same kind of adhesion and increased surface area, the stages of aggregate formation were similar but produced varied shapes and sizes.

4.2 Using Spheroids and Scratch Wound Healing Assays to Screen Natural Compounds

Botanical complex mixtures have been used since ancient times to heal minor wounds and scratches. Despite many hurdles to the research and development of natural compounds, there is renewed interest in their therapeutic potential, especially as sources for new antimicrobial treatments. Important considerations when characterizing natural compounds are their difficult solubility and availability to the cells in culture, the inherent difficulties in bioactive compound isolation and in elucidating its cellular target (Atanasov et al. 2021), and the fact that synergisms between various components with beneficial therapeutic outcomes (Schmidt et al. 2007) may obscure the contribution of a component in the mixture to the overall bioactivity of a natural compound.

We start characterization of spheroid-supporting or spheroid-disrupting compounds by measuring cell viability with the colorimetric MTT assay. The use of this method is not without controversy due to low reproducibility (Stepanenko and Dmitrenko 2015), but, in our experience, it serves didactic purposes. Our students visualize cellular activity under the microscope after a few minutes' incubation, and they also learn to collect colorimetric data and, more importantly, learn to appreciate the differences between monolayers and spheroids (Fig. 2a). Even though it is currently accepted that 3D in vitro models are superior to monolayers in drug screening, it is harder to find the most relevant end point to assess bioactivity (White et al. 2019). In our case, since we have characterized the formation of the spheroids, preventing formation via oxidative stress and supporting it with compounds such as MH, we are now in the position to set up screening for compounds that facilitate pulling and rolling of sheet edges to aid spheroid formation and support its stability.

Scratch assays are relatively simple assays that have provided important information about cell motility and have been used for drug screening. However, more information can be gained from more sophisticated image analysis as others have recently shown (Kauanova et al. 2021). Our image analysis focuses on measuring the clearly different region near and within the wound that is covered by highly motile cells. This phenomenon is useful as a unit of study because its speed of formation and cellular components can be altered by the presence of natural compounds such as MH.

There is some controversy in the literature as to whether or not spheroids may be fibrosis models. Some argue they are, because they can contain activated myofibroblasts (Kisseleva 2017), whereas others (Granato et al. 2017; Avagliano et al. 2019) suggest that spheroids are not fibrosis models because the myofibroblasts within them become deactivated and that their formation closely resembles the physiological modes of skin wound healing. Spheroids from dysregulated transcriptional coactivator with PDZ-binding motif (TAZ) grow more than normal (Jorgenson et al. 2017), and these could be considered a model of fibrosis. Our results with spheroids formed with NIH 3T3 cells support Granato's conclusions suggesting our spheroids are physiologically stable since (a) acidification of the media does not happen in spheroid containing wells while it happens in actively growing monolayers (our unpublished observations) and (b) there is no significant change in spheroid size after 2 weeks in culture. Though we have yet to ascertain the reason for some increase in size of the spheroids with MH, this change does not seem to be due to increased ECM production (collagen) because in our scratch wound assays, visually the amount of collagen accumulated in cells nearing the wound was less in MH treated wounds in accordance with a recent report of scarless wound healing with MH (Singh et al. 2018). Our analyses of the scratch wounds also uncovered the possible contribution of flat giant cells to the increased loose coverage with MH treatment. It is possible that MH increases the number and proliferation of these cells which may be proto myofibroblasts (Tomasek et al. 2002; Avagliano et al. 2019) or fibrocytes (Reilkoff et al. 2011). We still do not know the nature of these large cells, just that they are part of the loose coverage, and our future work will address this.

4.3 Usefulness of Our Proposed Workflow

With the workflow presented here, we can get insights on cellular events that may be modified by a wound healing compound. This can be useful for further characterization of the compound; similar to MH, other compounds that produce a moderate increase in viability can then be taken to the scratch assays in 2D to verify their ability to increase the cell motility needed to close the wound. In our analysis of scratch wounds, we look at the traditional ratio of wound closure, but we also include the analysis of the loose space around the wound, providing insight on potential effects of the compound on cellular subpopulations affected by wound. We added collagen staining of the 48-h wounds to our workflow to demonstrate whether the compound increases or decreases collagen production as a readout for fibrosis.

There are many case reports or anecdotal reports about natural compounds used in wound healing, many of which are based on animal testing, while our workflow uses in vitro systems that do not rely on animal testing.

Finally, we use these methods to teach cell biology and introduction to research to undergraduate students with very limited experience in the lab. The workflow was established as a collaborative effort among the students in a low budget setting. We hope these methods can be adopted and expanded in similar settings.

Acknowledgments This work has been supported by Montgomery College, SCIR 297 program, and by support from Schoenberg Fellowship to V.V. We are thankful to Lauren Kimlin and to Greta Babakhanova for their help with the ImageJ analyses and for the constant assistance and support of Arifur Rahman, Ya Yu Shao, and Chris Standing, our lab staff.

Contributions G. V. performed experiments, analyzed data, and reviewed the manuscript; M. A. analyzed data and prepared figures; L. P., S. A., M. V., D. C., W. T., and N. T. performed experiments and analyzed data; and V. V. conceived the studies, performed experiments, analyzed data, and wrote the manuscript. All authors read and approved the manuscript.

References

Atanasov AG, Zotchev SB, Dirsch VM et al (2021) Natural products in drug discovery: advances and opportunities. Nat Rev Drug Discov 20:200–216. https://doi.org/10.1038/s41573-020-00114-z

Avagliano A, Ruocco MR, Nasso R et al (2019) Development of a stromal microenvironment experimental model containing proto-myofibroblast like cells and analysis of its crosstalk with melanoma cells: a new tool to potentiate and stabilize tumor suppressor phenotype of dermal myofibroblasts. Cell 8. https://doi.org/10.3390/cells8111435

Beaune G, Blanch-Mercader C, Douezan S et al (2018) Spontaneous migration of cellular aggregates from giant keratocytes to running spheroids. Proc Natl Acad Sci U S A 115:12926–12931. https://doi.org/10.1073/pnas.1811348115

Berridge MV, Tan AS (1993) Characterization of the cellular reduction of 3-(4,5-dimethylthiazol-2-yl)-2,5-diphenyltetrazolium bromide (MTT): subcellular localization, substrate dependence, and involvement of mitochondrial electron transport in MTT reduction. Arch Biochem Biophys 303:474–482. https://doi.org/10.1006/abbi.1993.1311

Brüningk SC, Rivens I, Box C et al (2020) 3d tumour spheroids for the prediction of the effects of radiation and hyperthermia treatments. Sci Rep 10:1653. https://doi.org/10.1038/s41598-020-58569-4

Buckley CD, Pilling D, Lord JM et al (2001) Fibroblasts regulate the switch from acute resolving to chronic persistent inflammation. Trends Immunol 22:199–204. https://doi.org/10.1016/s1471-4906(01)01863-4

Bulman SEL, Tronci G, Goswami P et al (2017) Antibacterial properties of nonwoven wound dressings coated with manuka honey or methylglyoxal. Materials (Basel) 10. https://doi.org/10.3390/ma10080954

Caliari SR, Burdick JA (2016) A practical guide to hydrogels for cell culture. Nat Methods 13:405–414. https://doi.org/10.1038/nmeth.3839

Carter DA, Blair SE, Cokcetin NN et al (2016) Therapeutic manuka honey: no longer so alternative. Front Microbiol 7:569. https://doi.org/10.3389/fmicb.2016.00569

Clevers H (2016) Modeling development and disease with organoids. Cell 165:1586–1597. https://doi.org/10.1016/j.cell.2016.05.082

Decarli MC, do Amaral, R. L. F., Dos Santos, D. P. et al (2021) Cell spheroids as a versatile research platform: formation mechanisms, high throughput production, characterization and applications. Biofabrication. https://doi.org/10.1088/1758-5090/abe6f2

Frydman GH, Olaleye D, Annamalai D et al (2020) Manuka honey microneedles for enhanced wound healing and the prevention and/or treatment of methicillin-resistant staphylococcus aureus (mrsa) surgical site infection. Sci Rep 10:13229. https://doi.org/10.1038/s41598-020-70186-9

Fukushima T, Tanaka Y, Hamey FK et al (2019) Discrimination of dormant and active hematopoietic stem cells by g0 marker reveals dormancy regulation by cytoplasmic calcium. Cell Rep 29(4144–4158):e4147. https://doi.org/10.1016/j.celrep.2019.11.061

Gangwar R, Meena AS, Shukla PK et al (2017) Calcium-mediated oxidative stress: a common mechanism in tight junction disruption by different types of cellular stress. Biochem J 474:731–749. https://doi.org/10.1042/BCJ20160679

Graham AD, Pandey R, Tsancheva VS et al (2019) The development of a high throughput drug-responsive model of white adipose tissue comprising adipogenic 3t3-l1 cells in a 3d matrix. Biofabrication 12:015018. https://doi.org/10.1088/1758-5090/ab56fe

Granato G, Ruocco MR, Iaccarino A et al (2017) Generation and analysis of spheroids from human primary skin myofibroblasts: an experimental system to study myofibroblasts deactivation. Cell Death Disc 3:17038. https://doi.org/10.1038/cddiscovery.2017.38

Jorgenson AJ, Choi KM, Sicard D et al (2017) Taz activation drives fibroblast spheroid growth, expression of profibrotic paracrine signals, and context-dependent ECM gene expression. Am J Physiol Cell Physiol 312:C277–C285. https://doi.org/10.1152/ajpcell.00205.2016

Kalluri R (2016) The biology and function of fibroblasts in cancer. Nat Rev Cancer 16:582–598. https://doi.org/10.1038/nrc.2016.73

Kauanova S, Urazbayev A, Vorobjev I (2021) The frequent sampling of wound scratch assay reveals the "opportunity" window for quantitative evaluation of cell motility-impeding drugs. Front Cell Dev Biol 9:640972. https://doi.org/10.3389/fcell.2021.640972

Kisseleva T (2017) The origin of fibrogenic myofibroblasts in fibrotic liver. Hepatology 65:1039–1043. https://doi.org/10.1002/hep.28948

Livoti CM, Morgan JR (2010) Self-assembly and tissue fusion of toroid-shaped minimal building units. Tissue Eng Part A 16:2051–2061. https://doi.org/10.1089/ten.TEA.2009.0607

Lundholt BK, Scudder KM, Pagliaro L (2003) A simple technique for reducing edge effect in cell-based assays. J Biomol Screen 8:566–570. https://doi.org/10.1177/1087057103256465

Mokhtar JA, McBain AJ, Ledder RG et al (2020) Exposure to a manuka honey wound gel is associated with changes in bacterial virulence and antimicrobial susceptibility. Front Microbiol 11:2036. https://doi.org/10.3389/fmicb.2020.02036

Mueller-Klieser W (2000) Tumor biology and experimental therapeutics. Crit Rev Oncol Hematol 36:123–139. https://doi.org/10.1016/s1040-8428(00)00082-2

Pinto BI, Cruz ND, Lujan OR et al (2019) In vitro scratch assay to demonstrate effects of arsenic on skin cell migration. J Vis Exp. https://doi.org/10.3791/58838

Reilkoff RA, Bucala R, Herzog EL (2011) Fibrocytes: emerging effector cells in chronic inflammation. Nat Rev Immunol 11:427–435. https://doi.org/10.1038/nri2990

Robey P (2017) "Mesenchymal stem cells": fact or fiction, and implications in their therapeutic use. F1000Res 6. https://doi.org/10.12688/f1000research.10955.1

Rodrigues M, Kosaric N, Bonham CA et al (2019) Wound healing: a cellular perspective. Physiol Rev 99:665–706. https://doi.org/10.1152/physrev.00067.2017

Schmidt BM, Ribnicky DM, Lipsky PE et al (2007) Revisiting the ancient concept of botanical therapeutics. Nat Chem Biol 3:360–366. https://doi.org/10.1038/nchembio0707-360

Sies H (2017) Hydrogen peroxide as a central redox signaling molecule in physiological oxidative stress: oxidative eustress. Redox Biol 11:613–619. https://doi.org/10.1016/j.redox.2016.12.035

Singh S, Gupta A, Gupta B (2018) Scar free healing mediated by the release of aloe vera and manuka honey from dextran bionanocomposite wound dressings. Int J Biol Macromol 120:1581–1590. https://doi.org/10.1016/j.ijbiomac.2018.09.124

Smyrek I, Mathew B, Fischer SC et al (2019) E-cadherin, actin, microtubules and fak dominate different spheroid formation phases and important elements of tissue integrity. Biol Open 8. https://doi.org/10.1242/bio.037051

Stepanenko AA, Dmitrenko VV (2015) Pitfalls of the MTT assay: direct and off-target effects of inhibitors can result in over/underestimation of cell viability. Gene 574:193–203. https://doi.org/10.1016/j.gene.2015.08.009

Suarez-Arnedo A, Torres Figueroa F, Clavijo C et al (2020) An image j plugin for the high throughput image analysis of in vitro scratch wound healing assays. PLoS One 15:e0232565. https://doi.org/10.1371/journal.pone.0232565

Tashkandi H (2021) Honey in wound healing: an updated review. Open Life Sci 16:1091–1100. https://doi.org/10.1515/biol-2021-0084

Todaro GJ, Green H (1963) Quantitative studies of the growth of mouse embryo cells in culture and their development into established lines. J Cell Biol 17:299–313. https://doi.org/10.1083/jcb.17.2.299

Tomasek JJ, Gabbiani G, Hinz B et al (2002) Myofibroblasts and mechano-regulation of connective tissue remodelling. Nat Rev Mol Cell Biol 3:349–363. https://doi.org/10.1038/nrm809

Virador GM, de Marcos L, Virador VM (2019) Skin wound healing: refractory wounds and novel solutions. Methods Mol Biol 1879:221–241. https://doi.org/10.1007/7651_2018_161

White R (2016) Manuka honey in wound management: greater than the sum of its parts? J Wound Care 25:539–543. https://doi.org/10.12968/jowc.2016.25.9.539

White JR, Abodeely M, Ahmed S et al (2019) Best practices in bioassay development to support registration of biopharmaceuticals. BioTechniques 67:126–137. https://doi.org/10.2144/btn-2019-0031

Adv Exp Med Biol - Cell Biology and Translational Medicine (2022) 17: 243–248
https://doi.org/10.1007/978-3-031-20514-9

Index

Milton Keynes UK
Ingram Content Group UK Ltd.
UKHW050743081123
432189UK00003B/13